Canned Foods

Canned Foods

THERMAL PROCESSING AND MICROBIOLOGY

A. C. Hersom
Research Director, Londreco Ltd.
A research and development subsidiary of
Nestlé Products Technical Assistance Co. Ltd., Switzerland

E. D. Hulland
Senior Microbiologist, Londreco Ltd.

SEVENTH EDITION

CHEMICAL PUBLISHING COMPANY, INC.

Canned Foods
Thermal Processing and Microbiology

ISBN: 978-0-8206-0142-7

Chemical Publishing Company:
www.chemical-publishing.com
www.chemicalpublishing.net

First American Edition: Chemical Publishing New York 1981

Image credit: Steven Depolo, used under Creative Commons License

Printed in the United States of America

Preface

Since publication of the last edition a number of developments have influenced the canning industry. As alternatives to the tinplate can new containers have been introduced, some for in-pack sterilisation, and others for aseptic filling. Increased recognition of the influence of water activity on microbial growth and a gradual quantification of its effect has led to a better understanding of preservation processes. A desire to reduce the level of nitrites in canned cured meats has led to an intensification of work aimed at elucidating their role in the preservation system. Finally, knowledge of the causes of food poisoning and the properties of the organisms responsible has significantly increased. All these changes have necessitated a major revision.

A new chapter on continuous flow processing and aseptic filling has replaced the one on HTST processing and the chapters on food poisoning and the outline of canning operations have been extensively revised. The sections on moisture limitation and the curing of meats have been rewritten. New information has been added, where appropriate, to cover new containers such as the flexible retort pouch. The text throughout has been updated and many new illustrations have been added.

We wish to acknowledge the help received from Mrs A. Hersom and some of the staff at Londreco in preparation of the manuscript and checking the proofs. We are also indebted to our colleague, Mr P. J. Goode, in connection with compilation of the bibliography.

1980

A.C.H.
E.D.H.

Contents

1

Bacteria

For many years all living organisms were subdivided into two main groups — the Animal Kingdom and the Plant Kingdom. In the eighteenth and nineteenth centuries, when the study of micro-organisms commenced, these relatively simple organisms were allocated to one or other of the two Kingdoms on the basis of characteristics such as locomotion or the ability to photosynthesise, etc. As more micro-organisms were discovered, some of which possessed both plant and animal characteristics, considerable confusion developed over their classification, some organisms being assigned to both Kingdoms by different workers. Gradually as knowledge increased, it was realised that micro-organisms are descendants of very ancient evolutionary lines pre-dating the lines which eventually gave rise to the higher plants and animals. The situation was resolved in 1866 when Haeckel, one of Darwin's disciples, proposed the establishment of a third biological Kingdom, the Protista, to include the major microbial groups, the protozoa, algae, fungi and bacteria. Protista, therefore, may show characteristics which are specific to both the traditional Kingdoms but they are distinguished from plants and animals by their relatively simple biological organisation.

More recently the development of the electron microscope has enabled detailed studies to be made of the fine structure of micro-organisms which have revealed the existence of two distinct types of cell called procaryotes and eucaryotes. Eucaryotic cells are characterised by the possession of a true nucleus with multiple chromosomes enclosed in a nuclear membrane and are the unit of structure in plants, animals, protozoa, fungi and most groups of algae. Procaryotic cells are smaller, less differentiated and the nucleus consists of a single, naked cyclic DNA molecule without a nuclear membrane; all bacteria and the blue-green algae consist of procaryotic cells. The Protista, therefore, contains both procaryotic and eucaryotic organisms. Based on cell type all living organisms can be grouped in three main categories: acellular forms, represented by the viruses;

procaryotes, represented by the bacteria and blue-green algae, and eucaryotes represented by the fungi, algae, protozoa, plants and animals. Only bacteria and fungi are significant in canning microbiology.

Bacteria

Form

Bacteria are minute organisms consisting of a single procaryotic cell enclosed within a strong rigid cell wall. The presence of this rigid structure gives shape to the bacterial cell, the three basic forms being spheroidal, cylindrical or rod-like and spiral. Bacterial multiplication takes place by division of the cells. The spheroidal forms — which are termed cocci — divide in any plane whilst the rod-shaped types divide transversely. With many types of bacteria the cells remain attached to each other following division and groups of cells are formed, the arrangement of which depends on the planes through which division occurs. This grouping of the cells is very important in the classification of bacteria. With the coccal forms several groupings are observed. Cocci which divide in one plane only form pairs of cells known as diplococci or chains termed streptococci. Those dividing through any plane form irregular groups or clusters known as staphylococci. Division in two planes at right angles to each other results in the formation of a group of four cocci as in the tetracocci; division in three planes forms cubes of eight cells, the name Sarcina being given to groups of this type. With the rod-shaped organisms, to which the general term bacillus is applied, transverse division limits the possible arrangements to pairs or chains of cells.

Size. The extreme smallness of bacteria is indicated by the fact that the unit of measurement is the micrometre (μm) which is equal to one-thousandth of a millimetre. Each species exhibits more or less characteristic dimensions although irregularity may occur due to changes in environmental conditions and other causes. Between the different types there is a very wide variation in size, some species being many times larger than others. For many spheroidal forms the diameter lies between 0.75 and 2μm. A medium-sized rod has a width of between 0.5 and 1μm and a length of about 3μm. Some rods develop filamentous forms which may measure 100μm or more in length.

Cytology. The detailed structure of the bacterial cell has for many years been the subject of numerous investigations and much conflicting evidence has resulted, due chiefly to differences in the methods adopted. In recent years, however, improved techniques and the application of electron and other forms of microscopy have resulted in advances in the study of the bacterial cell, but many problems remain to be clarified.

From what is known at the present time, it may be stated that the bacterial cell comprises a rigid semi-permeable outer wall within which is the plasma membrane, also semi-permeable, enclosing a soft gel substance known as the cytoplasm. In addition to various organic and inorganic solutes, the cytoplasm also contains a nucleus and numerous small particles called ribosomes which are associated with protein synthesis. Other structures, including plasma membrane intrusions called mesosomes, have been reported, but some workers believe these to be artefacts arising during the chemical fixing of the cell when it is prepared for the electron microscope since they are not seen in freeze-etch preparations.

Cell wall. The bacterial cell is enclosed within a thick, strong and relatively rigid cell wall which is semi-permeable. It supports the weak plasma membrane against the internal osmotic pressure of the cytoplasm and maintains the characteristic shape of the bacterium. If the wall is damaged or weakened the cytoplasm swells from osmotic imbibition of water, bursts the weak plasma membrane and escapes into the environment, the cell disintegrating and dying. This process is termed lysis.

The rigidity of the cell wall is due to the arrangement of the macromolecules which form an open meshwork. Mucopeptide complexes are linked with polysaccharide chains and the structure is braced with peptide bridges. This open meshwork contains various matrix materials, of which the main constituent is a polysaccharide. The chemical composition of the cell wall varies considerably but remains fairly constant within a species or group of organisms, and is therefore a useful diagnostic feature. One of the most important subdivisions of bacteria depends upon their reaction to the Gram stain, those which retain the stain being termed Gram-positive and those from which it can be eluted, Gram-negative. The method involves treating the cells with a basic dye such as crystal violet, mordanting with iodine and then washing with an organic solvent, normally ethanol or acetone. The crystal violet and iodine react to form a complex which is retained by Gram-positive bacteria, which remain fully stained, and lost by Gram-negative bacteria, which are decolourised. A counter-stain such as carbol fuchsin or neutral red can be used as a final treatment to give the Gram-negative organisms a contrasting colour. The exact modus operandi of the Gram stain is not fully understood but there are considerable differences between the cell walls of the two groups of micro-organisms. Gram-positive bacteria generally have thicker walls which contain only three or four amino acids, a small quantity of lipid material and large amounts of polysaccharides and techoic acids, while the walls of Gram-negative cells are generally thin but chemically more complex, containing protein and a considerable amount of lipid material but less polysaccharide and no techoic acid.

The cell wall plays an important role in cell division. A transverse partition of cell wall material (and plasma membrane) grows inwards, like a closing iris diaphragm, from the lateral wall at the equator of the cell and forms a complete cross wall separating two daughter cells. Splitting of the cross wall may occur immediately, or it may be delayed so that chains or groups of cells are formed.

Plasma membrane. The bacterial protoplast is bounded by a thin elastic membrane consisting mainly of lipoprotein. It is separate from the cell wall and forms an osmotic barrier permitting the passive diffusion inwards and outwards of water and other small molecular substances. The active selection and transport of specific nutrient solutes into the cell and of waste products out of it are important functions of the plasma membrane. In addition to permeases responsible for nutrient uptake, it is associated with many other enzymes concerned with the various cell functions and may be regarded as the main organelle of the procaryotic cell. The membrane itself has little mechanical strength and is supported by the external cell wall.

Cytoplasm. The cytoplasm of the bacterial cell is a viscous watery solution, or soft gel, containing a variety of organic and inorganic solutes; it does not exhibit streaming. There are no membrane bound organelles in bacteria but the cytoplasm is filled with many thousands of tiny particles called ribosomes. These measure about 20nm in diameter, consist chiefly of RNA, and are the sole sites of protein synthesis in the cell. Much of this synthesis occurs in the region of the plasma membrane and long threads of integrated ribosomes, called polysomes, are loosely attached to the membrane.

Nucleus. The bacterial nucleus is not enclosed within a membrane and in ultra thin sections under the electron microscope it appears as a skein of parallel fibres. It is not visible with ordinary light microscopes unless special staining techniques are employed. The nucleus or nucleoid consists of a large cyclic DNA molecule containing all the genetic information essential to the bacterium. Multiple replication forks appear in fast growing cells to enable nuclear division to keep pace with cellular growth.

Flagella. Many species of rod-shaped bacteria and a few cocci are capable of movement in liquids. All these species possess appendages called flagella which consist of helical assemblies of the protein, flagellin (Murray, 1978). Each flagellum arises from a basal complex which is embedded in the plasma membrane. They are unbranched, of uniform thickness and their length may be several times that of the cell. The flagella appear to be in the form of a semi-rigid helix and it is thought that propulsion is achieved by each filament rotating in its insertion complex. A single bacterium may possess up to 100 flagella, the number and location being characteristic for a given species. If they are arranged round the whole organism, they are said to be peritrichous and, if at one or both ends of the cell, polar. Several polar flagella may aggregate into a bundle and function as a single unit or 'tail'.

They are not normally visible under the light microscope unless special stains are used but their motion can be observed by dark field microscopy. Flagella proteins constitute the H-antigens of motile bacteria.

Fimbriae or pili. Certain Gram-negative organisms possess very fine, hair-like appendages known as fimbriae. Like the flagella, they arise from the plasma membrane and project through the cell wall but are much smaller and more or less straight. They are arranged peritrichously and up to 500 may be present on a single cell. Fimbriae or common pili have adhesive properties enabling the organisms to attach themselves to a substratum, to each other, or in the case of pathogenic bacteria, to specific host cells and erythrocytes.

Several types of pili can be recognised but the functions of most are unknown. Certain Gram-negative bacteria, however, may possess up to four hollow sex pili which are believed to be concerned in conjugation. The donor cell forms a hollow sex pilus which passes through the cell wall, the distal end of the pilus locates and becomes attached to the recipient cell and a molecule of DNA passes down the hollow pilus from donor to recipient.

Slime layer. The cells of most species of bacteria are surrounded by a viscid slime layer which, depending on the species, may take the form of an extremely thin film or a relatively thick, well-defined coat, termed a capsule. This slime material consists mainly of water and contains about 2 per cent solids, usually a complex polysaccharide, although in some species it may be polypeptide or protein. Microscopically capsules can be seen in stained preparations, or wet preparations mounted in Indian ink, as clear zones or halos surrounding the cells; a special technique is necessary to stain the capsular material.

When slime-forming bacteria are grown on a solid medium the individual cells become embedded in a matrix of this substance and the colony appears sticky and mucoid in nature. Such colonies are able to adhere to surfaces in environments where they would otherwise be swept away and dispersed. The growth of capsule forming bacteria is responsible for the spoilage of milk, cream and bread due to 'ropiness'.

There is some evidence that the capsular material may be protective in nature: thus although the slime layer can be removed from a cell with no apparent effect on the viability or metabolism of the organism, the presence of this viscid layer affords some protection against enzyme attack. In the case of capsulate pathogenic bacteria, the capsule appears to protect the organism from phagocytosis by the white blood cells of the host since mutant strains which have become non-capsulate are no longer virulent.

Spores. A number of bacterial genera are now known to produce a spore or more correctly, an endospore, including the coccus *Sporosarcina*. An endospore, the term means formed within the cell, is a resting stage and is in no way concerned with reproduction since usually one cell produces one spore, which eventually germinates to produce one new bacterium, the old parent cell dying during the sporulation process. Endospores are much more resistant to heat, chemicals and desiccation than the vegetative forms and they may remain viable for very long periods. An example of this last characteristic has been given by Wilson & Shipp (1938) who found living spore-forming bacilli in meat preserved in a can for 114 years. Although usually considered to be a resting or dormant stage, spores are not completely inactive and may bring about changes in their immediate environment by means of enzymes. For example, Galloway (1946) has reported the slow liquefaction of gelatin by spores of the *Bacillus subtilis* type in the presence of disinfectant amply sufficient to inhibit germination and growth.

Until recently only the species of two genera of bacteria, *Bacillus* and *Clostridium* were known to produce endospores and most of our knowledge has been gained from these two organisms, the majority of the studies relating to the aerobic genus *Bacillus*. The spore consists of a central core of condensed cytoplasm containing the nuclear body; the cytoplasm is bounded by the thin plasma membrane and outside this is a thicker layer, known as the cortex. Finally the cortex is enclosed by several layers forming a relatively impermeable spore coat.

The resistance of spores to chemicals is due in part to their impermeable outer layers, as demonstrated in their resistance to the penetration of biological stains. Their resistance to heat, however, is less clearly understood. At one time, it was thought that a substance called dipicolinic acid, which is found only in the spore cortex, was intimately involved but the discovery of heat-resistant mutant spores containing no dipicolinic acid suggests that this is not the case.

Slepecky (1978) believes that dipicolinic acid, which may constitute up to 15 per cent of the dry weight of the spore, may possibly be involved with the maintenance of dormancy or the initiation of germination, or both. The most likely explanation of heat resistance has arisen from studies of the density and refractive index of spores which have shown that their water content is extremely low, varying from 5 to 15 per cent. In the dehydrated state proteins and nucleic acids are known to be far more resistant to thermal denaturation.

Several theories have been advanced to account for the process of sporulation, some workers believing that it is a normal process in the life-cycle of the cell. However, the generally accepted view is that spores are formed by healthy cells facing starvation or unfavourable conditions when the growth rate is slowing down. It is known to follow periods of rapid growth. The sporulation process is under genetic control and the fact that an elaborate sequence of special syntheses is involved indicates that the bacterial cell contains a considerable amount of genetic information concerning the process.

Chemical composition of bacteria

The chemical constituents of bacterial cells must obviously be related to the environment in which they were produced and this, no doubt, accounts for some of the widely discrepant analyses recorded in the literature. Bacteria contain a high proportion of water, for most species the water content of vegetative cells lying between 70 and 90 per cent: Anderson (1946) quotes figures of 73.3 per cent for *Escherichia coli* and 98.3 per cent for *Acetobacter aceti*. Although the core of bacterial endospores may contain only 5 to 15 per cent of water the outer cortex layer may contain up to 70 per cent moisture.

Bisset (1962) gives the following figures for the analysis of dry bacteria: carbon approximately 50 per cent, nitrogen 8 to 15 per cent and ash 2 to 15 per cent. According to Meadow (1978), of the 27 elements found in most living things, 4 — hydrogen, carbon, oxygen and nitrogen — constitute 99 per cent of their mass and only 12 others — sulphur, phosphorus, magnesium, calcium, sodium, potassium, chlorine, iron, manganese, copper, zinc, and molybdenum — occur in all species. Most of the nitrogen and carbon is contained in amino acids which are linked to form proteins, which in turn may constitute up to 50 per cent of the dry weight of the cell. Other macromolecules in descending amounts are nucleic acids, carbohydrates and lipids. The four main macromolecular types are usually found in combinations with each other as, for example, lipoproteins, lipopolysaccharides, glycopeptides, and nucleoproteins. Other constituents of bacterial cells may include alcohols, waxes, pigments and vitamins.

Nutrition

Before bacteria can proliferate they need to be supplied with materials from which they can derive energy and synthesise the protein, fat, carbohydrate and other constituents of which the bacterial cell is composed. In a monograph reviewing the subject of bacterial nutrition Knight (1936) states:

> The nutrients must supply:
> (a) Material which is ultimately transformed into living multiplying organisms.
> (b) The necessary energy which is required for this transformation and for other less energetically important purposes such as the maintenance of the new organisms and the performance of external work (motile organisms), etc.

Bacteria can be separated into two broad groups based on the form in which the necessary nutrients can be utilised. Those which can synthesise their protoplasm from simple inorganic compounds are termed autotrophic. These organisms assimilate carbon from carbon dioxide and nitrogen from ammonia, nitrate or nitrite. They obtain their energy for growth from oxidation reactions (chemosynthetic) or utilise the radiant energy of sunlight (photosynthetic). The autotrophs, which include the nitrifying bacteria important in soil microbiology, have no significance in the spoilage of foodstuffs.

The heterotrophic organisms are those for which organic materials are necessary as a source of carbon and energy. This group includes the pathogenic or disease-producing organisms and the saprophytic types which are responsible for the spoilage of foods. The simpler heterotrophs can assimilate nitrogen from inorganic sources, but the more exacting types require amino acids of varying complexity and in some cases specific amino acids such as tryptophane are essential. In the usual laboratory media used for the cultivation of bacteria, peptone and meat-extract broths serve as a source of carbon and nitrogen.

With regard to the mineral requirements of bacteria, Knight (1936) points out that the inorganic compounds (in nutrient media) perform several functions. Their effects may be physical, for example as buffer salts to oppose pH changes in the medium, and as salts for controlling osmotic and permeability effects. They may be required as nutrient salts for use in synthesising protoplasm or they may be utilised for catalytic or enzymic purposes. To satisfy the mineral requirements of bacteria, K, Na, Ca, Mg, Fe, phosphate, sulphate, carbonate and chloride ions appear to be necessary. In ordinary nutrient media these ions are normally present in sufficient concentration.

For some organisms specific growth-promoting substances are essential; their nature is not fully understood in all cases. These substances, which are often referred to as bacterial vitamins, are active in very low concentrations. The essential growth factors for some organisms can be synthesised by other types and Knight has suggested that these growth factors may be of wider importance in the growth of bacteria than is generally realised. If an organism is capable of synthesising the factor for itself the need for it is not apparent; the necessity for growth factors only becomes obvious for those organisms which have lost the ability to synthesise them. The subject of accessory growth factors for bacteria has been reviewed by Berkeley & Campbell (1971).

Bacteria absorb their nutrients through the permeable cell wall. If the material available is in a non-assimilable form it must first be broken down into simpler compounds which can pass into the cell. This conversion is accomplished by enzymes elaborated by the bacteria. After passing into the cell the material undergoes further enzymic changes and may be directly utilised for building cellular substance or broken down into yet more simple compounds. Bacteria vary widely in their enzymic activity; some species, for example, readily utilising a wide range of carbohydrates while others are inactive or attack only a few. Other species are particularly active against proteins, and fats may be decomposed by certain types. The end products resulting from these decompositions also show great variation with the different species. Hydrogen, carbon dioxide, alcohol, lactic and other acids are some of the products which may result from the decomposition of sugars (usually referred to as fermentation). The splitting of proteins, termed putrefaction, results in the formation of amino acids which may eventually be decomposed to give ammonia and carbon dioxide. Other products of protein decomposition include hydrogen sulphide, indole and other foul-smelling compounds. Pure proteins are not attacked by bacteria unless there is some other nitrogen source such as amino acid present to permit growth and the formation of proteolytic enzymes. Detailed information concerning bacterial nutrition and biochemical activity is given by Berkeley & Campbell (1971) and Whittenbury (1978).

Oxygen relationships

In relation to the influence which molecular oxygen of the atmosphere has on their development, bacteria are usually classified as follows:

1. Obligate aerobes — requiring molecular oxygen for growth
2. Obligate anaerobes — unable to develop in the presence of molecular oxygen
3. Facultative anaerobes — growing in the absence or presence of molecular oxygen
4. Micro-aerophiles — growing under an oxygen tension less than that in air.

The fact that an organism will develop in the presence of air does not necessarily mean that it has the ability to utilise molecular oxygen. Thus certain species of clostridia can reduce molecular oxygen but are unable to make use of this process in their metabolism. In recent years the terms oxybiontic and anoxybiontic respectively, have been proposed to describe the ability or inability of organisms to utilise molecular oxygen.

For details of the various theories concerning the role of molecular oxygen in bacterial respiration, Wilson & Miles (1975), Hewitt (1950), and Lamanna & Mallette (1959) should be consulted. As far as the anaerobic bacteria are concerned it is established that they are unable to develop in media with oxidation-reduction potentials (Eh) above certain limits, the inhibitory effect of oxygen on this class of organism being due to its action in raising the Eh of the media above the minimal limit. Reed & Orr (1943) have shown that the optimum Eh for 15 species of spore-forming anaerobes is in the region of −0.20 volt. It has been suggested that in the presence of oxygen, a strict anaerobe may produce toxic peroxides which it cannot destroy owing to its lack of catalase — an enzyme present in most aerobes and facultative anaerobes. The growth of anaerobic bacteria, therefore, requires that molecular oxygen be excluded from the medium or the medium be so compounded that its Eh is posed at a suitably low level. In the laboratory cultivation of anaerobic bacteria both of these procedures are used. Cultures in the ordinary type of nutrient broth or agar are stored in an inert atmosphere of hydrogen or nitrogen or, in the case of liquid cultures, the surface is sealed with paraffin oil or wax. Such liquid media must be boiled and rapidly cooled before inoculation, in order to remove dissolved oxygen. In the second method, strongly reducing conditions are maintained in the medium by the addition of such substances as glutathione, cysteine, ascorbic acid or cooked meat which are effective even when the medium is exposed to air; this method is applicable only to liquid media and for surface growth on solid media storage in an oxygen-free atmosphere is essential.

Bacterial toxins

Food-poisoning bacteria are characterised by the production of substances causing disease or illness as a result of their injurious effects on the body tissues. These substances, known as toxins, are usually antigenic, that is they stimulate the body tissues in the production of various specific neutralising substances which are collectively termed antibodies.

The ability to stimulate antibody production is not confined to living bacteria. Dead bacteria are antigenic as are all non-living proteins of colloidal dimensions provided they are foreign to the animal or host into which they are injected. Most toxins are protein or have protein components, but non-protein materials such as polysaccharide-phosphatide complexes which are toxic and antigenic have been isolated from bacteria.

Two main types of bacterial toxins are generally recognised but it should be noted that absolute distinction is not always possible because of the fact that some toxins exhibit properties common to both types. The classification is based on the following characteristics:

1. *Exotoxins.* Toxins of this class are extracellular, that is they diffuse freely into the medium in which the bacteria are growing and are easily separated from the cells by filtration. Exotoxins, which are active in very small amounts, stimulate the formation of the antitoxin type of antibody which specifically neutralises exotoxin and renders it harmless. The symptoms or lesions produced in the body by exotoxins are highly characteristic so that it is usually possible to recognise an exotoxin by the effects it produces. Toxins of this type are generally heat-labile, being inactivated at 60°C in an hour. Typical exotoxins are those produced by the botulism, tetanus and diphtheria organisms.

2. *Endotoxins.* This type of toxin appears to be an integral part of the cell and cannot be separated from the organisms by filtration. The symptoms produced in the body by endotoxins are less characteristic than those caused by exotoxins and can only be produced by the actual cells or cell extracts. Endotoxins do not usually stimulate antitoxin formation, the antibodies produced being of the antibacterial type such as the bacteriolysins which directly destroy bacteria or the agglutinins which immobilise them and cause them to aggregate into clumps. In comparison with typical exotoxins, endotoxins are relatively heat-stable.

The reactions which occur between antibodies and antigens show a relatively high degree of specificity and since many of these reactions are readily demonstrated *in vitro* they are of great value in the identification of various antigens such as bacteria and proteins. For such purposes the serum of an animal or person is used in which antibodies have developed as the result of injection or infection with the antigen substance. In particular, the agglutination test is widely used for identifying and separating species of bacteria which are morphologically and biochemically indistinguishable. The subject of bacterial toxins and immunity reactions is treated in great detail by Wilson & Miles (1975) and Ajl et al, (1970/1).

Reproduction and growth

Multiplication of bacteria normally takes place by division of the cell, a simple method known as binary fission. The process normally involves growth of the bacterial cell, division of the nuclear body, the formation of a transverse septum and separation of the resulting daughter cells. Thus, although bacterial growth usually refers to an increase in the number of viable cells present, it also involves the formation of new cellular material. Occasionally, the processes lose their synchronisation and for a time multinucleate cells may be seen in the culture. The rate at which binary fission occurs varies with different organisms and is affected by numerous factors such as the age of the culture, the temperature, available nutrients and the genetic make up of the organism. The rate is maximal at the optimum temperature.

The growth of a bacterial population shows four distinct phases as follows:

1. *Lag phase.* During the first phase which may last some hours, although there is little or no actual increase in numbers, growth or enlargement of some of the cells takes place and there is an increase in respiratory and other metabolic processes. The length of this phase varies according to the condition and number of cells in the inoculum; it may be regarded as a period of adaptation to the new environment. For example, if the new environment contains different nutrients to those present in the previous environment in which the cells had been growing, then it may be necessary for the cells to synthesise new enzymes, or for a process of selection of mutants to occur.

2. *Logarithmic phase.* This is the period of most rapid growth. It is the period in which the cells divide at a constant rate and, as a result of growth by binary fission, there is a linear relationship between the logarithm of the number of cells and time. Under the most favourable conditions the more rapidly growing organisms may undergo division in about 10 minutes. Cells in the log phase understandably have a very high rate of metabolism and, as a result of their adaptation to the maximum rate of synthesis and growth, may be very sensitive to the effects of heat and antimicrobial agents. The length of the logarithmic phase varies with different species and is influenced by temperature and other factors but rarely persists for more than a few hours.

3. *Stationary phase.* When, at the conclusion of the logarithmic phase, the number of viable organisms reaches a maximum, the rate of division gradually becomes slower and for a period is just sufficient to balance the number of cells dying, thus keeping the number of viable cells stationary. This slowing down of growth is usually caused by one of two factors (or both), the exhaustion of essential nutrients from the medium, or the accumulation of toxic waste products.

4. *Phase of decline.* After a variable time in the stationary phase, cell

division ceases completely and the cells remaining begin to die in large numbers. Death can be defined as the loss of ability to grow and multiply when the cells are transferred to a fresh nutrient medium and the cause of death is usually related to the factors which were responsible for the reduced rate of growth in the stationary phase. When the accumulation of toxic products is the main factor involved, there may be a very short stationary phase followed by a rapid death. In some cases there may also be a rapid fall in the total count if the dead cells undergo a process known as autolysis, in which the cells are broken down by their own cellular enzymes. Bacteria in this phase often take on unusual shapes and may change their staining characteristics, Gram-positive organisms becoming Gram-negative.

The type of growth discussed above in which the organisms pass through four distinct phases, occurs when an inoculum is placed in a culture vessel containing a finite amount of nutrient medium and is known as batch culture. In this form of growth the environment is restricted and as a result suffers changes due to the growth of the organisms in it. If, however, a fresh supply of nutrients is continuously fed into the vessel and, at the same time, organisms and spent culture are continuously removed by means of a constant level device, the growth rate and the concentration of viable organisms can be kept constant under controlled conditions for considerable periods of time. This system is known as continuous culture and is used industrially for the production of vinegar and antibiotics, etc.

Temperature range of growth

For all bacteria there is a minimum and maximum temperature at which development occurs. Between these points there is an optimum temperature — this usually being defined as that temperature at which cell reproduction is most rapid. The optimum temperature for cell reproduction does not necessarily coincide with that for other processes such as rate of fermentation or for the production of the largest total cell population. Dorn & Rahn (1939) have shown that *Streptococcus lactis* in milk gave the most rapid growth at 34°C; the largest cell population at 25° to 30°C; the most rapid fermentation at 40°C and the highest final acidity at 30°C.

With respect to their growth temperature requirements microorganisms are commonly separated into three broad groups. Such grouping, however, must be quite arbitrary since there is no sharp demarcation between the groups which are as follows:

Psychrophilic group. The organisms of this group develop readily at 0°C but often have optimum temperatures above 20°C. Ingraham & Stokes (1959) appear to favour a definition based on the ability of the organism to show appreciable growth within two weeks at 0°C. However, Eddy (1960) proposes the term *psychrotrophic* for these organisms and reserves the term *psychrophilic* for the bacteria with maximum growth temperature below 35°C. The *psychrotrophic* organisms are of practical significance in food preservation since they are capable of causing spoilage of foods in cold store.

Mesophilic group. Organisms with an optimum temperature between 25° and 40°C are termed mesophilic; members of this group do not as a rule develop below 5°C and the maximum growth temperature is usually about 40° to 45°C although certain thermo-tolerant spore-forming mesophiles will grow at 55°C. The mesophilic group includes all the pathogenic bacteria.

Thermophilic group. Much confusion exists concerning the meaning of the term 'thermophile' as applied to micro-organisms. It has been most commonly used to denote organisms with an optimum temperature of about 55°C but the present tendency appears to be to refer to any organism capable of growth at 55°C as a thermophile (*see* Ch. 7).

Temperatures outside the range for growth have a lethal or inhibitory action on bacteria. These effects of temperature, together with other controlling influences are dealt with later.

Identification

The morphological features of the individual cells are only sufficient to permit a primary grouping of bacteria and in order to distinguish the different species additional criteria have to be employed. Classification is, therefore, based on the determination of a variety of characteristics briefly indicated below:

Morphology. Size, shape and arrangement of the cells. Motility (arrangement of flagella). Spores (size, shape and position). Capsule formation.

Staining reaction. Gram's method. By means of the Gram staining method all bacteria can be placed into one of two groups — Gram-positive or Gram-negative. The essential feature of this important staining procedure is that after staining with one of certain of the aniline dyes and treatment with iodine solution some organisms (Gram-negative) can be decolourised by alcohol while others (Gram-positive) retain the dye.

Cultural characteristics. The form of the growth produced in solid and liquid culture media has diagnostic value. The size, shape, structure and pigmentation of discrete cell-masses or colonies on solid media are noted. In liquid media, pellicle formation, degree of turbidity and type of sediment (e.g. granular, viscid, etc.) are important. Other cultural tests such as growth in gelatin stab and agar streak cultures are also used.

Oxygen requirements. Anaerobic, aerobic, facultative anaerobic, micro-aerophilic.

Temperature requirements. Optimum temperature. Range for growth.

Biochemical reactions. Action on various carbohydrates, alcohols, proteins, inorganic salts, etc., are very important for identification purposes. Tests for the presence of certain end-products (e.g. indole, acetylmethylcarbinol) are also made.

Pathogenicity and serological tests. Although not normally used for the saprophytic types, injection of laboratory animals and serological tests are important in the study of pathogenic bacteria.

For details concerning the methods for carrying out all these tests reference should be made to *Identification Methods for Microbiologists*, Technical Series 1 and 2, published by the Society for Applied Bacteriology (1966 and 1968) or the numerous textbooks on practical bacteriology.

In recent years more sophisticated tests have been used to gain further information on certain microbial groups. However, some of these techniques are complex and only suitable for laboratories specialising in this type of work. Among the various methods which have been investigated are comparisons of gel electrophoresis patterns of cell protein extracts, ultraviolet and infrared spectrometry of organisms or their extracts and pyrolysis gas-liquid chromatography of micro-organisms. For information concerning new approaches to the problems of identification and taxonomy, including the use of computers, Sneath (1978) should be consulted.

Classification

A number of different systems have been advanced for the classification of bacteria and each has had supporters with the result that some confusion has existed. The problem of bacterial classification is difficult due to the lack of fundamental knowledge and adequate descriptions of many species so that no perfect scheme has been devised. Further information gained during the last decade and a basic re-thinking of the problems led Buchanan & Gibbons (1975) to propose a radically new scheme in the eighth edition of *Bergey's Manual of Determinative Bacteriology*. In this scheme the bacteria and the blue-green algae have been placed together in their own Kingdom, the Procaryotae, on the basis of their cellular organisation (procaryotic). The higher taxa such as Orders are no longer recognised and the bacterial genera are arranged in 19 different groups. A discussion on the theories involved is outside the scope of this book and the reader is advised to consult Buchanan & Gibbons (1975).

REFERENCES

Ajl S J, Ciegler A, Kadis S, Montie T C, Weinbaum G 1970/71 Microbial toxins. Academic Press, London, vols 1-5

Anderson C G 1946 Introduction to bacteriological chemistry. Livingstone, Edinburgh

Berkeley R C W, Campbell R 1971 In: Hawker L E, Linton A H (eds) Micro-organisms: function, form and environment. Arnold, London, p 143

Bisset K A 1962 The bacteria. Livingstone, Edinburgh

Buchanan R E, Gibbons N E 1975 Bergey's manual of determinative bacteriology, 8th edn. Williams & Wilkins, Baltimore

Dorn F L, Rahn O 1939 Arch Mikrobiol 10: 6

Eddy B P 1960 J Appl Bact 23: 189

Galloway L D 1946 Manuf Chem 17: 535

Hewitt L F 1950 Oxidation-reduction potentials in bacteriology and biochemistry. Livingstone, Edinburgh

Ingraham J L, Stokes J L 1959 Bact Rev 23: 97

Knight B C J G 1936 Bacterial nutrition. Med Res Council Spec Rept No 210. HMSO, London

Lamanna C, Mallette M F 1959 Basic bacteriology. Bailliere, Tindall & Cox, London

Meadow P 1978 In: Norris J R, Richmond M H (eds) Essays in microbiology. Wiley, Chichester, ch 6

Murray R G E 1978 In: Norris J R, Richmond M H (eds) Essays in microbiology. Wiley, Chichester, ch 2

Reed G B, Orr J H 1943 J Bact 45: 309

Slepecky R 1978 In: Norris J R, Richmond M H (eds) Essays in microbiology. Wiley, Chichester, ch 14

Sneath P H A 1978 In: Norris J R, Richmond M H (eds) Essays in microbiology. Wiley, Chichester, chs 9 and 10

Society For Applied Bacteriology 1966/1968 In: Gibbs B M, Shapton D A, Skinner F A Technical series 1, 2, Identification methods for microbiologists. Academic Press, London

Whittenbury R 1978 In: Norris J R, Richmond M H (eds) Essays in microbiology. Wiley, Chichester, ch 16

Wilson G S, Miles A A 1975 Principles of bacteriology and immunity (Topley & Wilson). Arnold, London

Wilson G S, Shipp H L 1938 Chem Ind, p 834

2

True fungi

MOULDS AND YEASTS

The True Fungi or *Eumycetes* differ from bacteria fundamentally in that they are composed of eucaryotic cells having the same basic organisation as those of plants and animals. In eucaryotic cells the nucleus is bounded by a nuclear membrane and the motile cytoplasm contains organelles such as mitochondria and chloroplasts. These structures are not present in the more simple procaryotic cell of bacteria.

Moulds and yeasts are frequently regarded as two distinct and separate groups of fungi. The distinction usually made is a morphological one, the moulds being defined as fungi which characteristically produce branched filaments (mycelium) and the yeasts as unicellular fungi. No true distinction exists, however, since some moulds assume a yeast-like form under certain conditions while a number of normally unicellular yeasts may give rise to mycelium. There is, in fact, a distinct relationship between the yeasts and the higher fungi, yeasts being regarded as previously filamentous forms which have degenerated into the unicellular state. For the present purpose, however, it is convenient to consider the moulds and yeasts separately.

Moulds

The common moulds consist of tubular filaments of microscopic size termed hyphae which, being profusely branched, intertwine to form a web-like structure known as mycelium. In many cases the mycelium can be differentiated into a vegetative portion which extends over and into the substrate on which it is growing and a fertile , spore-bearing portion extending into the air. Hyphae may be septate or non-septate. In the former case each hypha is divided by transverse walls or septa into a number of cells each containing one or more nuclei, while non-septate hyphae have the appearance of continuous, multi-nucleated tubes. The primary division of the fungi is based on the presence or absence of septa in the hyphae. Extension of the hyphae and growth of mycelium takes place either by growth at the tips (apical growth) or by division of the cells in the hyphae (intercalary growth).

Reproduction

Moulds normally reproduce by means of spores which are usually minute bodies a few micrometres in diameter and are formed in large numbers by a single organism. Sporulation in moulds is a truly reproductive process, each spore being capable of giving rise to a new growth by emitting one or more germ tubes which extend and branch to form fresh mycelium. Under favourable conditions the cycle from spore to spore may be completed fairly rapidly (24 to 48 hours). Mould spores resemble bacterial spores in that they are resistant to adverse conditions.

The identification of moulds is largely based on the characteristics of the reproductive spore systems which may be sexual or asexual in origin. In many species both types of spore formation occur, but if sexual spores are present they are regarded as the 'perfect' form and are primarily important for classification purposes; asexual spores are termed 'imperfect' or 'accessory' but their characteristics and mode of formation are important in the differentiation of the species, even in those types which form sexual spores.

Sexual sporulation is always preceded by the fusion of two separate hyphal cells and their nuclei but in asexual spore-formation there is no preliminary fusion of cells, the spores being produced at the ends of the fertile hyphae. In some species asexual spores arise singly or in groups as a result of budding or segmentation of the fertile hyphae. Such spores are termed conidia and the hyphae which bear them are conidiophores. Conidia may be attached to fertile hyphae by short stalks or sterigmata which, in some species, are branched. Asexual spores may also be produced within a sac or sporangium by the non-septate moulds. Such spores are sporangiospores and the hyphae bearing them are termed sporangiophores. The tip of the sporangiophore which projects into the sporangium is known as the columella and its shape and size are important for identification purposes. Sporangia are never produced by the septate moulds. In some species of fungi the hyphae break up into short segments, each of which is regarded as a spore since it is capable of giving rise to a new growth. The name oidia is given to these segments.

Asexual spores known as chlamydospores are formed by a thickening and swelling of the hyphal cells. They occur in any portion of the mycelium and are not confined to special areas. Such spores are very resistant to unfavourable conditions. In some species hyphae combine to form hard compact masses termed sclerotia which are also able to survive adverse conditions.

Principal groups

There are many thousands of species of fungi and their classification is a difficult matter. A number of different systems exist and for details or references to these the reader is advised to consult Henrici (1947) and Smith (1969). The latter gives full descriptions of the species commonly occurring in industry. Within the scope of this book it is only possible to refer to the principal groups and to indicate important genera or species associated with food spoilage.

True Fungi (*Eumycetes*) are usually divided into four main classes: the *Phycomycetes*, the *Ascomycetes*, the *Basidiomycetes*, and the *Fungi Imperfecti*. The *Phycomycetes* usually possess non-septate mycelia, while the *Ascomycetes, Basidiomycetes* and *Fungi Imperfecti* possess septate mycelia. The latter three groups may be differentiated according to their formation of sexual spores, the *Ascomycetes* producing sexual spores endogenously in sacs, while the *Basidiomycetes* produce sexual spores exogenously on stalks. The *Fungi Imperfecti*, so far as is known, produce no sexual spores. However, there is now considerable evidence that many of the *Fungi Imperfecti* have lost the ability to form sexual spores and the size of this group is being reduced as its members are gradually assigned to the *Ascomycetes*, or occasionally to the *Basidiomycetes*, by comparing the asexual states of species in the three groups.

More recent classifications divide the *Phycomycetes* into two distinct groups, thus producing five major groups of True Fungi (Booth, 1978).

Phycomycetes

Asexual spores are produced by cleavage of a multinucleate mass within a sporangium, and the sexual spores are formed by a simple direct union of gametes. The group is separated into three sub-classes, the *Oomycetes*, the *Zygomycetes* and the *Archimycetes*. Members of the first sub-class are unimportant in food spoilage but a number are responsible for diseases of plants and fish.

In the *Zygomycetes*, the sexual spores are zygospores. They are formed as the result of two hyphae coming into contact. At the point of contact, protuberances or progametangia develop, each of which becomes separated from its parent hypha by the formation of a septum. The cut-off portions which remain in contact fuse and the wall of the resulting cell becomes thick and warty. This group is divided into two orders:

Mucorales: asexual spores in sporangia.

Entomophthorales: asexual spores borne on conidiophores and shot off at maturity.

The order *Entomophthorales* includes a number of species parasitic on insects, but they are not found in food spoilage. The majority of the *Mucorales* species are saprophytic, members of the genera *Mucor*, *Rhizopus* and *Thamnidium* being commonly associated with food spoilage.

Ascomycetes

This group is characterised by the formation of sexual ascospores within a sac or ascus. The term endogenous is applied to this type of spore formation. There are usually eight ascospores within each ascus, many of which may be formed in a single growth. There is great diversity in the manner in which asci are formed by the various species. In the simplest forms, as in yeasts, there first occurs a fusion of two adjacent cells and their nuclei which proceed to divide into daughter nuclei, each of which becomes a spore. In this instance, the yeast cell forms the ascus. In higher members of the group, a fusion of two cells occurs and from the resulting cell new filaments or ascogenous hyphae arise in which asci containing ascospores are developed. In some species groups of asci become surrounded by a dense, protective mass of filaments to form a fruiting body known as an ascocarp. If the ascocarp is globular and closed it is known as a perithecium, if cup-shaped and wide open it is an apothecium.

The *Ascomycetes*, which include some 15 000 known species, form the largest class of Perfect Fungi and are divided as follows:

Plectomycetes: ascocarp absent or primitive and irregular with asci irregularly arranged
Discomycetes: asci in parallel series in apothecia
Pyrenomycetes: asci in parallel series in perithecia.

The *Plectomycetes* includes the yeasts and other fungi of industrial importance. Among the *Discomycetes* are plant parasites and saprophytic species; the large fungal growths found on tree-trunks, walls, etc., belong to this group. The *Pyrenomycetes* includes a number of plant pathogens, insect parasites and a few common moulds.

In relation to canned food spoilage, probably the most important of the *Ascomycetes* apart from the yeasts is the mould *Byssochlamys fulva*. The ascospores of this species are unusually heat-resistant.

Basidiomycetes

The characteristic spores from which the group derives its name are the basidiospores. The group includes the mushrooms, toadstools and parasitic rusts. These are not important in food microbiology and need not be considered here.

Fungi imperfecti

This class includes all the species in which sexual or 'perfect' sporulation is unkown and which reproduce by means of asexual spores. The great majority of the common moulds belong to the *Fungi Imperfecti* the number of which exceeds that of the *Ascomycetes*. Three main groups are recognised:

Sphaeropsidales: conidia borne within flask-shaped pycnidia
Melanconiales: conidiophores formed in a saucer-shaped depression in the surface of the substratum
Hyphomycetales: free conidiophores arising irregularly from mycelium.

The *Sphaeropsidales* includes many species parasitic on plants. The group *Melanconiales* is chiefly parasitic on plants and has no practical interest where canned foods are concerned.

The great majority of the saprophytic moulds which are of economic interest belong to the *Hyphomycetales*. Important genera associated with food spoilage include, *Aspergillus*, *Penicillium*, *Sporotrichum*, *Cladosporium*, *Alternaria*, *Fusarium*, *Oidium* and *Monilia*. Stolk & Dakin (1966) proposed the formation of a new genus, *Moniliella*, containing some species formerly classified as *Monilia* and a new species *Moniliella acetoabutans*. The latter exhibits a high tolerance to acetic acid, utilising the acid as a source of nutrient and causing the spoilage of products preserved by this means.

Many species of mould are of major industrial importance. The production of citric acid by *Aspergillus niger* and the ripening of cheeses by species of *Penicillium* are examples. *Penicillium notatum* is of interest as the original species observed to produce the antibiotic penicillin. Mutant strains of *P. chrysogenum* are now used for this purpose.

Yeasts

Yeasts cannot be clearly separated from the fungi which form a typical mycelium, but Henrici (1947) states that a fairly satisfactory definition of yeasts is that they are 'true fungi whose usual and dominant growth form is unicellular'.

A yeast cell consists of a permeable cell-wall or membrane enclosing a cytoplasm containing a nucleus and various granules. In young cells, the cell wall is thin and the cytoplasm relatively clear, but as the cell ages, thickening of the wall occurs and granules and vacuoles appear in the cytoplasm. The nucleus is not ordinarily visible except in stained cells. Yeast cells are mostly detached from each other except when actively growing and are of various shapes, the usual forms being round, oval or elliptical. In size they measure up to 8μm in diameter and 10μm in length.

Reproduction

Reproduction in yeasts is accomplished by budding, fission or spore formation. The normal method of vegetative multiplication in the majority of the industrially important yeasts is by budding. In this process a bulge or bud appears in the cell wall, and the bud, becoming constricted at its base, enlarges to approximately the same shape and size as the parent cell before finally becoming separated from it. Frequently, the bud cells give rise to new buds before they become detached from the parent cell so that a group of connected cells of varying sizes is formed. In multiplication by fission, the cell divides into two daughter cells, the process being similar to that occurring in bacteria.

In addition to these vegetative processes, reproduction may take place by means of spore formation and the method and type of spore formed divides the yeasts into three primary groupings. In the first group sexual reproduction occurs in which the cell becomes an ascus containing up to eight ascospores. Such yeasts are related to the *Ascomycetes* class of fungi and are known as ascosporogenous yeasts. The second group form ballistospores which are produced on small projections put out by the vegetative cell and discharged in a manner suggesting a relationship with the *Basidiomycetes* class of fungi. These yeasts are known as ballistosporogenous yeasts. The third group of yeasts produce neither ascospores nor ballistospores and are considered to be imperfect or asexual forms, and as such are classified with the *Fungi Imperfecti*.

The spores of yeasts are more resistant to unfavourable conditions than the vegetative cell, but as in the case of mould spores, the difference is not so marked as between bacteria and their spores.

Classification of yeasts

Differentiation of yeasts is principally based on their methods of reproduction, but this alone is insufficient basis for separation of the species and biochemical reactions, cultural characteristics and other features are of diagnostic importance. Details of the technical procedures and culture media required for the identification of yeasts are given by Smith (1969) and van der Walt (1970).

Although natural relationships can be postulated between certain yeasts to form small groups, insufficient information concerning morphology, physiology and genetics etc., plus in some instances the lack of clearly defined genera, make the establishment of a phylogenetic classification impossible at the present time. The classification proposed by Lodder (1970) is generally accepted, but it should be borne in mind that changes will be almost inevitable as further information becomes available.

Lodder recognises a total of 349 species of yeasts divided among 39 genera, which in turn are arranged into four major groups as follows:

Group I All the yeasts producing ascospores are placed in this group. They belong to the Order *Endomycetales* which is divided into four families:

- Family 1. *Dipodascaceae*: these are not considered as yeasts
- Family 2. *Endomycetaceae*: contains three genera, only one of which *Endomycopsis* is considered to be a yeast
- Family 3. *Saccharomycetaceae*: this is divided into four sub-families:
  - Subfamily 1. *Schizosaccharomycoideae* contains one genus, *Schizosaccharomyces*
  - Subfamily 2. *Nadsonioideae*: contains four genera, *Nadsonia*, *Saccharomycodes*, *Hanseniaspora*, and *Wickerhamia*
  - Subfamily 3. *Saccharomycoideae*: contains most of the species used in brewing and many spoilage organisms. There are at least 12 genera: *Saccharomyces*, *Kluyveromyces*, *Lodderomyces*, *Wingea*, *Pichia*, *Hansenula*, *Pachysolen*, *Citeromyces*, *Debaryomyces*, *Schwanniomyces*, *Dekkera* and *Saccharomycopsis*
  - Subfamily 4. *Lipomycetoideae:* contains only one genus, *Lipomyces*
- Family 4. *Spermophthoraceae*: contains seven genera, only three of which are regarded as yeasts, *Nematospora*, *Metschnikowia* and *Coccidiascus*

Group II Only two genera, *Leucosporidium* and *Rhodosporidium* are placed in this group. Both of these genera belong to the order *Ustilaginales* of the Class *Basidiomycetes*

Group III Contains one family, the *Sporobolomycetaceae*. This family comprises five genera, *Bullera, Sporobolomyces, Tilletiopsis, Itersonilia,* and *Sporidiobolus*. All the species form ballistospores which arise on sterigmata and are ejaculated by a drop-excretion mechanism

Group IV Contains 12 genera in which sexual reproduction has not been observed, and which do not form ballistospores. The genera are: *Brettanomyces, Candida, Cryptococcus, Kloeckera, Oosporidium, Pityrosporum, Rhodotorula, Schizoblastosporion, Sterigmatomyces, Torulopsis, Trichosporon* and *Trigonopsis*.

Important genera

Saccharomyces. This genus contains a number of important species of which *Saccharomyces cerevisiae* (Hansen) is the most notable. It has marked ability to produce alcohol and carbon dioxide in quantity from sugar and is used on a large scale in the brewing and baking industry. Reproduction takes place by budding and asexual sporulation. Other important yeasts in this genus are: *Saccharomyces cervisiae* var. *ellipsoideus* (commonly known as *S. ellipsoideus*). This yeast is used for the fermentation of wine. It commonly occurs on grapes and in the soil of vineyards. The cells are oval or pyriform and it has a higher alcohol tolerance but it is otherwise similar to *S. cerevisiae* (Hansen).

Saccharomyces pastorianus. This species is chiefly noted as a contaminant producing off flavours in beer and is a source of trouble in the brewing industry.

Other species of the genus are utilised in the production of various alcoholic liquors (cider, perry, etc.) and fermented milk liquors (Koumys, Kefer, etc.).

Saccharomyces rouxii and *Saccharomyces mellis.* These yeasts are important in food preservation because of their ability to develop in highly concentrated sugar products such as jams, honey, fondant cream, etc.

Saccharomyces bailii. This species which was formerly known as *Saccharomyces acidifaciens* also has a high tolerance to organic acids, such as acetic acid, which are used in food preservation. It is able to cause spoilage in such products as tomato ketchup, mayonnaise and various pickles.

Pichia and *Hansenula.* These genera are important as contaminants in the brewing industry. They form esters (ethyl acetate, amyl acetate, etc.) from sugar but do not produce alcohol; in alcoholic liquors most species utilise alcohol as a source of carbon and hence are responsible for serious spoilage of fermented liquors. When grown on liquids these yeasts form dry wrinkled pellicles. Such films are frequently found on the surface of pickle brines.

Debaryomyces. The species of this genus have a high salt tolerance and are commonly found in pickle brines.

Torulopsis. Species of this genus frequently occur as contaminants of food products. Some species are osmophilic and have been noted as spoilage agents in honey.

Brettanomyces. This genus contains yeasts of importance in the brewing industry where they are concerned in the 'after-fermentation' of beers.

Candida. Various species previously classified in the genus *Mycoderma*, and now considered to belong to a single variable species — *Candida mycoderma*, form dry, wrinkled pellicles on liquid media. Alcohol is used as a carbon source and these yeasts frequently give rise to spoilage of fermented beverages. They are also found on pickle brines. *Candida utilis*, previously known as *Torula utilis* and *C. tropicalis* grow readily in the process effluents from paper mills, distilleries, etc. and are used to produce animal fodder since they are rich in protein and vitamins. *C. lipolytica* and *C. intermedia* can be grown in an aqueous emulsion of crude petroleum; the process yields a yeast preparation which after solvent extraction, is of high quality. During their growth the yeasts remove normal alkanes in the oil which is a necessary and usually costly step in the refining of crude oil. They may also be used as an aid in the decomposition of oil slicks.

Rhodotorula. All the species produce red, pink, orange or yellow pigment of a carotinoid nature. Their fermentation abilities are low, and they are chiefly important as airborne contaminants, giving rise to the spoilage of food products. They have been noted as a cause of pink colouration of sauerkraut (Pederson & Kelly, 1938).

Biological characteristics of fungi

The fungi are devoid of chlorophyll and cannot synthesise their food solely from inorganic material. All are heterotrophic, therefore, and require an organic source of carbon for energy purposes and a few appear to require organic nitrogen. In addition to these and oxygen, other elements known to be essential in the nutrition of fungi are potassium, phosphorus, sulphur, magnesium and, in some cases, iron. Usually inorganic sources of these are utilisable. Moulds are distinguished for their ability to thrive on minute quantities of food materials and are frequently found growing in inorganic salt solutions and on fabrics, metal and glass surfaces, etc., where the amount of available nutrients is very low.

Fungi obtain their nutrients by absorption through the cell wall, as do the bacteria. They have a wide range of enzymic activity, carbohydrates of varying complexity, proteins, fats and organic acids being readily attacked by various species. The principal products of fermentation by fungi are non-volatile acids and alcohol. Carbon dioxide may be produced but not hydrogen or methane, and in this respect the fungi differ from bacteria, many species of which produce hydrogen, and less frequently, methane.

With regard to their oxygen requirements, the majority of the common moulds are strict aerobes but some species are capable of growth under markedly reduced oxygen tension. Many yeasts are facultative anaerobes, but according to Smith (1969), no truly anaerobic fungus exists.

The temperature relationships of the fungi vary widely. The optimum temperature for growth of most fungi lies between 25° and 30°C. Many species fail to develop above 30°C, but a few have their optimum above this temperature. The mould *Aspergillus fumigatus* can exist at temperatures above 50°C, although spore formation and optimum growth occur at 40°C. Castell (1944) has observed the growth of two *Mucor* species at 55°C. The optimum temperature for mycelial development does not necessarily coincide with that for other purposes, such as sexual spore production. The lower limit for growth varies but development of some species such as *Cladosporium herbarum* takes place at about -6°C.

The fungi also show considerable variation in their resistance to heat. Spores are generally slightly more resistant to heat than vegetative cells but are not nearly so resistant as bacterial spores. Some species are destroyed by long exposure to temperatures over 30°C and most species are destroyed by 30 mintues' heating at 65°C. An exception is the mould *Byssochlamys fulva*, which causes spoilage in canned fruit. The spores of this species survived heating for 30 minutes at 87° to 88°C. Williams et al (1941) have reported that the sclerotia of a *Penicillium* species required 300 minutes' exposure to 85°C for destruction.

REFERENCES

Booth C 1978 In: Norris J R, Richmond M H (eds) Essays in microbiology. Wiley, Chichester, ch 4

Castell C H 1944 Food Res 9: 410

Henrici A T 1947 Moulds, yeasts and actinomycetes. Chapman & Hall, London
Lodder J 1970 In: Lodder J (ed) The yeasts. North Holland Publ Co, Amsterdam, ch 1
Pederson C S & Kelly C D 1938 Food Res 3: 583
Smith G 1969 Introduction to industrial mycology, 6th edn. Arnold, London
Stolk A C, Dakin J C 1966 Antonie van Leeuwenhoek, 32: 399
van der Walt J P 1970 In: Lodder J (ed) The yeasts. North Holland Publ Co, Amsterdam, ch 2
Williams C C, Cameron E J, Williams O B 1941 Food Res 6: 69

3

Control of spoilage micro-organisms

Two general procedures are adopted in safeguarding food against spoilage by micro-organisms. The first, on which the canning operation is based, is sterilisation. This involves exposing the food to heat in order to destroy spoilage organisms, the food being protected against subsequent contamination by being enclosed in an airtight container. The second procedure aims at preservation by inhibiting the development of spoilage organisms and this may be achieved by various methods, the food being so treated that the activities of the organisms are inhibited or retarded. Preservation by such methods does not necessarily imply the destruction of the organism (i.e. a germicidal or fungicidal effect) and on removal or reduction of the inhibiting influence the food will undergo spoilage.

In reviewing the relationship of micro-organisms to canning it is necessary to give some consideration to the more frequently used methods of preservation other than heat, since food preserved by such methods is largely used as raw material for canning; furthermore, some canned and bottled products such as fruits, jams, sauces and pickles are preserved by the use of both heat and inhibitory substances. The chief methods which are applied commercially include refrigeration, gas storage, dehydration, filtration, pickling or curing, fermentation, smoking, irradiation and the addition of the so-called natural preservatives — sugar, salt, acid and spices — or of chemical preservatives such as sulphur dioxide and benzoic acid. Some of these methods are applied jointly, their effects being additive.

Refrigeration

Low temperatures preserve foods by retarding or preventing the growth of spoilage organisms and, in the case of fresh foods, by inhibiting the action of natural autolytic enzymes.

Ingraham & Stokes (1959) and Stokes (1960) drew attention to the fact that bacteria capable of development at low temperatures, in the past referred to as 'psychrophiles', usually grow best at temperatures between 20°C and 30°C and some at even higher temperatures.

Mossel (see Eddy, 1960) proposed that micro-organisms growing at below 5°C, but with higher optimum temperatures should be described as 'psychrotrophic'. This term has been adopted here to denote the micro-organisms growing at cold store temperatures and thus capable of causing spoilage in refrigerated foods. The bacteria primarily concerned belong to the genus *Pseudomonas* but others are found amongst the *Achromobacter, Flavobacterium, Micrococcus* and *Alcaligenes*. Yeasts of the *Torulopsis* type and mould species of the genera *Penicillium, Cladosporium, Mucor*, and *Thamnidium* are also involved. A number of psychrotrophic spore-formers have been isolated and identified. Larkin & Stokes (1967) investigated four new species of aerobic spore-formers capable of growth at refrigeration temperatures and named them *Bacillus psychrosaccharolyticus, B. insolitus, B. globisporus* and *B. psychrophilus*. These organisms had an optimum growth temperature of 20° to 25°C but grew well in 1 week at 0°C (Larkin & Stokes, 1968). A spore former named *B. macquariensis* growing at 0°C was isolated by Marshall & Ohye (1966). Psychrotrophic aerobic spore-formers are often isolated from raw milk and some strains assume importance as spoilage agents of pasteurised milk (Mourgues & Auclair, 1973; Grosskopf & Harper, 1974; Washam et al, 1977). Bhadsavle et al (1972) isolated strains of *Clostridium* from milk. These were identified as *C. carnis* and *C. hastiforme* both able to grow at 4°C. Psychrotrophs are often capable of comparatively rapid growth at 0°C, and although the rate of growth is slower than at higher temperatures, the total number of viable cells may be greater (Hess, 1934).

Michels & Visser (1976), because of their interest in chilled pasteurised foods, investigated the occurrence and heat resistance of psychrotrophic spore-formers in soils and in foods. They found a high number of psychrotrophic spores in soil. The heat resistance of these spores was generally low ($D_{90\,°C}$ from 1 to 11 mins).

The lower limit at which the growth of micro-organisms occurs in food is not determined by temperature alone, a very important factor being the amount of water frozen out of the substrate (Haines, 1934a; Ingram, 1951). When there is actual ice formation the growth of bacteria is retarded, whilst moulds and yeasts preponderate because they are better able to withstand the higher osmotic pressure which results from the concentration of solutes following the separation of the water as ice. For this same reason, the growth of bacteria on supercooled media occurs at a lower temperature than growth on frozen media. Jensen (1954) states that bacterial growth on supercooled media may occur at -7°C whereas the limiting temperature on frozen media is in the region of -3°C. Ingram (1951) has reported that organisms capable of resisting a high solute concentration may be exceptionally resistant to low temperature; he has observed the growth of halophilic bacteria on bacon and osmophilic yeasts in concentrated orange juice at temperatures down to −10°C. According to Haines (1934a), the limiting temperature for growth of the psychrotrophic group of organisms including bacteria, yeasts and moulds, lies between −5° and −10°C and is probably near −7°C. Lea (1931) found that storage at −5°C failed to prevent the development of yeasts and moulds on frozen meat, colonies appearing after 7 weeks. Berry & Magoon (1934) reported the growth of *Pseudomonas, Lactobacillus, Monilia* and *Penicillium* species at −4°C and *Cladosporium* and *Sporotrichum* species at −6.7°C. Data from numerous sources indicate that the minimum temperature for the growth of bacteria, in most foods, is about −10°C although enzymes within the cell may be active at lower temperatures.

Spoilage of fresh foods in cold stores takes place predominantly at the surface of the food. Thus considerable control over the extent of spoilage may be exercised by regulating the air flow and relative humidity within the chamber so that some surface drying occurs. A compromise between losses due to spoilage and weight loss due to evaporation is usually necessary.

The growth of most food-poisoning bacteria is prevented by storage at temperatures below 4.4°C. However, Schmidt et al (1961) showed that toxic outgrowth of several strains of heat shocked spores of Type E *Clostridium botulinum* took place in 31 to 45 days at an incubation temperature of 3.3°C. Admittedly high levels of inoculum (4 to 12 million spores per tube) were used but nevertheless the results indicated that a temperature of 3.3°C is not necessarily safe for extended storage of food products capable of supporting growth of *C. botulinum*; storage temperatures of 2.2°C and 1.1°C were satisfactory. Abrahamsson et al (1966) found that toxin was produced in chopped-meat medium and fish dialysate at temperatures between 3°C and 30°C but not at 1°C. In these experiments the inhibitory effect of salt was studied and found to be increasingly effective as the temperature was reduced. Segner et al (1966) found that 4 per cent sodium chloride, which had little effect at 30°C, led to long lag periods at 8°C and 4.5 per cent sodium chloride inhibited growth completely at 8°C but not at 16°C. Thus it seems clear that salt, in sufficient concentration, will raise the minimum growth temperature of *C. botulinum* type E. A reduction in the pH also inhibits growth at low temperatures. Emodi & Lechowich (1969) found that the strains of *C. botulinum* type E studied by them were inhibited at a temperature of 15.6°C if the pH was adjusted to 5.2. Angelotte et al (1961) carried out experiments with prepared food inoculated with salmonellae and staphylococci and concluded that the growth of both organisms was inhibited at temperatures below 5.6°C. The presence of salt raised the minimum growth temperature of salmonellae, for example the presence of 2 per cent raised the minimum temperature for *Salmonella heidelberg* to 8°C (Matches & Liston, 1972a). Matches & Liston (1972b) investigated also the effect of pH on the low temperature growth of three serotypes of salmonellae and found that growth only occurred over a narrow range of pH at low temperatures. They recorded the lowest temperature of growth at pH 7.0. Values of 5.2°C, 6.0°C and 9.0°C were obtained for *Salm. heidelberg, Salm. typhimurium*, and *Salm. derby* respectively. Michener & Elliot (1964) reviewed the literature and reported on the minimum growth temperatures for food poisoning and faecal indicator micro-organisms.

In recent years *Yersinia enterocolitica*, a human pathogen capable of growth at temperatures of 4°C, has assumed importance as a potential food poisoning organism. According to Hanna et al (1977a) the organism was readily destroyed by heat treatment and failed to survive an internal temperature 60° to 62°C in beef roasts. Extensive destruction during frozen storage on beef was reported by Hanna and colleagues although Leistner et al (1975) found only a slight decrease in numbers on chicken broilers after freezing for 90 days. There seems little doubt that it is capable of growth at 0° to 4°C and Hanna et al (1977b) reported growth on raw beef at temperatures as low as 0° to 1°C.

The act of freezing invariably results in a rapid initial decrease in the numbers of viable organisms. Depending on the temperature, the nature of the substrate, the type of organism and other factors, the survivors may then undergo further slow reduction or, in the case of psychrotrophic organisms, the initial decrease may be followed by a lag period and development of the survivors. Extremes of pH increase the susceptibility of micro-organisms to cold, whilst the presence of sugars, glycerol and colloids affords protection. These observations do not apply to bacterial spores which, according to Haines (1938), are practically unaffected by freezing or storage in the frozen state.

There is no general agreement as to the cause of bacterial death following cold treatment, the explanations advanced including mechanical damage by extracellular and intracellular ice crystals and alteration of cellular proteins. Evidence of metabolic injury, manifest as an increase in the nutritional requirements of the cell, has been presented by Straka & Stokes (1959). Various workers (Haines, 1934b; McFarlane, 1941; Sulzbacher, 1950) have noted that death of bacteria does not increase with decreasing temperatures once the substrate is frozen. According to Ray & Speck (1973) death during storage in the frozen condition is a function of time and temperature and may also be related to pH and solute ion concentration. For details of the various theories concerning bacterial destruction at low temperatures Rahn (1945), Weiser & Osterud (1945) and Ray & Speck (1973) should be consulted.

In general, micro-organisms are remarkably resistant to low temperatures, even some pathogenic types and faecal indicators surviving for long periods. *Escherichia coli* does not survive as well as the enterococci. Kereluk & Gunderson (1959), working with inoculated chicken gravy stored at -21.1°C, demonstrated that coliforms were reduced from 5 600 000 per ml to 66 per ml and enterococci from 15 000 000 per ml to 5 000 000 per ml in the same period. Ienistea & Pleceas (1967) also found that at -20°C, in contrast to other micro-organisms that decreased steadily in numbers, enterococci counts in samples of minced pork showed little change.

McCleskey & Christopher (1941) investigating pathogenic bacteria in 'quick-frozen' strawberries (-18°C) found that *Eberthella* (*Salmonella*) *typhosa* survived for 6 months and *Staphylococcus aureus* for 5 months. The survival of various pathogens in brain heart infusion broth, which had been frozen at -34.4°C and stored at -17.8°C was investigated by Raj & Liston (1961). After 393 days *Salmonella typhimurium* was virtually eliminated, *Escherichia coli* and *Staphylococcus aureus* were greatly reduced, *Streptococcus pyogenes* slightly reduced, and *Strep. bovis* and *Clostridium perfringens* little affected. In sterilised fish homogenate only *Salmonella typhimurium* and *Staphylococcus aureus* were significantly reduced. Zawadzki & Pogorzelska (1975) carried out studies of the effects of freezing, storage and defrosting on salmonellae inoculated into minced meat. The numbers decreased by 10.7 to 15.2 per cent after freezing, by 49.5 to 54.7 per cent after storage at -23°C for 10 days and by 88.1 to 91.2 per cent after 10 months.

A valuable review of factors affecting the growth of psychrotrophic micro-organisms in foods has been provided by Elliot & Michener (1965) and Ingram & Mackey (1976).

Gas storage

Considerable control of spoilage micro-organisms is attainable by modification of the atmosphere in which food materials are stored. Inhibition of growth of obligate aerobic organisms such as moulds may be achieved by storage under completely anaerobic conditions, but it is well known that some moulds are able to tolerate very low oxygen tensions and Miller & Golding (1949) have found that the moulds tested by them varied considerably in their oxygen requirements. The commercial procedures such as vacuum packaging and gas packing in which air is replaced by inert gas have great practical value in as much as the development of rancidity and other oxidative changes are prevented, but such packaging techniques cannot be relied upon to completely inhibit the growth of mould.

The introduction of gases into cold stores has been used to delay or prevent spoilage of fresh foods. For this purpose carbon dioxide has been most widely used although ozone, sulphur dioxide and nitrogen trichloride have been investigated and used experimentally.

According to Moran et al (1932) germination of mould spores is delayed by air containing as little as 4 per cent carbon dioxide. In 20 per cent carbon dioxide, the rate of growth is one-half to one-fifth that in air, the inhibition being more marked the lower the temperature. For the prevention of mould and bacterial growth on meat, Moran (1938) states that 40 per cent concentration is the ideal, but this amount has a deleterious effect on the quality of the meat (loss of bloom). With 20 per cent concentration the effect on the bloom of the meat is negligible over moderate storage periods while the growth of spoilage organisms is still retarded to a marked extent. In practice a concentration of 10 per cent carbon dioxide is used and under such conditions chilled beef remains free from microbial spoilage for 60 to 70 days. Moran (1938) states that the use of low concentrations of carbon dioxide also enables the storage life of chilled pork, mutton and lamb to be extended. Experimental work on the storage of eggs in carbon dioxide revealed a need for balancing good and bad effects. These are reviewed in Moran's paper.

The respiration and ripening of fruit can be retarded by controlling the carbon dioxide and oxygen contents of their atmosphere. This reduces decay since over-ripe fruit is liable to microbiological spoilage. In conjunction with low temperatures, gas storage of this type is known as controlled atmosphere (CA) storage and is used on vegetables and fruits, particularly apples and pears. The concentration of carbon dioxide required varies with the fruit and even for different varieties of the same fruit. To avoid abnormal (that is extreme anaerobic) respiration the oxygen concentration must not be allowed to fall below a certain critical level. The level is dependent on the time of exposure and temperature. In cool storage the critical level of oxygen in the atmosphere is about 2 per cent for apples and pears and probably for many other fruits. The carbon dioxide level is normally maintained at between 5 and 10 per cent. Most fruits reduce their respiration and rate of ripening as the oxygen tension is reduced but the response to carbon dioxide is variable. Thus Young et al (1962) demonstrated that the avocado, the banana and the lemon reacted differently to increased carbon dioxide. Respiration and ripening was reduced in the avocado; ripening was delayed but respiration not reduced in the banana; and respiration was actually increased in the lemon. According to Smith (1957) high concentrations of carbon dioxide (50 per cent) and temperatures of 4.4°C will suppress the fungal rotting of blackcurrants.

The recent use of liquid nitrogen as a refrigerant during transport has led to the interest in the effect of high nitrogen and low oxygen contents on fresh produce. In general, storage in 100 per cent nitrogen leads to damage but oxygen levels as low as 1 per cent are acceptable for most kinds of fruit and vegetable for short periods at temperatures of about 15°C, whilst at higher temperatures the oxygen level needs to be increased. The oxygen requirement for most fruit rotting fungi is very low and the growth of moulds in 1 per cent oxygen is not much different from that in air. However their growth is inhibited by carbon dioxide and some fruits such as cherries are remarkably tolerant to both high carbon dioxide and low oxygen levels. The subject of atmosphere control in storage and transport of fresh fruit and vegetables has been reviewed by Hall (1968).

There are reports in the literature of the use of ozone as a preservative. There are obvious objections to its use the foremost and most obvious being that it has been found to be a toxic agent and exposure of mammals to critical concentrations may lead to haemorrhaging and oedema of the lungs. Furthermore, since it is a powerful oxidising agent it gives rise to rancidity of fatty foods making it unsuitable for storage of meat, bacon, butter and similar products. Thus it would appear that ozone has limited applications in food preservation.

Moisture limitation

Under this heading, dehydration and the addition of solutes such as sugar may be considered jointly, since both methods of preservation are dependent upon the reduction of available moisture to a level where the development of micro-organisms is prevented. The availability of water has been defined by Scott (1957) as $a_w = p/p_o$ where p is the vapour pressure of the solution and p_o is the vapour pressure of the solvent (water). This is conveniently expressed in terms of the relative humidity of the atmosphere in equilibrium with the product (R.H.). This is numerically equal to a_w.

The level of a_w tolerated by micro-organisms varies with the properties of the cells concerned. Differences in cellular response to osmotic effects makes it possible to draw up tables relating the a_w to microbial growth. Such tables are often seen in the literature. However, this is a somewhat simplistic approach and may even be misleading since other factors such as pH, Eh, types of solutes used and the presence or absence of antimicrobial substances in the food have an effect as do extrinsic factors such as temperature of storage and composition of the headspace gas. In general terms the most sensitive bacteria are the Gram-negative rods (minima for growth in the a_w range 0.94 to 0.96) followed by *Clostridium* spp. (minima in the a_w range of 0.93 to 0.95) and then the *Bacillus* spp. (minima in the a_w range 0.90 to 0.93). Apart from the halophilic bacteria the Gram-positive cocci show the greatest tolerances to moisture limitation with reported minima at about 0.85. Yeasts and moulds, in general, grow at lower levels of a_w than do the bacteria. Those growing under the driest conditions have been termed osmophilic yeasts and xerophilic moulds respectively. Pitt (1975) suggested that arbitrarily the term xerophilic might be used for all species of fungi capable of growth at a water activity below 0.85.

The adjustment of a_w may be achieved either by water binding through addition of a solute or mixture of solutes or by removal of water which effectively concentrates the solutes already present. In practice, a combination of the two methods is often used to produce the so-called intermediate moisture foods. At the same molecular concentrations the water binding capacity decreases in the order (1) ionic substances such as sodium or potassium salts, (2) low molecular weight non-ionic compounds such as sugar and polyols, (3) high molecular weight compounds containing ionic groups such as proteins and peptides and (4) high molecular weight compounds containing polar groups such as starch, dextrins, pectins, hemicellulose and cellulose.

The nature of the solute has an effect on the pattern of growth of micro-organisms. Marshall et al (1971) compared the glycerol and sodium chloride tolerances of 16 species of non-halophilic bacteria at similar levels of a_w and found that most were more tolerant to the effects of reduced a_w using glycerol than when using salt. On the other hand, some, notably the cocci and salt tolerant bacteria, showed the reverse effect. Christian (1978), discussing specific solute effects, points out that either the minimum a_w for growth may be affected by the solute choice, or, in systems where other factors, none by themselves totally inhibiting, are involved, then the effect on growth rate may be important.

The water activity of the substrates also has a profound effect on the heat resistance of micro-organisms, even the enhanced thermal resistance of bacteria entrapped in fat or oil being capable of explanation by this means. The effect on heat resistance is dealt with in Chapter 8.

With the exception of the osmophilic yeasts, which are a special problem, moulds are able to cause spoilage of foods with lower values of a_w than other micro-organisms. Tomkins (1929) studied the germination of several types of mould spores at various temperatures and values of a_w. The results showed that a reduction in a_w caused a fall in the rate of germination, and for each mould there was an optimum temperature at which the range of a_w permitting germination was greatest. Heintzeler (1939) and Snow (1949) have studied the germination of various mould spores at different values of a_w. These workers have noted that, at the limiting minimum value of a_w for each fungus studied, the period required for germination may be very long. Several members of the *Aspergillus glaucus* group may germinate within a few weeks at 25°C at a_w values of 0.73 to 0.75 but at lower values of a_w germination may be considerably delayed.

Barton-Wright & Tomkins (1940) state that the critical humidity for mould growth in flour is 75 per cent. Their experiments show that the critical humidity increases as the temperature decreases. At 20°C mould was inhibited at 79 per cent R.H. (moisture 16 per cent). At 15°C 82.5 per cent R.H. (moisture 16.5 per cent) was inhibitory, and at 5°C 85 per cent R.H. (moisture 17.4 per cent) sufficed. These workers give 72 per cent R.H. as the lowest point at which mould growth has been noted on any material. Macara's (1943) experiments with dried meat indicate that the minimum humidity for mould growth on this material is slightly less than 75 per cent. Block (1953), who tested many materials for their liability to mould growth, found that at a relative humidity of 76 per cent cheese developed only slight mould growth after 1 year's storage. He concluded that the water-absorbing properties of the substance play an important part in determining the limiting humidity at which mould growth occurs, and postulated that, except at 100 per cent relative humidity, fungi are not capable of obtaining moisture for mycelial development direct from the atmosphere. The obligate xerophilic mould *Xeromyces bisporus* is reported to grow at a_w values as low as 0.6 (Pitt & Christian, 1968). Dallyn & Everton (1969) cite spoilage outbreaks due to its growth on table jellies, dried fruit and chocolate sauce.

Factors such as the presence of toxic substances, the pH of the media, and the nutritional value of the product for moulds can be expected to influence the limiting humidity but it may be said that foods in which the relative humidity lies below 70 per cent are generally resistant to mould spoilage. Peas, cereals and similar dry goods should, therefore, be dehydrated until the humidity of the atmosphere in equilibrium with the products is below this limit. Similarly, in sugar preserves the solutes must be present in sufficient concentration to reduce the water activity to the inhibitory level. It should be noted that fluctuating temperatures during storage can affect the liability to mould growth of products packed in sealed containers since sudden cooling may cause temporary localised condensation or humidities in excess of the equilibrium value for the pack.

For equal concentrations, the osmotic pressure of sugars in solution is highest with those sugars of low molecular weight. Since the water activity of a solution decreases with increase in osmotic pressure, monosaccharides such as dextrose and fructose are more effective in reducing a_w than sucrose. Thus, a preserve containing 65 per cent sugar as sucrose will be more liable to mould attack than a similar product with 65 per cent total sugar of which a proportion is present as invert sugar. Erickson & Fabian (1942) investigated the preserving action of various sugars and found that for bacteria the order of effectiveness was fructose > dextrose > sucrose > lactose. Thermophiles were more susceptible to the action of sugars than streptococcal species. For yeasts, fructose and dextrose were equally effective in 5 to 15 per cent lower concentration than sucrose. Fabian & Graham (1953) give the order of effectiveness of sugars on flat-sour thermophiles as dextrose > fructose > sucrose. Tarkow et al (1942) reported that in equal concentrations glucose was markedly more inhibitory than sucrose for yeast and mould. Mixtures of equal parts of the sugars were inhibitory to a degree intermediate to that of either sugar alone.

The osmophilic yeasts are able to tolerate very high concentrations of sugar and cause spoilage in honey, chocolate centres, jam, molasses and similar products in which the sugar content may be as high as 80 per cent. According to the nomenclature now generally accepted osmophilic yeasts are usually placed in *Saccharomyces* but sometimes in *Hansenula, Debaryomyces, Pichia* and *Citeromyces*. Many tolerant strains have been named as *Saccharomyces rouxii*, for example that isolated by Mossel (1951) which spoiled a sugared fruit product with an a_w of 0.81 and that studied by English (1953) for which the minimum a_w for fermentation was about 0.73. According to Pouncy & Summers (1939), confectionery products with a relative humidity of less than 69 per cent are resistant to spoilage by osmophilic yeasts. These authors devised an ingenious method of estimating the relative humidity of confectionery by means of the deliquescing of various crystals when exposed to air in equilibrium with the material. With material poor in protein the critical humidity for fermentation was considerably higher than that in protein-rich material. They point out that when products have a humidity above critical point this can in many cases be effectively lowered by the addition of 10 per cent invert sugar. Money & Born (1951) tabulate the equilibrium humidities of various sugar solutions and give an empirical formula by means of which the equilibrium humidity of jams, fondant cream, boiled sweets, etc., can be calculated. For further information on the role of osmophilic yeasts in food spoilage, publications by von Schelhorn (1951) and Tilbury (1976) should be consulted.

With the majority of canned foods, preservation by control of the moisture content is not possible. There are, however, some canned and glass-packed products for which moisture control is applied. These include cereals such as oatmeal, semolina and such sugar products as jam, candied fruit peel, confectionery and sweetened condensed milk. This latter product is invariably unsterile but the organisms normally present are unable to develop. With some types of jams and marmalade (as in some American products) where the sugar content is low (that is about 60 per cent) it is necessary to give heat treatment to prevent spoilage. Canned bread and canned chocolate nut roll are unsterile products in which the moisture content has a bearing on suppression of toxin production by *Clostridium botulinum.* Wagenaar & Dack (1954) found a limiting water content of 36 per cent in bread with a pH of 5.8. In chocolate nut roll the corresponding value was 39.75 per cent (Wagenaar & Dack, 1960).

The limiting a_w for germination and growth of clostridial spores is of particular significance to the canning microbiologist. Unfortunately most studies have been made in laboratory media and the results may not be directly applicable to foods. However, it appears that germination of the spores occurs at a lower value than growth of the vegetative cell, the latter being the important value for *Clostridium botulinum* since no significant toxin is produced until the cells multiply (Roberts & Smart, 1976). The nature of the added solute is also very important, glycerol being less effective than salt. Baird-Parker & Freame (1967) found that germination of *C. botulinum* spores types A, B and E occurred at a lower a_w with glycerol (minimum of 0.89) than with salt (minimum 0.93). The minimum values of a_w for vegetative growth were 0.93 and 0.96 respectively. Type E was generally more sensitive to water limitation than types A and B. Ohye et al (1967) studied the growth rates of four strains of *C. botulinum* type E in a basal N.Y.G. medium adjusted to varving a_w values with either sodium chloride or a mixture of sodium chloride, potassium chloride and sodium sulphate. They found a limiting value of 0.96 for the most tolerant strain (at 30°C). Emodi & Lechowich (1969) studied inhibition of *C. botulinum* type E strains by moisture limitation using various solutes. With sodium chloride they report a limiting value of 0.975. For types A and B Ohye & Christian (1967) report minimum values supporting growth as 0.95 for type A and 0.94 for type B the adjustments in a_w having been made by addition of sodium chloride, potassium chloride and sodium sulphate mixture.

Kang et al (1969) studied the growth of *Clostridium perfringens* in thioglycollate medium adjusted with glycerol, sodium chloride and sucrose and reported limiting growth values of 0.93, 0.95 and 0.97 respectively. Roberts & Smart (1976) discussed the control of clostridia by water activity and related factors, and concluded that there was a complex interaction of many factors including pH, incubation temperature and a_w and type of solute used.

The water relations of food-borne bacterial pathogens has been adequately reviewed by Troller (1973) and a useful summary of the limiting a_w for growth of food poisoning bacteria in complex media adjusted with various solutes has been given by Christian (1978). This is reproduced in Table 1. Of special significance is the position of

Table 1 Limiting a_w values for growth of food-poisoning bacteria in complex media adjusted with various solutes (courtesy of J.H.B. Christian)

Species	Solute	Limiting a_w	Species	Solute	Limiting a_w
Bacillus cereus	NaCl	0.94[4]	*Clostridium*	NaCl	0.97[5]
	Glucose	0.95[4]	*perfringens*	Sucrose	0.97[5]
	Glycerol	0.92[4]		Glycerol	0.95[5]
Clostridium			*Salmonella*	Salts	0.95[3]
botulinum			*oranienburg*	Sucrose	0.95[3]
type A	NaCl	0.96[1]		Glycerol	0.94[6]
	Salts	0.95[7]			
	Glycerol	0.92[1]	*Staphylococcus*	Salts	0.86[8]
type B	NaCl	0.96[1]	*aureus*	Sucrose	0.88[8]
	Salts	0.94[7]		Glycerol	0.89[6]
	Glycerol	0.92[1]			
type E	NaCl	0.97[1]	*Vibrio*	NaCl	0.95[2]
	Salts	0.97[7]	*parahaemolyticus*	Sucrose	0.96[2]
	Glycerol	0.94[1]		Glycerol	0.94[2]

References

1 Baird-Parker & Freame (1967)
2 Beuchat (1974)
3 Christian & Scott (1953)
4 Jakobsen & Murrell (1977)
5 Kang et al (1969)
6 Marshall et al (1971)
7 Ohye & Christian (1967)
8 Scott (1953)

Staphylococcus aureus which holds the distinction of being the most tolerant of the pathogens to restriction of moisture with a limiting a_w for growth at 0.86 in a salts solution or a sucrose plus salts solution (Scott, 1953). The minimal a_w for enterotoxin A formation was between 0.864 and 0.867 at 30°C and between 0.870 and 0.887 at 25°C (Lotter & Leistner, 1978).

Salt

The action of salt in preserving food is not fully understood, but it does not appear to act solely by virtue of its osmotic effect. According to Spiegelberg (1944), the osmotic pressure at which growth of bacteria ceases is much lower for salts than for sugars. The concentration of salt necessary to inhibit the growth of micro-organisms in food is related to many factors, including pH, temperature, protein content and presence of inhibitory substances such as acids. The water content is obviously of major importance, since it is the concentration of salt in the water phase and not the amount in the food as a whole which is significant. According to Labrie & Gibbons (1937), the inhibitory action of salt on bacteria increases with decreasing temperature from 21° to 10°C. Callow (1929) also noted that the amount of salt required to inhibit mould growth decreased with decreasing temperature, 8 per cent sufficing at 0°C, whereas 12 per cent was necessary at room temperature. A similar effect has been observed with sporing bacteria; for example, the maximum levels of salt permitting growth of *Clostridium botulinum* type E are 5.8 per cent at 30°C, 5.1 per cent at 20°C and 4.3 per cent at 15°C (Ohye et al, 1967). The influence of the composition of the medium on the salt tolerance of micro-organisms has been repeatedly demonstrated. Garrard & Lochhead (1939) observed that organisms exhibited greater salt-tolerance in curing pickle than in broths of similar salt content. Stuart (1940) showed that growth of halophilic bacteria may be either stimulated or retarded by varying the protein content of the medium. The effect of pH on salt tolerance was studied by Joslyn & Cruess (1929), who found that lowering the pH caused a marked decrease in the salt tolerance of various yeast and mould species.

Micro-organisms have been divided into halophiles and non-halophiles based upon their response to salt concentration. However, such a demarcation may be misleading since many intermediate types exist and there are reports of so-called obligate halophiles growing on low salt media (Stuart & James, 1938). Stuart (1938) has also shown that, contrary to the general view, halophilic bacteria are not exclusively associated with salt environment such as solar salt, sea-water and fish. They are in fact widely distributed in nature and can be isolated in 25 per cent salt media from non-salt environments including stagnant water, sulphur springs, rat-dung and soil, provided an incubation period up to 90 days is permitted.

The extreme halophiles, that is those requiring above 12 per cent of sodium chloride for growth, have been classified as *Halobacteriaceae* and comprise those micro-organisms now generally regarded as true obligate halophiles.

Flannery (1956) has suggested a classification, based on tolerance to 2 per cent salt solution, as follows:

1. Non-halophiles
   a. *Salt-sensitive*. Grow only in media containing less than 2 per cent salt
   b. *Salt-tolerant*. Grow best in media containing less than 2 per cent salt, but will grow in media containing more than 2 per cent salt
2. Halophiles
   a. *Facultative*. Will grow in media containing less than 2 per cent salt, but will grow best in media containing more than 2 per cent salt
   b. *Obligate*. Grow only in media containing more than 2 per cent salt.

From the wide diversity of the types reported in the literature, it is evident that a typical halophilic flora does not exist, many organisms of a widely varying morphological and biochemical character being represented. Micro-organisms show such variation in their ability to tolerate salt that growth of one species or another occurs throughout the range up to saturation. Pathogenic organisms generally are more susceptible to the action of strong salt solution than are the saprophytic types and rod forms are usually more sensitive than cocci. Tanner & Evans (1933) reported that growth of *Clostridium botulinum* was prevented by 6.5 to 12 per cent salt, the critical concentration depending on the medium. Oxhøj (1943) states that *C. perfringens* and *C. sporogenes* are suppressed by 5.7 to 7.4 per cent salt, the amount varying with the medium. In canned cured meat, for example, Greenberg et al (1959) state that the salt level should reach 9 per cent to inhibit toxin production. Wagenaar & Dack (1958) showed that growth and toxin production by *C. botulinum*, Type A, in a surface-ripened cheese were not inhibited until the salt concentration was above 8 per cent; Type B spores tolerated a maximum concentration of 6 per cent. Abrahamsson et al (1966) working with Type E, demonstrated that the salt concentration necessary to inhibit toxin production varied with the incubation temperature but that no toxin was detectable within 3 months at 30°C in Robertson's chopped meat medium containing 4.5 per cent sodium chloride. Boyd & Southcott (1971) investigated the salt tolerance of various strains of *C. botulinum* Type E in heat processed cod homogenates. No toxin was produced by the Minnesota strain in 3.7 per cent salt although outgrowth and cell multiplication was not inhibited; toxin of low potency was produced in 4.4 per cent by the Tennessee strain and 3.84 per cent by the Saratoga strain. A brine concentration of 5.5 per cent was sporostatic to the Saratoga and Tennessee strains but not to the Minnesota strain. Iandolo & Ordal (1963) found that the growth rate of *Staphylococcus aureus* was stimulated by addition of sodium chloride up to 4 per cent but at greater concentrations the growth rate and total population was reduced. Nunheimer & Fabian (1940) state that sodium chloride in 15 to 20 per cent concentration prevents the growth of some strains of food-poisoning staphylococci, while 20 to 25 per cent has a definite lethal effect. Salt is capable of limiting the growth of salmonellae in foods the levels required being temperature dependent. For example, Matches & Liston (1972a) found that three serotypes were able to increase in numbers in concentrations of 5 to 8 per cent salt at 22°C but at 12°C growth occured only in the 0 to 4 per cent range.

Livingstone (quoted by Buchanan & Fulmer, 1930) pointed out that the spherical form presents the least surface for water exchange and is, therefore, advantageous in concentrated solution, and it is interesting to note that, as a group, micrococci generally exhibit a high salt-tolerance, many types developing freely in the presence of 25 per cent salt.

Many of the bacterial species growing in strong salt solutions are chromogenic and cause spoilage by producing discoloration on salted fish and hides. Such spoilage has been reported by Robertson (1931) and others. A non-sporing anaerobic rod isolated and described by Baumgartner (1937) developed in media saturated with salt. This organism is the cause of gaseous spoilage in unsterilised salted fish products such as pastes and fish sauce. It can be effectively controlled by reducing the pH of such products to below 5.5.

Film-forming yeasts examined by Mrak & Bonar (1939) grew in brines containing 24 per cent salt. Yeasts of this type grow on the surface of pickled vegetable brines and by oxidising the lactic acid produced during fermentation of the vegetables, impair their keeping quality. Moulds may also show this undesirable activity. According to Tanner (1944), mould growth may occur in the presence of 20 to 30 per cent salt. Under anaerobic conditions, in moderate salt concentrations, lactobacilli usually predominate lowering the pH and rendering the product less likely to spoilage by undesirable micro-organisms. This condition exists in the desirable and normal fermentation of brined vegetables such as in the production of sauerkraut.

In connection with the brining of meats, Jensen (1954) has called attention to the fact that many organisms can tolerate high salt concentrations in brines containing large pieces of meat; growth appears to take place at the interfaces of brine and tissues and is very slow in the clear brine. At present, little is known concerning such interfacial growth.

A review of the role of salt as a preservative for foods has been made by Ingram & Kitchell (1967).

Acid preservatives

The chemical preservatives most widely used in foods as antimicrobial agents are the benzoates (as sodium benzoate), sorbates (as sorbic acid or potassium sorbate), and the propionates (as sodium or calcium propionate). In addition acetic and lactic acids are widely used for preservation of acid foods such as pickles and sauces.

The effect of acids in preventing the development of micro-organisms may be due to the hydrogen ion concentration or to the toxicity of the undissociated molecule or the anion. With mineral acids the toxic effect is related to the hydrogen ion concentration, but the toxicity of the organic acids is disproportionate to the degree of dissociation and is due mainly to the undissociated molecule or the anion.

The yeasts and moulds are very much less susceptible to the effect of high hydrogen ion concentration than bacteria. Most species of bacteria have their optimum pH in the region of neutrality and are unable to develop below about pH 4.5. Among the most acid-tolerant bacteria are those of the *Lactobacillus* and *Clostridium butyricum* groups growing down to about pH 3.5, whereas moulds and yeasts which develop best in the slightly acid range (pH 5.0 to 6.0) may tolerate pH 2.0 or even lower.

In the preservation of foods, acetic and lactic acids are those most commonly employed. Fabian & Wadsworth (1939) found acetic acid to be a better preservative than lactic acid for pickles, and Levine & Fellers (1940a) have also reported that acetic is more toxic than lactic acid for bacteria, yeast and mould species. When the reaction of the medium was adjusted with acetic acid, bacteria were inhibited at pH 4.9, *Saccharomyces cerevisiae* at pH 3.9 and *Aspergillus niger* at pH 4.1, the titratable acidities being 0.04 per cent, 0.59 per cent and 0.27 per cent respectively. These acidities relate to the inhibition of a few species in laboratory media. In commercial practice higher concentrations of acetic acid (for example, up to 3.6 per cent calculated on the aqueous phase) are used to prevent spoilage in products such as sauces and pickles. In a later report, Levine & Fellers (1940b) noted that the addition of 5 per cent salt or 20 per cent sugar did not appreciably reduce the amount of acid required to prevent growth. In non-toxic concentration acetic acid stimulated the growth of the mould by serving as a source of energy. Erickson & Fabian (1942) found that when based on pH values, the order of preserving and germicidal power for bacteria was acetic > citric > lactic acid. When judged by the amount of acid, the order was lactic > acetic > citric. For yeasts, the order was acetic > lactic > citric, irrespective of whether it was based on pH value or concentration of acid. They also observed that a combination of sugar with the preserving quantity of acid rendered the mixture germicidal. Working with flat-sour thermophiles Fabian & Graham (1953) concluded that the order of germicidal action of acids at pH 5.5 was citric > acetic > lactic.

Nunheimer & Fabian (1940) found that the amount of dextrose required to exert a germicidal action on staphylococcal strains could be reduced by 50 per cent when used in the presence of half the inhibitory concentration of acid. Salt and sucrose could be reduced only by 30 and 20 per cent respectively to bring about a germicidal effect. Shillinglaw & Levine (1943) examined the germicidal effects of a number of edible acids, their work having reference to the possibility of water-borne diseases being carried by carbonated beverages. In a concentration of 0.02 N (the approximate strength used in many beverages) the order of effectiveness for destroying *Escherichia coli* at 30°C was tartaric > glycolic > phosphoric > lactic > acetic > citric. Temperature coefficients for rates of destruction varied with the different acids with the result that the order of effectiveness at 30°C (tartaric > phosphoric > lactic > citric) became phosphoric > lactic > tartaric > citric at 0.6°C. The toxicity of 0.02 N lactic and citric acids was increased by the addition of 10 per cent sucrose or 2.5 volumes of carbon dioxide. Mosell (1963), in a review of the survival of salmonellae in foods, states that a pH value of about 4.5 limits growth. He also reproduces a table from earlier work showing decimal reduction times of enterobacteriacae in various foods, containing different organic acids, at lower pH values. Bell & Etchells (1952) investigated the acetic acid tolerance of spoilage yeasts isolated from commercially prepared sweet pickles and found that the addition of sugar or sodium benzoate reduced the amount of acetic acid required for preservation. Bell and Etchells provide a chart from which it is possible to predict from the sugar and acid content whether a given pickle will support the growth of spoilage yeasts.

The fungistatic action of the fatty acids was studied by Hoffman et al (1939). Over the pH range of 2 to 8 many of these acids were effective in preventing mould growth. Below pH 5.0 acetic acid was very effective, the amount required decreasing with pH. At pH 2.0, less than 0.04 M acetic acid was sufficient, whilst at pH 5.0 between 0.08 M and 0.12 M concentration was necessary. At the same pH propionic acid was effective in lower concentrations than acetic acid and was active up to pH 6.0 to 7.0.

The survival of staphylococci in acid media was studied by Campanini et al (1976). With hydrochloric, citric, lactic or acetic acid the rate of inactivation increased with a fall in pH from 3.5 to 2.5 and increased with increase in the pK of the acid. Minor & Marth (1972) found that acetic, lactic and phosphoric acids were more active against *Staphylococcus aureus* than was hydrochloric acid.

Propionic acid and its salts have been recommended for prevention of spoilage in foodstuffs. They have a limited spectrum of activity being effective against moulds but with limited activity against bacteria and practically no effect on yeasts. Its value as a mould inhibitor in flour products is summarised by Seiler (1964). Olsen & Mady (1940) reported that propionic acid inhibited surface mould growth on butter. The acid was more effective than the calcium salt and the latter was more effective than the sodium salt. The pH of the substrate was an important factor. Calcium propionate in 0.1 to 0.5 per cent concentration has been found to be effective in preventing mould growth on fruit jellies, glazed jellies and similar products. Propionate is used in bread doughs as a bacterial spore inhibitor, particularly those of *Bacillus mesentericus* (*subtilis*) the causative organism in rope formation. The UK regulations permit the inclusion of propionic acid (or the appropriate salt) in bread and certain flour confectionery (Ministry of Agriculture, Fisheries and Food, 1972).

Benzoic and sulphurous acids. These acids and their salts are widely used preservatives; both are permitted in specified foods in the United Kingdom. They are weak acids mainly effective in the undissociated form as shown by Rahn & Conn (1944) and Scardovi (cited by Ingram, 1959). Von Schelhorn (1953) introduced the term 'absolute efficacy' to describe the efficiency of the undissociated fraction. A comparison of the 'absolute efficacy' of various preservatives showed undissociated sulphurous acid to be 100 to 500 times more effective than undissociated benzoic acid against the organisms studied by von Schelhorn. Since the effective unit is the undissociated molecule it follows that the lower the pH the greater the preservative action. For this reason benzoates and sulphites are used at pH values of less than 4.0. Sulphurous acid suffers from the disadvantage that it enters into combination with aldehydic sugars in which form it is inactive. In addition a large proportion is normally in the undissociated form. Green (1976) has reviewed the use of sulphur dioxide as a food preservative.

Sorbic acid. Gooding (1945) first drew attention to the fungistatic properties of sorbic acid and many later investigations have confirmed these findings. Wright (1961) states that sorbic acid has an inhibitory action on catalase positive bacteria, moulds and yeasts but catalase negative bacteria such as lactobacilli and clostridia are usually unaffected. Wallhäuser & Lück (1972) state that the antibacterial activity of sorbic acid is not of such a universal character as its antifungal activity although in some cases it was significant. Tompkin et al (1974) investigated the use of sorbic acid (0.1 per cent w/w) against various pathogens associated with cooked uncured sausage. Growth of salmonellae was markedly retarded. Growth of *Staphylococcus aureus* was delayed by one day but total growth eventually reached the same level as the non-sorbate controls. Growth of *Clostridium botulinum* and formation of toxin was retarded but not prevented. Although more effective in the acid range sorbic acid is extremely useful in mildly acid conditions at pH values close to 5.0, a region in which benzoic acid becomes ineffective. Sorbic acid is of very low toxicity and is metabolised to water and carbon dioxide in the same way as the natural fatty acid analogue caproic acid (Barnes, 1959). A number of investigations have demonstrated the value of sorbic acid for inhibiting mould spoilage in cheese (e.g. Smith & Rollin, 1954). Its value as a fungistatic agent in bakery goods has also been established (Melnick et al, 1956; Ottaway, 1956) and it has been advocated for inhibition of scum yeasts in cucumber fermentations (Phillips & Mundt, 1950). Its efficiency as a preservative in acetic acid preserves has been assessed by Dakin & Crawford (1966). *Saccharomyces bailii*, *Pichia membranaefaciens*, *Candida krusei* and *Torulopsis holmii* have been found to be particularly resistant to acid preservatives (Pitt & Richardson, 1973). Pitt (1974) found that in a medium containing 10 per cent glucose with a pH of 3.5, *S. bailii* is not inhibited by 600mg per kg of either sorbic or benzoic acid, or 2 per cent acetic acid.

The antimicrobial spectrum, pH dependence and solubility of permitted preservatives is summarised in Table 2 reproduced from a paper by Jarvis & Burke (1976). The same paper gives useful information on the legal status of the use of preservatives in various countries.

The action of preservatives is influenced to a marked degree by a number of factors, a full discussion of which is outside the scope of this book, but a brief statement may indicate their practical importance.

Table 2 Antimicrobial spectrum, pH dependence and solubility of commonly permitted perservatives. Reproduced by permission of the Society of Applied Bacteriology from Jarvis & Burke (1976)

Preservative	Spectrum[3]	pH Dependence	Solubility in water g/100 at 25°C
Sorbates	S,B,b,*Y*,*M*	±	139 (potassium salt)
Parabens	S,*B*,b,*Y*,*M*	—	0.25-0.015[1]
Diacetate	S,B,b,Y,M	±	100 (sodium salt)
Sulphites	*B*,*b*,Y,M	±	V.S.
Benzoates	*Y*,M,(B)	+	50 (sodium salt)
Propionates	S,*M*	+	150 (sodium salt)
Nisin	S,B	±	[1,2]
Nitrite	*S*,*B*,*b*	±	66 (sodium salt)

1 Solubility depends on length of alkyl chain.
2 Very soluble in dilute acid; solubility in water improved by admixture with carrier proteins (e.g. skim milk powder).
3 S, spores; B, Gram positive bacteria; b, Gram negative bacteria; Y, yeasts; M, moulds; ( ), some strains; Italics, major effects.

The activity of a preservative is primarily dependent on its concentration. If present in adequate concentration the effect on micro-organisms is lethal. In lower concentration, inhibition of growth but not death results, while in very low concentration the toxic effect is lost completely and development of the organisms may even be stimulated. The degree of dilution required to produce these changes in activity varies with each preservative so that if two different preservatives are diluted in the same proportion the effect on their toxicity may be quite different. A numerical expression, the concentration coefficient, is used to denote the effect of dilution on the activity of a preservative.

Temperature has an important effect on the activity of preservatives. In general, toxicity is markedly increased by rise in temperature, the increase varying with different substances. The degree to which toxicity is increased by a given rise of temperature is expressed as the temperature coefficient. In addition to the effect of temperature on the activity of the preservative, the effect on the organisms has also to be considered. Where the concentration of preservative is only sufficient to cause inhibition of growth the stimulatory effect on the organisms of a slight rise in temperature may outweigh the effect on the activity of the preservative. At temperatures in excess of the maximum for growth, however, very small amounts of preservatives may exert a pronounced lethal effect.

The type of organism and the number present are factors requiring consideration. As with other harmful agencies, the spores of micro-organisms are more resistant to toxic chemicals than are vegetative cells. A given preservative cannot be relied upon to be equally effective against all types of organisms, even different strains of the same species varying considerably in their resistance to the same preservative. The number of cells present may influence the efficacy of a preservative and a concentration which is adequate to control a slight infection may be inadequate to deal with a large number of organisms. In this connection it is clearly necessary to keep contamination of preserved foods to a minimum.

In addition to these factors the nature of the food to which the preservative is added is very important. As mentioned earlier the hydrogen ion concentration has a marked effect, the toxicity of the weak acid preservatives being increased in acid medium. In the presence of organic matter the action of most preservatives is retarded, sulphur dioxide particularly since it enters into combination with sugar and other substances to form inactive compounds.

A final cautionary note on the use of preservatives concerns the possibility of induced development of resistance in micro-organisms to certain preservatives. Many strains of osmophilic yeasts initially sensitive to benzoate and sorbate develop resistance to these preservatives (Harman et al, 1974).

For further information concerning the preservation of foods by chemicals the reader may consult Ingram (1959), Kimble (1977) and Jarvis & Burke (1976).

The curing of meat

In addition to imparting desirable colour and flavour, the curing of meat has an appreciable preservative effect. The reactions underlying the formation of the characteristic red colour of cooked cured meats are the combination of muscle pigment myoglobin with nitric oxide to form nitric oxide myoglobin which is converted by heat to the more stable compound, nitric oxide haemochromogen. The source of the nitric oxide is the nitrite present in the curing solution or pickle. In the presence of ascorbate, as little as 30 p.p.m. of nitrite is sufficient to impart the characteristic colour to cured meat, but rather more, about 50 p.p.m. is required to give a characteristic flavour. Details of the chemistry of the curing process are given by Fox (1966) and Bard & Townsend (1971).

The traditional method of curing meat is to use a salt/nitrate brine containing from 20 to 28 per cent salt and sodium or potassium nitrate at about one-tenth the weight of salt. It is customary to inject the pickle into the meat by pumping in order to introduce the salt into the meat as rapidly and as evenly as possible. After pumping the meat is immersed in the pickle in which salt tolerant bacteria develop and convert nitrate to nitrite. A wide variety of organisms are present and, in order to control the activities of spoilage types, the curing process is carried out at a low temperature — usually about 5°C.

It has been known for many years that the nitrite is the essential component in producing the colour and flavour in cured meat, but because of the good safety record for these products, there was a reluctance to replace nitrate by nitrite. Recently, however, it has been shown that nitrates give little inhibition to clostridia at the concentrations present in cured meats and many curers have stopped using nitrate and now add nitrite direct. The trend away from nitrate usage is clearly shown in a survey of American products reported by Binkerd & Kolari (1975). Other additives such as ascorbic acid and polyphosphates are sometimes used. Phosphates do not appear to reduce the effectiveness of nitrite and indeed Jarvis et al (1977) have shown that some phosphates may enhance the preservative effect of curing salts in meat systems in respect of growth and toxin production by *Clostridium botulinum*. Regarding the addition of ascorbic acid the evidence is conflicting. It is known to diminish the formation of nitrosamines but whether it reduces the antimicrobial effect of nitrite is less clear. In unheated systems there are indications that it has little effect (Baird-Parker & Baillie, 1974; Crowther et al, 1976). Bowen et al (1974) also reported that sodium ascorbate does not appreciably alter the inhibitory action of sodium nitrite against toxin production by *C. botulinum* in heated vacuum-packed wieners with nitrite levels above 50 p.p.m. In a further paper Bowen & Deibel (1974) report, however, that levels of ascorbate above 500 p.p.m. decreased the efficacy of nitrite in inhibiting botulinal growth in cured bacon. Tompkin et al (1978) report that the addition of sodium isoascorbate to perishable canned cured meat appreciably increased the efficacy of nitrite against *C. botulinum* and in a later paper suggest that this may be attributed to its ability to retard repair of nitrite injured cells by withholding an essential cation.

Investigations into the reasons for the preservative action of curing salts have been made by numerous workers but in many cases these studies have been carried out on model systems and the results may not be directly applicable to practical situations. Undoubtedly the effectiveness of the curing salts is influenced by other factors, particularly the pH of the meat, the storage temperature and the number and types of contaminating micro-organisms.

It is now generally accepted that in mildly acid conditions, as in cured meat, the nitrite component accounts for a large part of the preservative action of curing salts. Tarr (1941a, 1941b, 1942) showed that the inhibitory effect was dependent on the pH of the medium and increased significantly when the pH dropped below 6.0. It has been suggested that the inhibition is due to the presence of undissociated nitrous acid. For a unit drop in pH, Castellani & Niven (1955) demonstrated a 10-fold diminution in the concentration of nitrite required to inhibit the growth of *Staphylococcus aureus*. Similar results were obtained by Eddy & Ingram (1956) using a species of *Bacillus* and by Perigo et al (1967) using vegetative *Clostridium sporogenes* P.A. 3679. According to Tarr, 200 p.p.m. of nitrite in media at a pH of 6.0 is inhibitory to the genera *Achromobacter, Aerobacter, Escherichia, Flavobacterium, Micrococcus* and *Pseudomonas*. Some bacteria are more resistant, notably *Salmonella* (Castellani & Niven, 1955); *Lactobacillus* (Spencer, 1971) and *Clostridium perfringens* (Perigo & Roberts, 1968). It appears that nitrite gives maximum inhibition at a pH of about 5.6 since at greater acidities it disappears too rapidly to be useful (Ingram, 1959).

Sodium chloride itself is capable of acting as an important bacteriostatic agent (see section on Salt p. 40). Spores are sensitised to the inhibitory action of salt by heat, for example, according to Jarvis et al (1976) heating at 85°, 90° or 95°C sensitised spores of *Clostridium botulinum* type B to 2.5 per cent and 3.5 per cent sodium chloride. Heating to 70° or 80°C required higher concentrations of 4.5 per cent and 5.5 per cent. In the context of cured meat its effect is best considered in conjunction with the nitrite content with which there is an interactive preservative effect. This has been demonstrated by Wood & Evans (1973) and Baird-Parker & Baillie (1974). Roberts & Ingram (1973) have shown triple interactive effects between salt, nitrite and pH.

There are other important factors affecting the preservative action of curing salts. According to Ingram (1974) the work of Roberts' group at the ARC Meat Research Institute in the UK has shown that the minimum inhibitory concentrations are markedly affected by the number of vegetative cells when their concentration fell below 100 per gram. Also the importance of incubation temperature is emphasised; the minimum inhibitory concentration being reduced at lower incubation temperatures. Henry et al (1954) noticed the importance of redox potential and it has been confirmed by Castellani & Niven (1955) and Eddy & Ingram (1956) that nitrite is more effective under anaerobic conditions.

The foregoing applies mainly to unheated systems. In considering canned foods the other major factor to be considered is the heat treatment. Perigo et al (1967) working with laboratory media, found the ability of nitrite to inhibit the vegetative growth of clostridia was enhanced if it was first autoclaved with the medium. They suggested the formation of an unidentified inhibitory substance at least 10 times more inhibitory than nitrite alone. The rate at which this substance is formed is greatest at about 110°C. At higher temperatures the substance appears to decompose or combine with other ingredients so that its inhibitory activity declines. This unidentified substance has become known as the Perigo Factor or Inhibitor (PF). Roberts & Garcia (1973) reported that an inhibitor of *Clostridium* spp., formed by heating sodium nitrite in a laboratory medium, also inhibited 9 out of 14 strains of *Bacillus* spp. but that *B. circulans, B. polymyxa, B. macerans, B. pantothenticus* and *Bacillus F* were resistant.

In heated systems most investigations have been concerned with the inhibitory effects on bacterial spores. Yesair & Cameron (1942) investigated the effect of curing salts on the growth and heat resistance of *Clostridium botulinum*. They showed that in meat-infusion agar more than 70 per cent reduction in spore germination was produced by 0.1 per cent sodium nitrate, 0.005 per cent sodium nitrite or 2 per cent salt. They suggested that the quantities used in commercial practice may result in complete inhibition. They also demonstrated an apparent reduction in the heat resistance of *C. botulinum* when heated in cured meat, but this effect was found to be due to the growth-inhibiting action of the curing salts. When heat-treated cured meats were cultured so that the inhibitory salts were diluted by the culture medium, the heat resistance proved to be unaffected.

Stumbo et al (1945a) investigated the effect of meat-curing agents on the thermal resistance of a putrefactive anaerobe in meat and concluded that the curing materials did not influence the heating processes necessary to sterilise the meat. They subsequently (1945b) examined the effect of curing agents on the growth of the organism in heat-processed meat and found salt (3.5 per cent) to be the principal inhibitory factor. Sodium nitrate (0.17 per cent) and sodium nitrite (0.015 per cent) were ineffective in preventing spoilage of the meat although sodium nitrite did appreciably delay spore germination. Salt and sodium nitrate, salt and sodium nitrite and a combination of all three proved little, if any, more effective than salt alone. These authors point out that some of the apparently contradictory views concerning the inhibitory effects of meat curing agents may be due to differences in the media in which the agents have been tested. Silliker et al (1958), investigated the effect of individual curing ingredients on the shelf stability of canned comminuted meats and stated that 'a combination of 78 p.p.m. of sodium nitrite, 3.5 per cent brine and a mild cook results in sufficient injury to the contaminating flora to prevent subsequent growth in the product'. Riemann (1963) refers to unpublished work in which the effect of heat processing and curing salts on putrefactive anaerobic spores was tested in a factorial experiment. The factors concerned were salt, nitrite, nitrate, pH and F_0 values and all were significant except nitrate.

Other workers have demonstrated significant interactions between factors. Pivnick et al (1969) showed that toxinogenesis from *Clostridium botulinum* types A and B was affected by the concentration of salt, the concentration of nitrite, the amount of thermal processing and the extent of contamination. Grever (1974) worked with canned sausage emulsions containing 3.5 per cent salt, 100 p.p.m. and 200 p.p.m. of sodium nitrite and processed to $F_0 = 0.05$ and $F_0 = 0.5$. At the lower heat process clostridia were not inhibited but at the higher value 100 p.p.m. of nitrite was sufficient provided that the salt content was 3.5 per cent. Grever, like Stumbo et al (1945b), remarks on the importance of salt as an inhibitor for heat treated spores. Spencer (1966) reviewing earlier work emphasised that the supplementary effect of heating, of the order of $F_0 = 0.1$ to 1.0 (*see* Ch. 9), is critical and that a low incidence of contaminating spores is necessary if spoilage is to be prevented.

Johnston et al (1969) considered four possibilities to account for the inhibitory properties of nitrite in cooked cured meats:

1. Enhanced destruction of spores by heat
2. Increased rate of germination of spores during the process, followed by death of the germinated spores during heating
3. Prevention of development of spores which have survived the heat process
4. Production of an inhibitory substance from reactions occurring during the process, e.g. Perigo Factor.

The first two possibilities have been discredited. The third possibility, namely inhibition of spore development, offers the most plausible explanation and it is supported by the work of Roberts & Ingram (1966), Duncan & Foster (1968) and Nordin et al (1975).

Johnston et al (1969) concluded that a Perigo inhibitor was not produced when meat was added to the medium, thus casting doubts on the significance of the Perigo Factor in the preservation of canned cured meat. However, other workers have reported the formation of inhibitory compounds during the heating of meat systems. Ashworth & Spencer (1972) worked with minced pork using *Clostridium sporogenes* and *C. botulinum* as inocula. Ashworth et al (1973) used a model pork system, simulating pasteurised hams, and challenged it with *C. sporogenes* P.A. 3679. Chang et al (1974) and Chang & Akhtar (1974) used commercially formulated pork luncheon meat processed at 110°C to an $F_0 = 0.4$. Using *C. botulinum* the processed luncheon meat was inoculated with washed spores. Even if the nitrite had been depleted before inoculation took place, inhibition was observed, the extent being greater if higher levels of nitrite had been present initially. However, when the formation of inhibitors in meat systems has been observed, the effect does not correspond in intensity to the Perigo Factor produced in culture media.

It is evident that the preservative action of curing salts is a complex matter but there can be no doubt that the nitrite component is of paramount importance. Unfortunately, because of the fear of the formation of carcinogenic nitrosamines, and recently the implication of nitrite itself, there is inevitable pressure for the reduction, or even removal of nitrite from canned meats. In view of the fact that the presence of *Clostridium botulinum* spores can be demonstrated in pork (Roberts & Ingram, 1976 quote a survey carried out in the UK showing about two cells per kilogram, a figure considerably higher than that found by American workers) this could result in growth and toxin production in canned cured meats leading to cases of botulism. There have been a number of inoculated pack studies designed to show the result of reducing nitrite levels on the formation of toxin production from *Clostridium botulinum* (Greenberg, 1972, 1973) but it is dangerous to draw general conclusions. Roberts & Ingram (1976) doubt whether it is possible to legislate for nitrite alone in view of the many factors involved.

Smoking

The process of smoking meat and fish is carried out after pickling or brining by exposing the products to the smoke of smouldering wood dust. Generally, hard woods such as oak, ash and elm are preferred for producing smoke, soft resinous woods being unsuitable since they contain volatile substances causing unpleasant flavours in the food. The smoking operation is usually carried out by suspending the food directly over the smouldering wood or by producing smoke in one chamber and blowing it by means of ducts and fans into another containing the food. The process needs careful control for the production of high-quality foodstuffs.

In addition to imparting desirable flavour, the smoking process has a marked preservative action which is in part due to the absorption of bactericidal substances from the smoke. Jensen (1954) stated that the view of the packing industry was that smoking had a definite preservative action, the effect being produced by aldehydes, phenols and aliphatic acids. During the operation, the surface layers of the product become impregnated with these bactericidal constituents of smoke and non-spore forming bacteria are very largely destroyed. Subsequent contaminants are to some extent controlled by the residual preservative action of the absorbed bactericidal substances, but the presence of salt and the dehydration which occurs during smoking also contribute to the keeping properties of smoked products. The mycostatic action of wood smoke constituents is not very pronounced, and smoked foods are more susceptible to mould than to bacterial spoilage. Kersken (1974) studied the fungistatic action of smoke on six species of mould frequently found on cured meat products, namely *Cladosporium herbarum, Penicillium chrysogenum, Mucor hiemalis, Aspergillus repens, A. amstelodami* and *A. chevalieri*. The fungistatic effect increased with the temperature of smoke production and varied with the species. The most sensitive was *C. herbarum* and the most resistant the three species of *Aspergillus*. Shewan (1949), discussing the smoking of fish, states that the pH of the superficial layers has been found to fall from 6.7 to about 5.9 during smoking. This is thought to be the result of absorption of acid smoke constituents and has the effect of increasing the susceptibility of micro-organisms on the fish to the bactericidal agents in the smoke. Gibbons et al (1954) investigated the bactericidal effects of smoking on bacon. The temperature reached in the smoke-house had an important effect on the bactericidal action of the smoking process; variation in humidity had little effect. The combined action of heavy smoke and high temperature — 60°C — reduced the number of bacteria 100 000-fold. Heiszler et al (1972) studied the effect of smoking temperature on the micro-organisms present on frankfurters and found, in general, that survival or growth of organisms during subsequent storage at 5°C was inversely related to the temperature or time of smoking.

Experiments carried out by Handford & Gibbs (1964) using a liquid smoke preparation, showed that lactic bacteria were more resistant than catalase positive cocci. Thus smoking tended to change the balance in the flora, results which are in agreement with the conclusions of Eddy & Ingram (1962) following examination of various commercial bacon samples.

The survival of salmonellae and *Clostridium botulinum* type E in 'hot' smoked fish was considered by a committee of the National Research Council (1975). It was concluded that in properly smoked fish non-sporing organisms should be largely eliminated at the temperature reached and any risk of salmonellosis would arise primarily from cross infection through raw fish. The spores of *C. botulinum* type E would survive the heat treatment but the spores would not normally germinate and produce toxin due to the salt content and the low temperature at which these products are kept (below 3.3°C). Failure to achieve the correct salt content combined with inadequate refrigeration could lead to a risk of botulism since the fish are normally eaten without further cooking.

Jensen's review provides a full summary of the literature and deals with the chemical and bacteriological aspects of the subject. Detailed information on methods of smoking are given by Jones (1942).

Spices

The preservative action of some of the spices and herbs used for flavouring foods is well established and, as Potter (1939) has remarked, the efficiency of spice oils as preservatives is often greater than that of some chemical preservatives.

In all cases the inhibitory or toxic effect of spices and herbs is due to the essential oils. Most workers have come to the conclusion that cloves, cinnamon and mustard exert a greater preservative action than other spices and herbs. Corran & Edgar (1933), who cite earlier work on the subject, investigated the action of various spices, herbs and their oils on yeast (*Saccharomyces cerevisiae*). They found that brown mustard flour was the most potent, followed by cloves and cinnamon. Cardamom, cummin, coriander, caraway, celery seed, pimento, nutmeg, ginger, marjoram and other spices and herbs had little or no effect. Similarly, volatile oil of mustard was found to have a greater preservative action than the essential oils of other spices and herbs. Volatile oil of mustard in a concentration of 200 parts per million or 5000 parts per million of brown mustard flour was more potent than sulphur dioxide and benzoic acid in concentrations of 350 and 600 parts per million respectively. Fabian et al (1939), using a number of bacterial species as test organisms, concluded that there was considerable variation in the resistance of different organisms to the same spice and of the same organism to different spices. Their results showed that ground cinnamon and cloves were the only spices which were inhibitory in low concentration for bacteria. Ground peppercorn and allspice were inhibitory in 1 per cent concentration, and mustard, mace, nutmeg and ginger in 5 per cent concentration. 0.1 per cent of a 50 per cent emulsion of oil of mustard gave some inhibition, while 1 per cent of a 50 per cent emulsion was completely inhibitory. Martinez et al (1973) carried out a detailed survey of the antimicrobial activity of the essential oils obtained from bay leaf, pimento berry and leaf and cloves. They found that these oils were effective against Gram-positive and Gram-negative bacteria, acid-fast bacteria and fungi.

The antimicrobial properties of onion and garlic have been observed and recorded for many years. The activity of onion is attributable to volatile aliphatic disulphides such as n-propyl allyl disulphide and di-n-propyl disulphide. The effect of these constituents present in reconstituted dehydrated onion was investigated by Johnson & Vaughn (1969) who showed that both onion and garlic exhibited activity towards *Salmonella typhimurium* and *Escherichia coli*. Garlic was more effective than onion against the latter organism. Drechsel (1965), investigating the sensitivity of organisms in salami to garlic found that the antimicrobial action was selective, the desirable organisms being resistant whereas organisms such as *Pseudomonas aeruginosa* were very sensitive.

Blum & Fabian (1943) examined a number of spice oils and their components for controlling microbial surface growth. The organisms tested were *Saccharomyces ellipsoideus, Saccharomyces cerevisiae, Mycoderma vini* and *Acetobacter aceti*. Results showed that the organisms varied in their resistance to the spices. Oil of mustard was, however, consistently the most germicidal, followed by cinnamon, cassia and clove. The decreasing order of toxicity of the spice components was found to be allyl isothiocyanate and carvacrol (equal), cinnamic aldehyde and cinnamyl acetate (equal), eugenol methyl ether and eucalyptol. Germicidal efficiency of the spice oils was not correlated with surface tension, and Blum and Fabian suggest that toxicity is due to chemical rather than physical causes.

Webb & Tanner (1945) found spice oils, because of greater concentration of active principle, to be more effective than whole or ground spices in preventing yeast growth in laboratory media. Oils of cinnamon, mustard, cloves, allspice, bay, wintergreen and peppermint in 0.1 per cent concentration retarded and in most cases completely inhibited growth. In concentrations over 1 per cent, mustard, cinnamon and clove oils were germicidal to yeasts in spice oil — dextrose agar plates. Tested by the cup-plate method, oils of allspice, almond and bay were also germicidal to yeasts. Anise, lemon and onion oils were classified as bacteriostatic. Anderson et al (1953) have tested a number of essential oils for inhibition of growth of acid food spoilage organisms (bacteria and yeasts) in dextrose broth. Mustard, garlic, onion and cinnamon oils were the most effective. In acidified broth, the inhibitory action of most of the oils was increased, but for one strain of yeast the amount of oil necessary to inhibit growth was as great or greater in the acidified broth than in broth at pH 7.2. According to Scheibe (1958), allicin, the active principle of garlic, is inhibitory to Gram-negative and Gram-positive bacteria at dilutions of 1:125 000 but not at dilutions of 1:250 000.

These and other investigations indicate that the preservative action of some spices may be of practical importance, but it must be recognised that the amounts which can be used in foods are often strictly limited by flavour considerations. The literature concerned with the antimicrobial action of essential oils has been reviewed by Koedam (1977). The effect of spice oils on the thermal resistance of food spoilage organisms is referred to in Chapter 8.

Antibiotics

Since the early experiments of Tarr and his associates on the use of antibiotics for the preservation of fish, numerous investigations have been made with various fresh foods. The tetracyclines, particularly chlortetracycline (Aureomycin) and oxytetracycline (Terramycin), have been used to lengthen the storage life of poultry and fish. At one time they were widely used in the USA and Canada for poultry and fish and in the UK for extending the life of fresh fish. The tolerance for antibiotic residues permitted under the US Food and Drug Administration has now been revoked. The tetracyclines are generally described as 'broad-spectrum' bacteriostats. They are effective against both Gram-positive and Gram-negative bacteria of diverse types; they are not effective against yeasts and moulds. Since residues are destroyed during cooking (Meredith et al, 1965) they have been considered acceptable in fresh foods subsequently heated. The main objection to their use is the possibility of inducing resistance in micro-organisms against substances used clinically in human or animal medicine (Jarvis & Morisetti, 1969).

Preservation of poultry. Investigations by workers such as Ayres et al (1956), Wells et al (1957) and McVicker et al (1958) have established that chlortetracycline (CTC) is superior to other antibiotics tested for the preservation of poultry. Kohler et al (1955), Shrimpton (1957), Walker & Taylor (1960) and many others have demonstrated that the storage life of eviscerated whole chickens or cut-up chickens is extended by an immersion treatment in water or slush ice containing 10 p.p.m. of this antibiotic. Abbey et al (1960), reporting on the effectiveness after 5 years of continued commercial use, state that treatment of this type gives, on average, a 75 per cent extension of shelf life. It has been shown by Wells et al (1957) and Shrimpton (1957), in comparative studies with untreated controls, that to be of practical value the treated birds must be held under refrigeration at a temperature between 0°C and 3°C. Although bacterial spoilage may be retarded, CTC is not effective against yeasts or moulds. Furthermore the existence of resistant strains of bacteria or their development (Vaughn et al, 1960) leads to hazards of selective bacterial growth during storage (Thatcher & Loit, 1961) and so imposes limitations on the usefulness of antibiotic treatments.

Preservation of fish and shellfish. A number of antibiotics have been tested for this purpose but the tetracyclines, particularly chlortetracycline and chloramphenicol (Southcott et al, 1958) and oxytetracycline (Lennon, 1972), have proved to be the most effective.

Various methods have been used to treat fish, but because best results are obtained by treatment as soon as possible after capture there has been special interest in application by means of the ice or chilled sea water in which the fish are held. Boyd et al (1953) found that ice containing 1 p.p.m. of CTC brought about a marked improvement in the keeping quality. This preservative effect has been demonstrated for salmon (Gillespie et al, 1955; Stern et al, 1958), white fish such as haddock and cod (Shewan, 1956), pacific round herring (Tomiyama et al, 1956) and numerous other types (Firman et al, 1956). Farber (1954) and Castell & Greenough (1958) observed that dipping of fish fillets in water containing chlortetracycline improved their keeping properties.

Studies on the preservation of crabmeat, shrimp and oysters with antibiotics have been made in several laboratories but sometimes show conflicting results. For example, Abbey et al (1957) found that an immersion of freshly shucked oysters in a CTC solution extended the storage life, but Benarde & Littleford (1957) reported that whilst the bacterial count of crabmeat and oysters is decreased the useful shelf life is not extended. Fieger et al (1956) found that ice containing 10 p.p.m. of CTC added a few days to the storage life of shrimp, and this is in agreement with the work of Farber (1959) and his colleagues, who showed that a five-minute dip in a 15 p.p.m. solution prolonged the freshness of both peeled and unpeeled shrimp, the peeled raw shrimp benefiting to a greater extent. Lerke & Farber (1957) screened 19 antibiotics for their ability to prolong the storage life of Mexican shrimps. The shelf life was extended by 4 days if they were first dipped in a 15 p.p.m. solution of chlortetracycline and held at 6°C. Continued use of the antibiotic led to the growth of resistant strains. It has been suggested that antibiotic resistant strains may be sensitive to nitrofurans (Paul et al, 1952). In a study of the microbial contamination of shrimps, Waters & Hamdy (1970) investigated the combined use of nitrofurans and chlortetracycline and found some combinations to be effective. Generally the results of the various investigations show retardation of bacterial spoilage if the concentration of antibiotic is sufficiently high but sometimes when judged organoleptically the results are disappointing.

It can be concluded from the many observations that the

tetracyclines, particularly chlortetracycline, will delay spoilage of fish and fish fillets, the increase in storage life depending on the initial freshness of the fish and the storage temperature. The use of antibiotics for extending the storage life of shellfish appears to be of less practical value.

Preservation of meat. Goldberg et al (1953) found that for ground fresh beef the most effective antibiotics were chlortetracycline, oxytetracycline and chloramphenicol. Weiser et al (1953), by infusing beef rounds with CTC, brought about a significant improvement in keeping quality at room and chill-room temperatures. Jay et al (1957a, 1957b, 1958) showed that spoilage organisms are often more sensitive to low concentrations of CTC when tested in meat than when tested in laboratory media. The development of a resistant bacterial flora which occurs in treated meat was shown to be due to a selective increase of the resistant strains.

Ordal & Brown (1957) reported an extension of about 10 days in the storage life of ground lean pork when it is treated with 10 p.p.m. of oxytetracycline combined with 3 per cent salt. The use of oxytetracycline in curing solutions for ham also resulted in a longer storage life.

The use of antibiotics for treatment of meat has been purely experimental and we cannot find any instance where their commercial use for this purpose is permitted.

Practical and public health aspects. Factors of public health significance which have to be taken into account if the widespread use of antibiotics is contemplated include their oral toxicity and the possibility of the development of a resistant pathogenic bacterial flora in the stored food. According to Hines (1956), who reviewed the effect of long-term administration of CTC, there is no evidence that oral administration at the levels normally used is harmful. In addition, it has been shown by Stern et al (1958) and Lerke & Farber (1960) that CTC used for preserving fish intended for canning is inactivated during the processing. The possibility of the selective development, during storage, of CTC resistant spoilage organisms, particularly psychrotrophic bacteria and yeasts, has been established by Thatcher & Loit (1961) although there was no evidence of additional hazard due to *Staphylococci*, *Enterococci*, *Coliforms* or pathogenic yeasts. The numbers of *Salmonellae* actually decreased in the presence of CTC. Boyd et al (1956) found that vegetative cells of *Clostridium botulinum* Type E tended to decrease in minced fish treated with CTC and held at 10°C; spores of the same organism were not affected.

In conclusion it may be stated that certain antibiotics, such as chlortetracycline, used in conjuction with storage at temperatures between 0°C and 3°C delay the spoilage of fresh fish and poultry. Chlortetracycline is not effective against yeasts and moulds and resistant strains of bacteria exist or may develop.

Some antibiotics known to inhibit fungal micro-organisms were evaluated by Klis et al (1959), and pimaricin (Myprozine) was found to be the most effective of those tested. In the paper by Ayres et al (1956) it was reported as being effective in controlling yeasts on dressed poultry when used in combination with CTC. For further information the reader is referred to Clark et al (1964). Nisin (*see also* Ch. 8) has proved effective for the control of *Clostridium* species in cheese and its use for this purpose is permitted in the UK. For further information on this subject the reader is referred to reviews by Farber (1959), Wrenshall (1959), Tarr (1960) and Kimble (1977).

Fermentation

Vegetable products used in the manufacture of sauerkraut and pickles are preserved by brining and fermentation. Micro-organisms play an important role in the fermentation of vegetables such as cabbage, olives and cucumbers. The vegetables are placed in brine, usually of about 5 to 10 per cent concentration, or lower in the case of sauerkraut, which suppresses Gram-negative bacteria, and allowed to undergo spontaneous lactic fermentation. The natural sugars are converted to lactic acid, which, in conjunction with the depletion of carbohydrates, exerts a preservative action. The process should be carried out in vessels designed to minimise contact with the atmosphere since the development of film-forming yeasts may lead to colonisation of the surface by undesirable micro-organisms. A full discussion of the process is outside the scope of this book, but for further information the reader is referred to Cruess (1958), Pederson (1960), Etchells et al (1975) and Vaughn (1975).

Certain types of sausages such as Salami, Thuringer and Cervelat undergo a lactic fermentation for flavour development. The process leads to a fall in pH and a reduction in the water holding capacity of the meat which has a certain preservative effect, particularly in conjunction with mild heating or smoking to which they may be subsequently subjected. Various organisms have been suggested as starter cultures, including lactobacilli and *Pediococcus cerevisiae*. For further information on this aspect Niinivaara et al (1964) should be consulted.

Air filtration

The mechanical removal of micro-organisms by filtration is the commonly used method of sterilising air for use in aseptic processing plants. The most common type of filter for handling large volumes of air is a fibrous depth type such as the ME filter element available from Dominick Hunter Engineers Limited. This element uses glass micro-fibres with a mean diameter of 0.5 μm laid down in a random manner. The gas passes through the matrix and micro-organisms are trapped both on the surface and in the depth of the filter. Whilst the actual voids may be comparatively large, the mechanism is such that its efficiency in removing small particles is high. Dwyer (1966) indicates that such filters with an effective diameter of 5 to 100 μm will remove more than 99 per cent of sub-micron particles. The ME filter element is claimed to remove particles down to 0.01 μm in size. The element is supported by inner and outer expanded stainless steel cylinders, the composite cylindrical section being permanently dip-sealed with silicone rubber into end caps. The assembly is held in a stainless steel housing and connected to the system so that it may be pre-sterilised by passing steam through it at a temperature of 121°C for 20 minutes.

REFERENCES

Abbey A, Kohler A R, Hopper P E 1960 Food Tech 14: 609
Abbey A, Kohler A R, Upham S D 1957 Food Tech 11: 265
Abrahamsson K, Gullmar B, Molin N 1966 Can Journal Microbiol 12 (2): 382
Anderson E E, Esselen Jr, W B, Handleman A R 1953 Food Res 18: 40
Angellote R, Foter M J, Lewis K H 1961 Amer J publ Hlth 51:76
Ashworth J, Hargreaves L L, Jarvis B 1973 J Food Technol 8:477
Ashworth J, Spencer R 1972 J Food Technol 7: 111
Ayres J C, Walker H W, Fanelli M J, King A W, Thomas F 1956 Food Tech 10: 563
Baird-Parker A C, Baillie M A H 1974 In: Krol B, Tinbergen B J (eds) Proc 1st Int Symp on Nitrites in Meat Products. Centre for Agr Pub and Doc, Wageningen, Netherlands
Baird-Parker A C, Freame B 1967 J appl Bact 30: 420
Bard J, Townsend C T 1971 In: Price J F, Schweigert B S (eds) The science of meat and meat products, 2nd edn. Freeman, San Francisco
Barnes J M 1959 Chem Ind p 557
Barton-Wright E C, Tomkins R G 1940 Cereal Chem 17: 332
Baumgartner J G 1937 Food Res 2: 321
Beauchat L R 1974 Appl Microbiol 27: 1075
Bell T A, Etchells J L 1952 Food Tech 6: 468
Benarde M A, Littleford R A 1957 Appl Microbiol 5: 368
Berry J A, Magoon C A 1934 Phytopath 24: 780
Bhadsavle C H, Shehata T E, Collins E B 1972 Appl Microbiol 24: 699
Binkerd E F, Kolari O E 1975 Food Cosmet Toxicol 13: 655
Block, S S 1953 Appl Microbiol 1: 287
Blum H B, Fabian F W 1943 Fruit Prod J 22: 326, 347
Bowen V G, Deibel R H 1974 Proc Meat Ind Res Conf. American Meat Science Association in cooperation with the American Meat Institute Research Foundation, p 63
Bowen V G, Cerveny J G, Deibel R H 1974 Appl Microbiol 27: 605
Boyd J W, Southcott B A 1971 J Fisheries Res Board Canada 28 (8): 1071
Boyd J, Brumwell C, Tarr H L A 1953 Prog Rep Pacific Coast Stns Fish Res Bd Canada No 96, 25
Boyd J W, Bluhm H M, Muirhead C R, Tarr H L A 1956 Amer J publ Hlth 46: 1531

Buchanan R E, Fulmer E I 1930 Physiology and biochemistry of bacteria Vol 2, Baillière, Tindall & Cox Ltd, London
Callow E H 1929 Food Invn Bd Ann Rept p 72. HMSO, London
Campanini M, Casolari A, Gola S 1976 Industria Conserve 51 (2): 88
Castell C H, Greenough M F 1958 Canad Fisherm 45 No 11: 20
Castellani A G, Niven Jr C F 1955 Appl Microbiol 3: 154
Chang P C, Akhtar S M 1974 Canad Inst Food Sci and Technol J 7 (2): 117
Chang P C, Akhtar S M, Burke T, Pivnick H 1974 Canad Inst Food Sci and Technol J 7 (3): 209
Christian J H B 1978 Proc 2nd Internat Symp on Properties of Water in Relation to Food Quality and Stability. in press
Christian J H B, Scott W J 1953 Aust J Biol Sci 6: 565
Clark W L, Shirk R J, Kline E F 1964 Pimaricin, a new food fungistat. In Microbial inhibitors in food, Proc of the 4th Int Symp on Food Microbiology, Göteborg. Almqvist and Wiksell, Uppsala
Corran J W, Edgar S H 1933 J Soc Chem Ind 52: 149T
Crowther J S, Holbrook R, Baird-Parker A C, Austin B C 1976 Tinbergen B J, Krol B (eds) Proc 2nd Int Symp on Nitrites in Meat Prods. Centre for Agr Pub and Doc, Wageningen, Netherlands
Cruess W V 1958 Commercial fruit and vegetable products, 4th edn. McGraw-Hill, New York
Dakin J C, Crawford C 1966 BFMIRA Tech Circ No 316
Dallyn H, Everton J R 1969 J Food Technol 4: 399
Drechsel H 1965 Thesis for Doctorate in Veterinary Medicine, Ludwig Maximilians University, Munich
Duncan C L, Foster E M 1968 Appl Microbiol 16: 401
Dwyer J L 1966 Contamination analysis and control. Reinhold, New York
Eddy B P 1960 J appl Bact 23: 189
Eddy B P, Ingram M 1956 J appl Bact 19: 62
Eddy B P, Ingram M 1962 Factors affecting the nitrite content of Wiltshire bacon, in Vol 2, Proc of the 1st Cong of Food Science and Technol, London. Gordon, Breach
Elliot R P, Michener H S 1965 US Dept Agric Tech Bull No 1320
Emodi A S, Lechowich R V 1969 J Food Sci 34: 78
English M P 1953 J Gen Microbiol 9: 15
Erickson F J, Fabian F W 1942 Food Res 7: 68
Etchells J L, Fleming H P, Bell T A 1975 In: Carr J G, Cutting C V, Whiting G C (eds) Lactic acid bacteria in beverages and food. Academic Press, London, p 281
Fabian F W, Graham H T 1953 Food Tech 7: 212
Fabian F W, Krehl C F, Little N W 1939 Food Res 4: 269
Fabian F W, Wadsworth C K 1939 Food Res 4: 511
Farber L 1954 Food Tech 8: 503
Farber L 1959 Annu Rev Microbiol 13: 125
Fieger E A, Bailey M E, Novak A F 1956 Food Tech 10: 578
Firman M C, Abbey A, Darken M A, Kohler A R, Upham S D 1956 Food Tech 10: 381
Flannery W L 1956 Bact Rev 20: 49
Fox J B 1966 Agr Food Chem 14: 207
Garrard E H, Lochhead A G 1939 Canad J Res Sect D 17 No 2: 45
Gibbons N E, Rose D, Hopkins J W 1954 Food Tech 8: 155
Gillespie D C, Boyd J W, Bisset H M, Tarr H L A 1955 Food Tech 9: 296
Goldberg H S, Weiser H H, Deatherage F E 1953 Food Tech 7: 165
Gooding C M 1945 US Patent 2 379 294
Green L F 1976 Food Chem 1 (2): 103
Greenberg R A 1972 Proc Amer Meat Inst Res Conf. American Meat Science Association in cooperation with the American Meat Institute Research Foundation, p 25
Greenberg R A 1973 Proc Amer Meat Inst Res Conf, 1973, p 69
Greenberg R A, Silliker J H, Fatta L D 1959 Food Tech 13: 509

Grever A B G 1974 Krol B, Tinbergen, B J (eds) Proc 1st Int Symp on Nitrites in Meat Products. Centre for Agr Pub and Doc, Wageningen, Netherlands
Grosskopf J C, Harper W J 1974 Milchwissenschaft 29 (8): 467
Haines R B 1934a J Hyg 34: 277
Haines R B 1934b Food Invest Bd, Dept Sci Ind Res, London. Ann Rept 44
Haines R B 1938 Proc Roy Soc Sect B 124: 451
Hall E G 1968 CSIRO Food Preserv Q 28 (1-2): 2
Handford R M, Gibbs B M 1964 Antibacterial effects of smoke constituents on bacteria isolated from bacon. In: Microbial inhibitors in foods. Proc of the 4th Int Symp on Food Microbiology, Göteborg. Almqvist and Wiksell, Uppsala
Hanna M O, Stewart J C, Carpenter Z L, Vanderzant C 1977a J Food Protection 40: 689
Hanna M O, Stewart J C, Zink D L, Carpenter Z L, Vanderzant C 1977b J Food Sci 42: 1 180
Harman D F, Hocking A D, Pitt J I, Warth A D 1974 CSIRO Division of Food Res Report 1973-74, p 9
Heintzeler I 1939 Arch Mikrobiol 10: 92
Heiszler M G, Kraft A A, Rey C R, Rust R E 1972 J Food Sci 37: 845
Henry M, Joubert L, Goret P 1954 C R Seance Soc Biol 148: 819
Hess E 1934 Contr Canadian Biol Fish 8: 491
Hines L R 1956 Antibiotics and Chemotherapy 6: 623
Hoffman C, Schweitzer T R, Dalby G 1939 Food Res 4: 539
Iandolo J J, Ordal Z J 1963 Bact Proc A27: 7
Ienistea C, Pleceas P 1967 Nahrung 11: 55
Ingraham J L, Stokes J L 1959 Bact Rev 23: 97
Ingram M 1951 Proc Soc appl Bact 14: 243
Ingram M 1959 Chem Ind p 552
Ingram M 1974 Krol, B, Tinbergen B J (eds) Proc 1st Int Symp on Nitrites in Meat Products. Centre for Agr Pub and Doc, Wageningen, Netherlands
Ingram M, Kitchell A G 1967 J Food Tech 2: 1
Ingram M, Mackey B M 1976 In: Skinner F A, Hugo W B (eds) Inhibition and inactivation of vegetative microbes. Soc for Appl Bact Symp Series No 5. Academic Press, London, p 111
Jakobsen M, Murrell W G 1977 In: Barker A N et al (eds) Spore research 1976 Academic Press, London. Vol 2 p 819
Jarvis B, Burke C S 1976 In: Skinner F A, Hugo W B (eds) Inhibition and inactivation of vegetative microbes. Soc for Appl Bact, Symp Series No 5. Academic Press, London, p 345
Jarvis B, Morisetti M D 1969 Internat Biodeterioration Bulletin 5: 39
Jarvis B, Patel A M, Rhodes A C 1977 J appl Bact 43 (3): p XV (Abstr)
Jarvis B, Rhodes A C, King S E, Patel A M 1976 J Food Technol 11: 41
Jay J M, Weiser H H, Deatherage F E 1957a Food Tech 11: 563
Jay J M, Weiser H H, Deatherage F E 1957b Antibiotics Annu, 1956-57 p 954
Jay J M, Weiser H H, Deatherage F E 1958 Appl Microbiol 6: 343
Jensen L B 1954 Microbiology of meats. Garrard Press, Illinois
Johnson M G, Vaughn R H 1969 Appl Microbiol 17: 903
Johnston M A, Pivnick H, Samson J M 1969 Canad Inst Food Sci Technol J 2: 52
Jones O 1942 Food 11: 191
Joslyn M A, Cruess W V 1929 Hilgardia 4: 201
Kang C K, Woodburn M, Pagenkopf A, Cheney R 1969 Appl Microbiol 18: 798
Kereluk K, Gunderson M F 1959 Appl Microbiol 7: 327
Kersken H 1974 Fleischwirtschaft 54 (8): 1341
Kimble C E 1977 In: Block S S (ed) Disinfection, sterilization and preservation, 2nd edn. Lea and Febiger, Philadelphia, p 834
Klis J S, Witter L D, Ordal Z L 1959 Food Tech 13: 124
Koedam A 1977 Riechstoffe, Aromen, Korperpflegemittel 27 (1): 6 and (2): 36
Kohler A R, Miller W H, Broquist H P 1955 Food Tech 9: 151
Labrie A, Gibbons N E 1937 J Biol Bd Canada 3: 439

Larkin J M, Stokes J L 1967 J Bacteriol 94: 889
Larkin J M, Stokes J L 1968 Canad J Microbiol 14 (2): 97
Lea C H 1931 J Soc Chem Ind 50: 207T
Leistner L, Hechelmann H, Kashiwazaki M, Albertz R 1975 Fleischwirtschaft 55: 1599
Lennon R E 1972 Fish Farming Ind 3: 15
Lerke P A, Farber L 1957 Antibiot Ann, 1956-57 966
Lerke P A, Farber L 1960 Food Tech 14: 217
Levine A S, Fellers C R 1940a J Bact 39: 499
Levine A S, Fellers C R 1940b J Bact 40: 255
Lotter L P, Leistner L 1978 Appl and Environmental Microbiol 36: 377
Macara T J R 1943 J Soc Chem Ind 62: 104T
Marshall B J, Ohye D F 1966 J Gen Microbiol 44: 41
Marshall B J, Ohye D F, Christian J H B 1971 Appl Microbiol 21: 363
Martinez Nadal N G, Montalvo A E, Seda M 1973 Cosmetics and Perfumery 88 (10): 37
Matches J R, Liston J 1972a J Milk Food Technol 35: 39
Matches J R, Liston J 1972b J Milk Food Technol 35: 49
McCleskey C S, Christopher W N 1941 J Bact 41: 98 (Abst)
McFarlane V H 1941 Food Res 6: 481
McVicker R J, Dawson L E, Mallman W L, Walters S, Jones E 1958 Food Tech 12: 147
Melnick D, Vahlteich H W, Hackett A 1956 Food Res 21: 133
Meredith W E, Weiser H H, Winter A R 1965 Appl Microbiol 13: 86
Michels M J M, Visser F M W 1976 J appl Bact 41 (1): 1
Michener H D, Elliot R P 1964 Advances in Food Res 13: 349
Miller D D, Golding N S 1949 J Dairy Sci 32: 101
Ministry of Agriculture, Fisheries and Food 1972 Food Additives and Contaminants Committee report on the review of the Preservatives in Food Regulations 1962. HMSO, London
Minor T E, Marth E H 1972 J Milk Technol 35 (4): 191
Money, R W, Born R 1951 J Sci Food Agric 2: 180
Moran T 1938 Food Res 3: 149
Moran T, Smith E C, Tomkins R G 1932 J Soc Chem Ind 51: 114T
Mossel D A A 1951 Leeuwenhock ned Tijdschr 17: 146
Mossel D A A 1963 Ann Inst Pasteur, Paris 104: 551
Mourgues R, Auclair J 1973 Lait 53 (528): 481
Mrak E M, Bonar L 1939 Zbl Bakt, Sect 2 100 (Nos 14/17): 289
National Research Council 1975 Committee on Food Protection. Prevention of microbial and parasitic hazards associated with processed foods
Niinivaara F P, Pohja M S, Komulainen S E 1964 Food technol 18 (2): 25
Nordin H R, Burke T, Webb G, Rubin L J, van Binnendyke D 1975 Canad Inst Food Sci and Technol J 8 (2): 58
Nunheimer T D, Fabian F W 1940 Amer J publ Hlth 30: 1040
Ohye D F, Christian J H B 1967 Ingram M, Roberts T A (eds) Proc 5th Int Symp Food Microbiol held Moscow, 1966. Chapman and Hall, London, p 217
Ohye D F, Christian J H B, Scott W J 1967 In: Ingram M, Roberts T A (eds) Proc 5th Int Symp Food Microbiol held Moscow, 1966. Chapman and Hall, London, p 136
Olsen C J, Mady, H 1940 J Dairy Sci 23: 509 (Abstr)
Ordal Z J, Brown W L 1957 Antibiotics Annu, 1956-57 p 860
Ottaway F J H 1956 Brit Bak Ind Res Assoc Bull No 54, 108
Oxhoj O 1943 Chem Zbl 11: 34
Paul H E, Harrington C M, Bender R C, Briggs W P 1952 Proc Soc Exp Biol Med 79: 199
Pederson C S 1960 Adv Food Res 10: 233
Perigo J A, Roberts T A 1968 J Food Technol 3: 91
Perigo J A, Whiting E, Bashford T E 1967 J Food Technol 2: 377
Phillips G F, Mundt J O 1950 Food Tech 4: 291
Pitt J I 1974 Food Technol in Australia 26 (6): 238
Pitt J I 1975 In: Duckworth R B (ed) Water relations of foods. Academic Press, London, p 273

Pitt J I, Christian J H B 1968 Appl Microbiol 16: 1853
Pitt J I, Richardson K C 1973 CSIRO Food Res Q 33 (4): 80
Pivnick H, Barnett H W, Nordin H R, Rubin L J 1969 Canad Inst Food Technol J 2 (3): 141
Potter R S 1939 Flavours 2 No 1: 7
Pouncy A E, Summers B C L 1939 J Soc Chem Ind 58: 162T
Rahn O 1945 Bact Reviews 9: 1
Rahn O, Conn J E 1944 Analyst 69: 222
Raj H, Liston J 1961 Food Tech 15 No 10: 429
Ray B, Speck M L 1973 CRC Critical Reviews in Clinical Laboratory Sciences 4: 161
Retzlaff N, Seydler B 1965 Archiv Lebenmittelhyg 16 (6): 128
Riemann H 1963 Food Technol 17: 39
Roberts T A, Garcia C E 1973 J Food Technol 8: 463
Roberts T A, Ingram M 1966 J Food Technol 1: 147
Roberts T A, Ingram M 1973 J Food Technol 8: 467
Roberts T A, Ingram M 1976 Tinbergen B J, Krol B (eds) Proc 2nd Int Symp on Nitrites in Meat Products. Centre for Agr Pub and Doc, Wageningen, Netherlands
Roberts T A, Smart J L 1976 In: Davies R, Birch G G, Parker K J (eds) Intermediate moisture foods. Applied Science, London, p 203
Robertson M E 1931 J Hyg 31: 84
Scheibe W 1958 Medizinische, 1633 (Abstracted in Chem Abs, 1959 53: 1538a)
Schmidt C F, Lechowich R V, Folinazzo J F 1961 J Food Sci 26 No 6: 626
Scott W J 1953 Aust J Biol Sci 6: 549
Scott W J 1957 Adv Food Res 7: 83
Segner W P, Schmidt C F, Boltz J K 1966 Appl Microbiol 14: 49
Seiler D A L 1964 Factors affecting the use of mould inhibitors in bread and cake. In: Microbial inhibitors in foods. Proc of the 4th Int Symp on Food Microbiology, Göteborg. Almqvist and Wiksell, Uppsala
Shewan J M 1949 Chem Ind 501
Shewan J M 1956 Mod Refrig 59: 423
Shillinglaw C A, Levine M 1943 Food Res 8: 464
Shrimpton D H 1957 J Sci Food Agr 8: 485
Silliker J H, Greenberg R A, Schack W R 1958 Food Tech 12: 551
Smith D P, Rollin N J 1954 Food Res 19: 59
Smith W H 1957 Nature, Lond 179: 876
Snow D 1949 Ann Appl Biol 36: 1
Southcott B A, Baker E G, Boyd J M, Tarr H L A 1958 Food Tech 12: 108
Spencer R 1966 Food Manuf 41 (Feb): 41
Spencer R 1971 Proc 17th Eur Meet Meat Res Workers Bristol, Sept 1971
Spiegelberg C H 1944 J Bact 48: 13
Stern J A, Liebman H L, Kudo G, Chapel J, Olsen R A, Farber L L, Grennan M 1958 Food Tech 12: 132
Stokes J L 1960 Proc of 12th Res Cong of Am Meat Inst Found, Circular No 61, July, 1960
Straka R P, Stokes J L 1959 J Bact 78: 181
Stuart L S 1938 Food Res 3: 417
Stuart L S 1940 Food Res 61: 267
Stuart L S, James L H 1938 J Bact 35: 381
Stumbo C R, Gross C E, Vinton C 1945a Food Res 10: 283
Stumbo C R, Gross C E, Vinton C 1945b Food Res 10: 293
Sulzbacher W L 1950 Food Tech 4: 386
Tanner F W 1944 Microbiology of foods. Twin City Printing Co, Champaign, Illinois
Tanner F W, Evans F L 1933 Zbl Bakt, Sect 2 88: 44
Tarkow L, Fellers C R, Levine A S 1942 J Bact 44: 367
Tarr H L A 1941a J Fish Res Bd Canada 5: 265
Tarr H L A 1941b Nature, London 147: 417
Tarr H L A 1942 J Fish Res Bd Canada 6: 74

Tarr H L A 1960 Canad Fd Ind 31 No 4: 39
Thatcher F S, Loit A 1961 Appl Microbiol 9: 39
Tilbury R G 1976 In: Davies R, Birch G G, Parker K J (eds) Intermediate moisture foods. Applied Science Publishers Ltd, London, p 138
Tomiyama T, Kuroki S, Maeda D, Hamada S, Honda A 1956 Food Tech 10: 215
Tomkins R G, 1929 Proc Roy Soc, Sect B 105: 375
Tompkin R B, Christiansen L N, Shaparis A B 1978 J Food Sci 43: 1368
Tompkin R B, Christiansen L N, Shaparis A B, Bolin H 1974 Appl Microbiol 28 (2): 262
Troller J A 1973 J Milk Food Technol 36 (5): 276
Vaughn R H 1975 In: Carr J G, Cutting C V, Whiting G C (eds) Lactic acid bacteria in beverages and food. Academic Press, London, p 307
Vaughn R H, Ng H, Stewart G F, Nagel C W, Simpson K L 1960 Appl Microbiol 8: 27
von Schelhorn M 1951 Adv Food Res 3: 429
von Schelhorn M 1953 Food Tech 7: 97
Wagenaar R O, Dack G M 1954 Food Res 19: 521
Wagenaar R O, Dack G M 1958 J Dairy Sci 41: 1182, 1191, 1196
Wagenaar R O, Dack G M 1960 Food Res 25: 646
Walker E A, Taylor J H 1960 J appl Bact 23: 155
Wallhäusser K H, Lück E 1972 Deutsche Lebensmittel Rundschau 68 (2): 39
Washam C J, Olson H C, Vedamuthu E R 1977 J Food Protection 40 (2): 101
Waters M E, Hamdy M K 1970 J Milk Food Technol 33: 221
Webb A H, Tanner F W 1945 Food Res 10: 273
Weiser R S, Osterud C M 1945 J Bact 50: 413
Weiser H H, Goldberg H S, Cahill V R, Kunkle L E, Deatherage F E 1953 Food Tech 7: 495
Wells F E, Fry J L, Marion W W, Stadelman W J 1957 Food Tech 11: 656
Wood J M, Evans G G 1973 IFST Proceedings 6 (3): 111
Wrenshall C L 1959 In: Goldbert H S (ed) Antibiotics, their chemistry and non-medical uses. van Nostrand, Princeton, p 449
Wright H B 1961 Food Tr Rev 31 No 4: 43
Yesair J, Cameron E J 1942 Canner (Convention No) 94: 89
Young R E, Romani R J, Biale J B 1962 Plant Physiol 37: 416
Zawadzki Z, Pogorzelska E 1975 Medycyna Weterynaryjna 31 (8) 488

4

Containers

For those who have an interest in the bacteriological condition of thermally processed foods, a working knowledge of the fundamentals of container construction and closure is essential since the maintenance of sterility is dependent on the efficiency of operations such as seaming and closing of cans or heat sealing of thermoplastic based containers.

Three-piece cans

The earlier types of metal containers were made exclusively by hand. All seams were jointed by solder and the food was filled into the can through a small hole in the end which was afterwards covered by a disc or cap which was soldered in place. The inherent disadvantages with these 'hole and cap' containers included the difficulties of packing large pieces of food through the relatively small hole and of efficient cleansing of the can before filling. Out of these disadvantages and the need for increased production arose by degrees the modern 'open top' or sanitary can which consists of a cylindrical body with a soldered lock seam and unsoldered double-seamed ends. Except for a few products, the 'open top' can has superseded other types of containers. For a history of the development of the modern can the reader should consult works by Cameron (1936) and May (1937).

The high speed of modern production and the function which the can has to fulfil in hermetically sealing the food necessitates that all the processes concerned in can fabrication and seaming are precision operations. The success of these operations depends to a great extent on the use of tinplate of uniform quality.

The hand-fed mills used for rolling the hot steel baseplate into sheets have been superseded by continuous strip mills in which the final reduction to the correct thickness is carried out by a cold rolling process. Cold-reduced tinplate has superior mechanical properties and is more uniform, particularly with respect to gauge thickness, giving rise to a higher quality standard in can making and bringing about economies by enabling the manufacturer to use thinner plate without loss in strength. The use of thin tinplate often necessitates a modification in the design and construction of the can in order to confer to it extra rigidity. Tinplate with a high tensile strength can be made by a process which involves subjecting the baseplate to a heavy reduction in thickness during the final stages of rolling. After tin coating, the resulting product, known as double-reduced tinplate, has a tensile strength up to twice that of conventional tinplate but because of its high strength and the direction of the grain, conversion to can components is more difficult than with ordinary tinplate. For example, the beading of the sidewall of can bodies is made difficult because of the low ductility of double-reduced tinplate. The problem was solved by adopting a design consisting of a central cluster of 15 beads with a shallow profile. Tinplate on which the tin coating has been deposited electrolytically is also being extensively used. The coating applied in this manner is more uniform than that obtained by the older hot-dipping process. A lower average thickness of tin coating is obtainable and it is customary to apply a protective film of lacquer when such grades of electrolytic tinplate are used for food can production. The requirements, as far as tin thickness is concerned, may be different for the internal and external surfaces of the can. To satisfy this need differentially coated plate, that is, plate with different densities of tin on the two surfaces, is produced. Substantial economies in the cost of can production and in the world usage of tin are being effected by the replacement of hot dipped tinplate by various grades of electrolytic tinplate. The weight of tin deposited is expressed as g per m^2. For electrolytic plate this is normally within the range 2.8 to 11.2, the actual amount required depending upon the conditions applying, and whether or not a protective lacquer is used. A common specification in the UK is D 11.2/5.6 indicating differential plate with a coating of 11.2 g per m^2 on one side and 5.6 g per m^2 on the other. This would normally be used with the high tin side in contact with the product and without lacquer application to either surface. Apart from acid products, such as fruit, a lower specification would probably be used if lacquering was applied. The international standard covering qualities of the steel base and the tin coating is ISO 1111: cold-reduced tinplate and cold-reduced blackplate; Part 1, Sheet.

After electro-deposition of the tin coating a heat treatment is applied in order to give a momentary fusion (flow-melting or flow-brightening). This leads to the formation of an intermetallic tin-iron compound the character of which influences the corrosion resistance of the tinplate. To protect the tin surface from oxidation during storage or the stoving of lacquer coatings, a chemical passivation treatment is applied by spraying or dipping in a chromic acid or sodium dichromate solution. This deposits a layer containing chromium to a weight of 0.1 to 1 μg per cm^2 of surface.

Tin is the widely used coating material but there has been interest in the use of chemically treated steel plate as a substitute for tinplate. The most favoured methods involve treatments which result in deposition of chromium or chromium compounds on the steel plate. These are well described by Smith (1967). Because of low abrasion resistance such tin-free steels need to be protected by a lacquer film if they are to be used for can making purposes. Furthermore since tin-free steel cannot be machine soldered it is not suitable for conventional can bodies but it could find application for drawn cans or for cans in which the side seam is adhesively bonded or welded. Otherwise its use is limited to can end manufacture. Since cans made from tin-free steel have to be lacquered externally the economics have not favoured a change from tinplate (Norman & Rees, 1972), but, of course, the situation is subject to change depending upon the price of tin and other factors. According to Norman & Rees (1972) tin-free steel has proved particularly suitable for some products where it showed technical advantages over tinplate. However there were difficulties with external protection, breaks occurring in the lacquer film during seaming and handling, leading to corrosion. Thus improved attention to equipment is necessary to enable tin-free steel cans to be used. However tin-free steel is being used for can ends on pet-foods and tin-free steel drawn cans are now being offered by manufacturers in the USA (Anon, 1978).

For certain foods such as some fruit, vegetables and meat products, the tinplate, whether electrolytic or hot-dipped, needs to be protected against chemical interaction of the contents and the metal by a lacquer film. This is usually applied to the sheets of tinplate before they are cut and made into cans. With this method, however, the lacquer film generally sustains some damage during the subsequent fabrication of the can, and methods designed to obviate such damage are used, e.g. spraying the finished can with lacquer. Typical operations involved in can fabrication and seaming are as follows:

Body Blanks

The blanks for the can bodies are cut by means of automatic slitters which cut the tinplate by the action of rotary cutters made from hardened steel. For high-speed production it is now customary to use Duplex slitters by which the sheets are cut in one direction and the strips so formed are fed automatically to the second section where transverse cuts produce body blanks of the correct size. It is essential that body blanks be cut absolutely square.

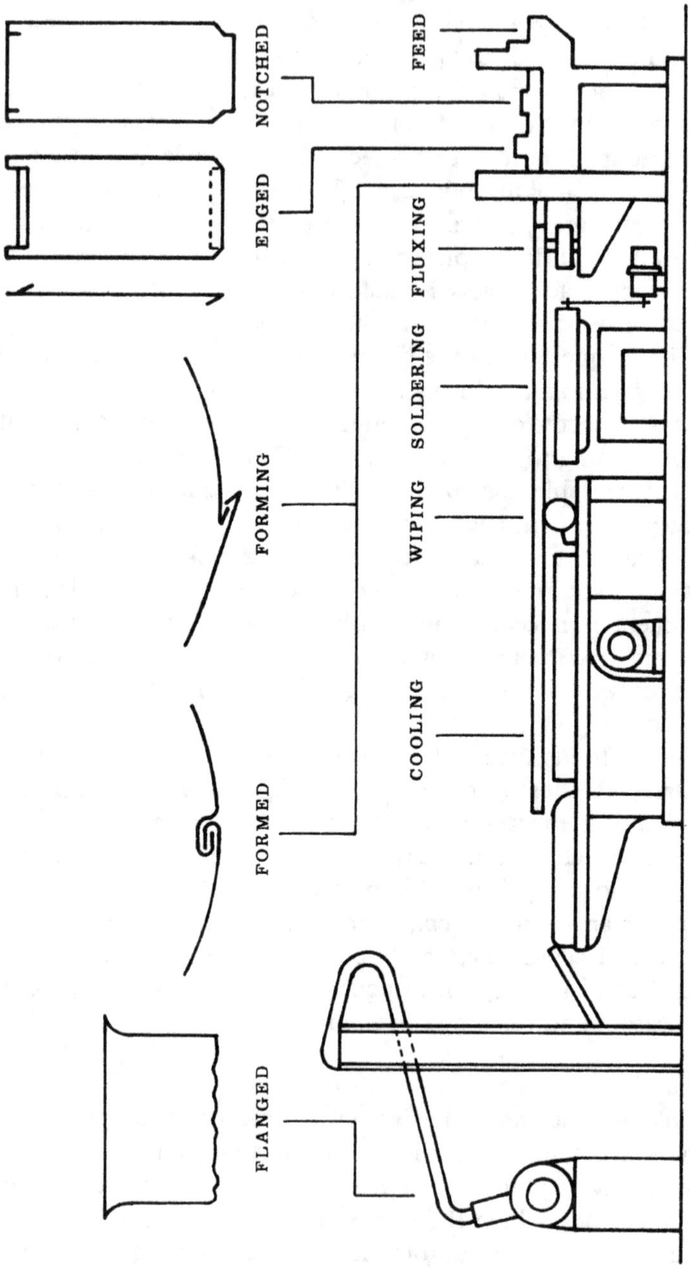

Fig. 1 Sequence of operations in body forming

The blanks are formed into cylindrical bodies by the can body-making machine which automatically performs several successive operations. First, each blank passes through a flexing unit which removes irregularities and work-hardens the plate slightly. The corners of each blank are then notched so that when the lock seam is formed the ends of the seam comprise only two thicknesses of metal. The significance of this notching operation becomes apparent in the formation of the double seams. After notching, the blank is edged, that is, a hook is formed on each of the two shorter edges of the blank by bending one edge downwards and the other upwards. The hooks are then coated with flux, after which the blank is formed into a cylindrical body by being wrapped by means of two 'wings' round an expanding horn or mandrel so that the hooks lock tightly together. The locked hooks are then hammered into the special grooves along the length of the horn. After a second application of flux to the outside of the lock seam the body is then carried over a revolving solder roll. The solder roll rotates in a bath of molten solder and is aligned in such a manner that solder is applied to the lock seam as the can body passes along it. Immediately after leaving the solder roll, excess solder is removed from the seam by a revolving brush. Before the can body finally leaves the Body Maker it receives a blast of cold air to solidify the solder in the lock seam.

Can making speeds have been increased by various methods. Now in common use is the technique of making a long cylinder which is later parted into two or three can cylinders, thus doubling or trebling the speed of the Body Maker.

Flanging. The can bodies are next fed to the Flanger, where they are centred between two flanging punches which fit into both ends of the can. These punches then move slightly towards each other and force the edges of the body outwards, thus simultaneously forming flanges on both ends of the can body.

Manufacture of can ends (Fig. 2)

The strips from which the ends are stamped out are cut from sheet tinplate by the Scroll Shears machine. From the point of view of wastage of tinplate it is more economical to stamp out the ends in a staggered or scroll pattern than in straight parallel rows, and the sheets are therefore cut into scrolled strips. Even so, there are a number of areas corresponding to half-ends and the edge of the scroll strip which are wasted and important economies are being made by the use of coiled plate for the stamping of can ends. The ends are stamped out by automatic presses, the concentric rings, termed expansion rings, which are formed on the ends by the press being designed to permit ready bulging of the ends during heat-processing and thus to relieve strain on the seams of the can. Another advantage arising out of the use of these rings is the readiness with which the ends bulge in case of spoilage. In addition, their springiness is an aid to good double seam formation.

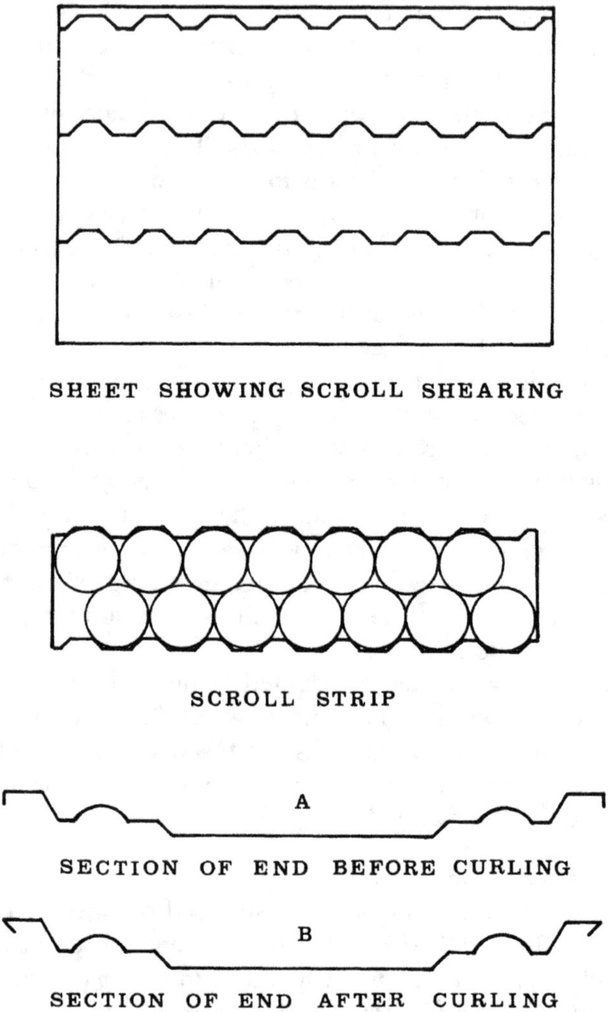

Fig. 2 Formation of can end

After the rim of the ends has been curled inwards by the Curling Machine, the inside of the curl is lined with a sealing compound which may be a rubber solution or emulsion or some other material with similar resilient properties. The sealing compound, which is dried by passing the ends through an oven, acts as a gasket and ensures an airtight sealing of the finished seam by filling in the small irregularities between the metal faces in the seam.

Over the last few years easy-opening features have been incorporated into cans, particularly beverage cans of various types. This usually takes the form of an easy-opening end made from aluminium which is scored so that the end may be removed, or a pouring aperture formed, by pulling a tab or ring. A serious technical limitation to the use of these ends on many processed foods is the corrosion which occurs, mainly bi-metallic, when they are seamed on to a tinplate can. The use of an all-aluminium can is often too expensive to be acceptable.

Double seaming

The can ends are attached to the body by a Double Seamer machine. Although there are a number of different seaming machines, fundamentally they consist of a base plate or table on which stands the can to be closed, a chuck which fits firmly into the countersink of the can end and a pair of seaming rolls or grooved wheels which operate in succession. As the can body is delivered to the Double Seamer it is automatically covered by an end, the curl of which fits over the flange of the body. The body with the end in place is carried to the base plate, which by lifting automatically, causes the chuck to fit into the end of the can. The first of the seaming rolls then comes into operation, and by pressing tightly and revolving at high speed round the curled rim of the end, rolls or tucks it under the flange of the body. The second wheel, which has a more shallow groove than the first, then comes into operation and tightly flattens together the hooked edges of the end and the body against the chuck.

The finished double seam, which should conform to certain standards of measurement (*see* Ch. 17), consists of five thicknesses of metal, except where it coincides with the body seam, where seven thicknesses are present. If the body-blanks were lock seamed right up to the edges, the double seam would, at the point where it meets the lock seam, comprise eleven thicknesses of metal. By notching the blanks four thicknesses of metal are eliminated. When correctly formed, a double seam gives an efficient, airtight seal, but slight malformation may result in the seam being unable to withstand moderate strain or damage which may cause leakage and subsequent spoilage of the can contents. The production of mechanically sound seams is only possible with mechanically sound and carefully adjusted machinery and the maintenance of seaming machines in first-class condition is essential for the avoidance of losses due to leakage. Furthermore, it is essential that can seams be examined at regular and frequent intervals whenever a seaming machine is in production; although adjustment may be carefully made at the commencement of a run, slight changes in the setting may, through a multitude of causes, arise at any time with serious results.

The various processes involved in can manufacture are shown diagrammatically in Figures 1 to 3.

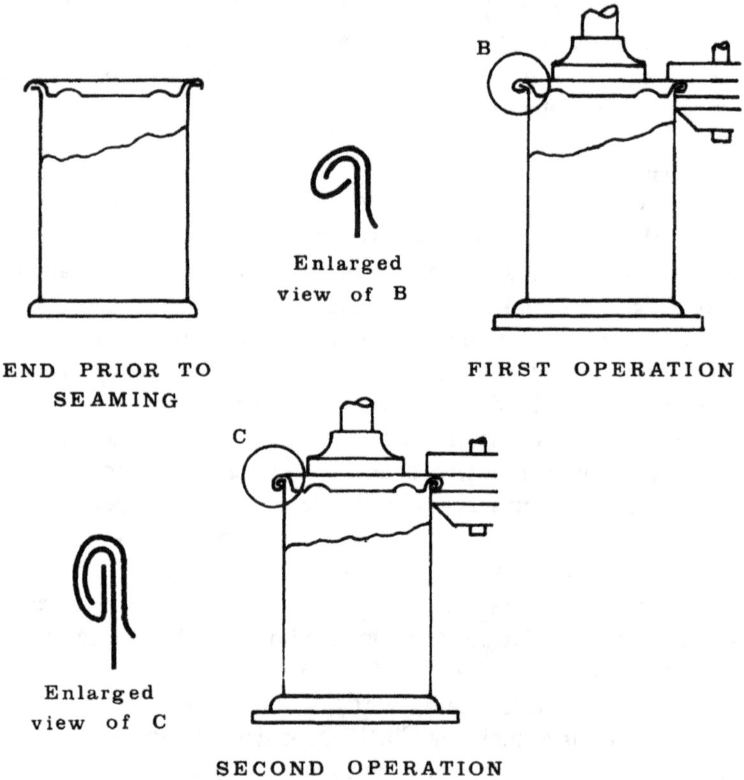

Fig. 3 Seaming operation

Whilst the soldered tinplate can is still by far the most widely used container for processed foods, cans made from aluminium alloys are available. Aluminium may readily be converted into containers by a variety of processes but the containers so formed have poor mechanical properties. By alloying with magnesium harder and stronger material can be achieved at the expense of corrosion resistance (Gotsch et al, 1958). The all-aluminium alloy container, generally with easy open end, offers advantages of convenience, lightness, and appearance. Apart from beverages, for which it is widely used, it has found application for the packaging of spreads, pâtés, fish products and desserts. Like tin-free steel it cannot be readily soldered and it is used for cans with welded or cemented side seams or alternatively for two-piece cans (see below).

Non-soldered cans

The canned beverage market has seen a move away from the use of the soldered side seam in favour of the adhesive bonded or welded side seam or to the two-piece drawn can. As with the easy-open end, the trend may be expected to carry over to the sterilisable container with many of the innovations pioneered in aluminium alloy being adopted for tinplate or tin-free steel cans.

Cemented side seams

The cemented side seam has been used for the manufacture of tin-free steel (TFS) cans. The nylon bonding material used is applied to the pre-lacquered body blanks and a lap joint is formed which is bonded by the application of heat followed by rapid cooling. An essential requisite is the coverage of the cut side seam by the nylon cement in order to avoid iron pick-up. The formed can is sprayed internally with a coat of lacquer as is the conventional beverage can.

Welded side seams

The development of the welded side seam is a more recent innovation and is typified by processes such as the Continental Can Company's Conweld process. As an example, the electronically controlled Conweld process uses tin-free material to produce a precision steel to steel weld with just a thin line to show the seam. The thin seam line makes full circle lithography possible (Furtado, 1975). In the UK, Metal Box Limited have used the Soudronic welding process for the production of 5 pint cans made from tinplate and intended for the packaging of low carbonation (draught) beer.

Two-piece cans

The production of a seamless body and end from a single disc of metal has technical, marketing and financial attractions. The technical advantages attributable to the elimination of the two seams are reduction in the possibility of leakage and removal of a potential source of lead contamination. From the marketing aspect the two-piece can undoubtably has aesthetic appeal presenting a smooth profile and streamline appearance. The absence of a side seam permits uninterrupted print decoration externally and the bottom of the can can be designed and formed for better stackability. Since less metal is used in its construction due to elimination of the overlaps at the two seams there are also financial and conservation reasons for its adoption.

Shallow profile cans in aluminium or tinplate such as oval top cans of the type used for packing fish have been available for many years. Cans of this type are usually made by cutting, drawing and flanging in one operation but because of the strain to which it is subjected the tinplate used is generally heavier and more ductile than that used for round cans.

In recent years there has been considerable progress in methods of producing drawn cans the improvements being directed towards increasing the can height, reducing the metal thickness and distributing the metal more evenly. Two types of can may be considered namely (1) the drawn and ironed (DWI) can and (2) the drawn/redrawn (DRD) can.

The DWI can was developed by the aluminium industry in the USA for the beer can market and was first produced commercially in 1964 (Maeder, 1975). Tinplate versions became available in 1972. The manufacturing process involves blanking out a disc of metal from sheet, drawing it into a cup and then forcing it by means of a punch through a series of dies, each slightly smaller than the preceding one, thus elongating the wall by a stretching or ironing action. The process is described in more detail by Ings (1976). The process of ironing the wall reduces its thickness and this can be exploited for carbonated beverages where the internal pressure imparts rigidity to the container. For processed foods, however, the internal vacuum necessitates thicker walls if panelling is to be avoided. In these circumstances either a beaded DWI can or the DRD process could be used.

In the DRD process a blank disc is drawn into a cup as before but in this case excess metal made available by reducing the circumference of the body is used to elongate the wall thus maintaining the thickness of the metal. For this process the plate must be pre-lacquered to assist in fabrication but this may be an advantage since it is cheaper to apply lacquer to sheet than to formed cans. The DWI and DRD processes are compared by Senzacqua (1976). In Europe a DRD line is in operation at Masilly, France, for pet-food cans and in the UK Metal Box Limited are developing the method for processed food cans.

A development worth noting is the 'necked-in' can in which the ends are the same diameter as the body. Such cans do not over-ride one another and this gives rise to high efficiences on can lines. Some material is saved, both in construction of the can by virtue of the smaller diameter end, and in the outer packaging material since the cans may be packed closer together.

Easy-open ends

The widespread consumer acceptance of the aluminium easy-open end (EOE) for the beverage can led to a search for a full aperture EOE suitable for processed foods. The first generation had a tear-off tongue with a spiral incision, the can end being removed as a ribbon of metal. The latest type has a circular incision which enables the whole central panel to be removed by pulling an attached ring.

The aluminium EOE must generally be used in conjuction with an aluminium can otherwise bi-metallic corrosion is liable to occur. The cost of an all aluminium construction plus the easy opening feature has mitigated against its general adoption although it has been used on cans for desserts, pâté and fish products. The German Hansa all-aluminium can with easy-open ends is widely used for some fish products in the United Kingdom. In order to extend the use of the EOE and to reduce the cost, tinplate versions have been developed (Anon, 1971). Difficulties of maintaining a controllable score on tinplate have given some problems and as yet this end is not in widespread use for thermally processed cans.

For fruit juices, which are processed by a hot-fill and hold technique, tinplate ends pre-punched with an aperture which is sealed by an adhesive strip of metallised polyester film or aluminium foil are used (Anon, 1967; Norman & Rees, 1972).

Semi-rigid and flexible packs

Aluminium has many attractions for construction of containers and apart from its use in the construction of cans described above it is also used in the form of thin sheet or foil in conjunction with a lacquer coating or as a laminate with thermoplastics.

The semi-rigid tray

Semi-rigid aluminium trays, heat sealed with an aluminium cover and capable of being thermally processed, are available and have been used in Europe and Japan for processed foods. One of the preferred constructions for the processable tray and lid is aluminium foil/polypropylene laminate. Such a container is heat sealable and may be processed at temperatures up to 120°C. The trays and lids are formed on deep draw presses from aluminium roll stock. This operation may be carried out on the users premises or the trays and ends may be supplied pre-formed. Heat processing of the filled and sealed container must be carried out either in super-heated water with an overpressure or in a steam/air mixture (see Ch. 5). The customer must remove the lid by cutting with a knife, scissors or a special tool, since a suitable peelable bond, adequate to withstand thermal processing, has not yet been developed. Compared with conventional cans, material costs are higher and line speeds are lower. Against this must be off-set the advantages of the more rapid heat penetration obtainable from a shallow tray and the consumer appeal of a different and lightweight container.

Formed foil containers

These are made by a process in which a special thin aluminium foil (20 to 50 μm), laminated to thermoplastic on either side, is formed into shallow dies either by applying air pressure or by the action of a forming tool. After filling and sealing the packs are punched from the web to give individual trays. The Bosch Packaging Group have developed a modified version of the Hofliger and Karg Servac platen blister sealing machine for this purpose. This method seems to be a better option than the air forming method which has been used by DRG Packaging in the United Kingdom. Bosch gives the maximum depth of draw as 45 mm but for all practical purposes trays with a steep wall gradient, thus retaining a reasonable size base, are limited to a 25 mm draw. The formed foil tray is probably best compared with the flexible pouch. It has several advantages over the latter, mainly the provision of a more rigid container, reduction in pack area for hard vacuum-packed products, higher line speeds and easier filling of some products such as sausages or slices of meat for which it should be ideally suited.

The flexible pouch

As an alternative to the rigid or semi-rigid container the construction and use of a flexible pouch made from thermoplastic and aluminium foil laminate has proved technically and commercially feasible. The origins date back to American work in the mid or late 1950s (Hu et al, 1955; Wallenburg & Jarnhell, 1957; Gould et al, 1962) although most commercial applications have taken place in Japan and Europe.

The commonest material in use is a three-ply laminate based on a sandwich of thin gauge aluminium foil (9 to 25 μm) between two thermoplastic films. The outer ply is generally made from polyester about 12 μm thick which serves to protect the aluminium from scuffing and to give strength to the pack. The composition of the inner ply has been the subject of considerable research since this determines the seal characteristics and furthermore it must withstand contact with fat, oil and other food components. Modern pouches generally employ either a modified high density polyethylene or a polypropylene copolymer 50 to 80 μm thick as the inner film. The adhesive used for bonding the plies together is conventionally a polyester-isocyanate

(Goldfarb, 1973). The retort pouch was designed for operating retort temperatures up to 121°C, the polypropylene based material being more suitable for higher temperatures than the modified high density polyethylenes. According to Tsutsumi, quoted by Lampi (1977), products are being processed by HTST methods in Japan, time temperature combinations of 2.7 to 9 minutes at temperatures up to 135°C being used for this purpose. The material used for these applications is described as 12-μm polyester/9-μm aluminium foil/15-μm oriented Nylon/6/50-μm polypropylene. In high temperature use failures, if experienced, are often attributable to delamination due to adhesive failure aided by film shrinkage (Lampi, 1977).

The pouches may be fabricated on-line by the packer using a form-fill-seal system, several of which are available and comprehensively described by Lampi (1977), or they may be purchased premade. The normal design is a flat rectangular shape with four fin seals about 1 cm wide. The size is determined by the pouch thickness that can be tolerated at the correct filling weight. According to Lampi the nominal thickness ranges from 1.58 cm for the 114 g size to 3.33 cm for the 909 g size. Notches are sometimes provided for easy tear opening. It is undesirable to leave a portion above the top seal open and secondary top seals, with or without trimming, should be made.

It should be possible to achieve a leakage failure rate no higher than that acceptable for cans (0.01 per cent) provided suitable precautions are taken. Due to the flexible nature of the pouch pressure changes or handling cause a pumping action which tends to promote leakage and for this reason, as well as for the physical protection afforded, pouches should be over-wrapped. Furthermore if leakage does occur it is desirable that the organisms involved should not be of public health significance and this risk is reduced if the naked pouch is not handled unnecessarily. Normally retort pouches are packed in rectangular flat cartons which provide a good surface for decorative and display purposes. Toyo Seikan Kaisha Ltd. (1973) recommend an outer carton of 300 g per m^2 base weight with glued flaps.

The suitability of the retort pouch for the packaging of food has been proved technically but its success in Europe and the USA will depend on consumer acceptance and so far, at least in Europe, it has not fulfilled many early expectations.

Glass containers and closures

A variety of heat-processed foods are packed in glass containers hermetically sealed by means of closures, usually of tinplate or aluminium, appropriately lacquered to prevent corrosion. These foods range from pasteurised or hot-filled products such as jams to fully processed packs such as infant foods and bottled vegetables.

It is not the intention here to discuss the use of glass containers but some comments on available closure systems are appropriate (Fig. 4). Moody (1977) divides closures into four groups.

1. *Normal seals.* These provide a good seal when the pressure inside the container is essentially the same as outside and varies only due to short-term fluctuations largely due to temperature effects
2. *Vacuum seals.* These give an airtight seal when the pressure inside is significantly lower than outside
3. *Pressure seals.* These are designed to contain high internal pressures of 5 to 10 bars and are mostly used for sealing pasteurised beer and carbonated beverages
4. *Venting seals.* These are designed to contain limited pressure but vent if the pressure increases further and then reseat as it falls.

The basis of sealing any closure is to seat the lid to the glass finish via a gasket of resilient material which is maintained in a compressed condition. The sealing surface may be on the top rim or wrapped around the finish to provide 'top-and-side' seal. Some vacuum closures rely entirely on an annular side seal. The resilient compound itself may be in the form of a disc faced with a material compatible with the food or, more often for processed foods, as a ring seal of 'flowed-in' compound.

Since heat processed foods are normally packaged in a manner designed to provide a vacuum most closures of interest are of the vacuum seal type held in place solely by the developed vacuum or by a combination of vacuum and mechanical retention.

A simple and well tried closure of the first type is the 'pry-off' cap, e.g. those developed by the White Cap Company and the Anchor Hocking Glass Corporation. The 'pry-off' cap is side sealing using a butyl rubber gasket crimped into the inside of the skirt. The vacuum is developed by steam flushing the headspace as the cap is applied and pushed in place by downward pressure. A more recent development is the PT (Press-on, Twist-off) cap developed and patented by the Continental Can Company and sold in the UK by Metal Box Packaging Limited. This is a smooth-skirted tinplate cap with a plastisol compound covering a top ring and the inside of the skirt. It is applied to a container with a multi-start thread finish by downward pressure in the same way as the prise-off cap. During the heat treatment, which must be applied, the plastisol on the skirt softens and takes up the profile of the glass thread. On cooling the cap can be renewed by the normal twist-off action.

Fig. 4 Various closures for glass containers

A screw closure with suitable liner makes a good vacuum closure but because of the frictional forces required and the retention attributable to the vacuum it is difficult to remove. The widely used 'twist-off' vacuum seal overcomes the problem by provision of lugs which mate with a four-start thread of coarse pitch on the glass finish. This provides a rapid lift on applying the opening torque which, in combination with low friction sealing compounds, enables the vacuum to be released. These caps are widely used on pickles, sauces, salad creams and jams many of which are hot-filled.

Another closure of the vacuum type with mechanical retention is the Omnia cap, made by Thomas Hunter. During application a number of indentations are made on the edge of the skirt to hold the cap in position. On releasing the vacuum by piercing the aluminium, the cap is pushed off by hand. The Omnia S type has a deeper skirt and is crimped to a screw thread finish connecting it to a twist-off closure. A vacuum type closure widely used in the UK for hot-filled products such as jam is the squeeze cap produced by Metal Box Packaging Ltd. This cap may be applied by the Whitecap machine. It is not suitable for products requiring in-bottle pasteurisation.

REFERENCES

Anon 1967 Modern packaging 40 (9): 94
Anon 1971 Canner/Packer 140 (3): 36
Anon 1978 Food Engng 50 (2): 57
Cameron E J 1936 Canner 82: No 20, 7; No 21, 13; No 22, 14
Furtado R 1975 Food in Canada 35 (6): 42
Goldfarb P L 1973 Flexible packaging for heat processed foods — package, process and product evaluation. Packaging Inst, St Louis
Gotsch L P, Pilcher R W, Lueck R H 1958 Food Technol 12: 487
Gould W A, Geisman J R, Weiser H H, Bash W D, Moore W J, Salzer R H, Long F E 1962 Ohio Agric Exp Stn Res Bull 905
Hu K H, Nelson A I, Legault R R, Steinberg M P 1955 Food Technol 9: 236
Ings J 1976 Metals Australia August: 150
Lampi R A 1977 Adv Food Res 23: 305
Maeder E G 1975 Tech Q, Master Brewers Assoc of America 12 (1): 63
May E C 1937 The Canning Clan. Macmillan, New York
Moody B 1977 Packaging in glass, 2nd edn. Hutchinson Benham, London
Norman G F, Rees J A G 1972 Food Manuf 47 (9): 27
Senzacqua A 1976 Imballagio 27 (241): 28
Smith E J 1967 Iron and Steel Engineer 44 (7): 127
Toyo Seikan Kaisha Ltd 1973 Brochure on RP-F Tokyo
Wallenberg E, Jarnhell B 1957 Mod Packag 31 (1): 165

5

Outline of canning operations

Canning may be loosely defined as a process for preserving food achieved by the application of a thermal sterilisation procedure to products packed in hermetically sealed containers. Thus the canning process may be carried out, either, (1) by filling the product into a container, sealing it hermetically and subsequently heating and cooling the container and contents until all micro-organisms capable of growth in the food are destroyed, or (2) by sterilising the product continuously in-line by heating and cooling it in heat exchangers and then filling it aseptically into pre-sterilised containers which are subsequently sealed under aseptic conditions.

The aseptic canning process involves special techniques and equipment which are discussed in Chapter 10. In this chapter the basic operations connected with the conventional process are outlined. These are:

1. Preparation of the food
2. Filling of the container
3. Exhausting
4. Sealing of container
5. Thermal processing
6. Cooling of the container and contents.

Preparation of the food

A variety of processes such as grading, trimming, washing, blending, blanching, pre-cooking, etc. are employed. A discussion of the detailed procedures is outside the scope of this book and the reader is referred to Lopez (1975) who gives general descriptions of various canning processes and Lock (1969) who deals with fruit and vegetables.

The question of microbial contamination derived from raw materials, containers, etc., is considered in Chapter 6. The pre-canning operations should be carried out effectively but rapidly since undue delay at this stage will permit the development of rapidly growing micro-organisms which may render the heat process inadequate. Even if the subsequent heat process does effectively sterilise the product, incipient gas formation can give rise to the liberation of gas during processing, reducing headspace vacuum, or even giving rise to sterile 'blown' cans. In any event delay in handling invariably leads to a loss of quality in the finished product.

The operations during preparation of the food which are of significance to the microbiologist are (1) washing, (2) peeling and (3) blanching, since all these operations should lead to a reduction in the numbers of contaminating organisms which are present on the surface of the produce, usually in the adhering soil. If these operations are incorrectly carried out, or carried out in badly designed equipment, this will lead to an increase in the initial bacterial load which could result in an inadequate process.

Washing

Washing processes are carried out in various ways and may involve machinery designed specially for the product. Usually the foodstuff is subjected to high pressure water sprays, or is immersed in streams of water whilst it is moved on a belt or tumbled on agitating or revolving screens. Flotation is frequently used as a means for removing stones, dirt and foreign matter, the product being flumed over collector trays into which the denser material settles out. Lightweight foreign matter is often removed by allowing it to float over weirs. Various industrial washing machines are described by Slade (1967).

Peeling

The peeling of fruits and vegetables is an important preliminary operation since, in conjunction with washing, it removes surface soiling and associated microbial contamination. Incorrect or inefficient peeling can lead to excessive wastage of sound raw material in addition to increasing the effluent problem. Various methods are used such as steam peeling, mechanical peeling, flame peeling, abrasive peeling and lye peeling or combinations thereof.

1. Steam peeling. The raw material, usually vegetables, is exposed for a short period of time to steam under pressure in a retort which revolves slowly, mixing the product. After steaming the pressure is released quickly causing a steam flash under the skin, thus breaking the surface, facilitating peeling. The product is discharged on to rubber-coated rollers running at different speeds so that the vegetables rotate whilst passing under high pressure water from jets. The skins are removed by this action and discharged into a hopper underneath.

2. Mechanical peeling. This is generally a knife cutting action applied to fruit such as apples and pears and may be carried out on equipment designed specially for each type of product. Modern mechanical peelers incorporate automatic feed and hydraulically controlled rotary knives.

3. Flame peeling. This method is rarely used but may be particularly suited to some products, e.g. pimentos. The produce is conveyed through a furnace where it is subjected to direct contact with live flames and the charred skins are subsequently removed by high-pressure water sprays.

4. Abrasive peeling. The product undergoes a tumbling action so that its surfaces are subjected to the action of an abrasive material such as a carborundum surface, thus loosening the skin which is removed by water sprays.

5. Lye peeling. This is generally carried out by passing the produce through a tank of heated sodium hydroxide solution and subsequently removing the skins by high pressure water sprays in a rotary washer. Sometimes an acid dip is used following the washing operation. A short lye treatment may be used in advance of an abrasive peeler to increase capacity and to reduce weight loss.

Details on various types of peeling machines are given by Slade (1967).

A new development is the so-called dry caustic peeling system designed to reduce or eliminate the need for large amounts of fresh water. After lye immersion, the skins are removed by flaming and a dry scrubbing action. Further details of the process are given by Farrall (1976).

Blanching

Many products such as vegetables, fruit and meat are scalded or blanched in hot water at 87.5° to 95°C, or exposed to live steam. Blanching has various objectives as follows: (1) removal of respiratory gases which would reduce the ultimate vacuum in the can if they were liberated during processing; (2) inhibition of enzymic reactions which might occur during the period prior to heat processing leading to an adverse effect on product quality and nutritive value; (3) shrinkage of the product thus permitting an adequate fill of the container; (4) further cleansing of the product; (5) hydration of certain dried products which are not fully rehydrated prior to the blanching operation.

There are generally two types of equipment used for water blanching. The most common is the rotary immersion blancher in which the product is conveyed by a screw on the inside of a rotating reel, the bottom part of which is immersed in hot water. Chain conveyors are also used for transporting products through the hot water zone. The other is the hydraulic type pipe blancher wherein the product is pumped with hot water through a continuous length of pipe, steam jets along the pipe length being used to heat the water and assist the product flow.

Continuous steam blanchers are mechanically more complex, the product being moved through a tank containing steam. Ray (1975) has described a steam blancher with hydrostatic seals which is claimed to use 50 per cent less energy than conventional steam blanchers.

Hot-water blanching requires the use of large volumes of water much of which runs to waste. Since there is direct contact with the product, some nutrients are lost and this, in turn, gives an effluent problem. Other methods of blanching have been suggested as a means of overcoming this problem: namely, hot-gas blanching and IQB (Individual Quick Blanch) but neither has yet been widely applied on a commercial basis.

Hot-gas blanching is described by Robe (1973) and various trials conducted by the National Food Processors Association are reported by Ralls et al (1973). The IQB method has been described by Lazar et al (1971), the principle of which is to subject each particle of food for a relatively short period to an atmosphere of steam, and then to accumulate the product in a deep bed without further application of heat until equilibration of temperature takes place. Field trials on the IQB method have been reported by Bomben et al (1973) and comparisons with other methods were made by Lund (1973). Generally the results obtained for hot-gas blanching and IQB blanching indicate that the products are of similar quality to that obtained from steam or water blanching but that nutrient loss is reduced and the volume of water effluent is reduced by up to 99 per cent of that obtained by the conventional methods.

Filling

The filling of containers, which is accomplished mechanically or by hand, requires to be carefully controlled. This applies not only to the gross weight of material filled, but, where the product is non-uniform, as in the case of vegetables in a liquid, to the amount of each material filled. Apart from the economic aspect to the packer or consumer, the filling of the correct weight of material has an important bearing on other operations in the canning process. For example, the efficiency of exhausting procedures is, in part, dependent on the amount of free space (i.e. headspace) above the surface of the food in a can, whilst the ratio of solid to liquid material markedly influences the rate of heat penetration into the container and thus affects the processing treatment. With sterilisation processes in which agitation is used for increasing the rate of heat penetration the headspace aids mixing of the contents and therefore its extent has a profound effect on the process (*see* Ch. 9).

In addition to controlling the weight of food filled, it is necessary, for certain types of pack, to give attention to the method of filling in order to prevent occlusion of relatively large volumes of air. It is, for example, inadvisable to fill thick sauce or gravy over pieces of meat, vegetables, etc., as this tends to result in the trapping of air at the base of the can. Special care is also necessary to avoid the trapping of air in the filling of semi-solid products such as meat and fish pastes.

Exhausting

An essential operation in the canning process is the removal of air from the container before it is closed. This procedure is necessary for the following reasons:

1. Minimisation of strain on the can seams or pouch seals through expansion of air during heat processing
2. Removal of oxygen which accelerates the internal corrosion of the container
3. Creation of a vacuum when the container is cooled. Cans with bulging ends being regarded as unsound, it is necessary to ensure that the can ends remain flat or slightly concave throughout moderate changes of storage temperature or barometric pressure.

Additional advantages which attend exhausting include prevention of oxidation and preservation of vitamin C content.

In commercial practice the procedures adopted for the removal of air from the can are:

1. Heat-exhaust
2. Mechanical exhaust
3. Steam injection.

Heat exhaust

In the heat exhaust method, the contents of the can are heated immediately before sealing. The effects of heating are to release air or gas trapped in the product, to expand the product and to displace air in the headspace by water vapour. Each of these factors contributes to the formation of a vacuum when the can is cooled to a temperature below that at which it was sealed. The final vacuum in the can generally increases with the closing temperature but the volume of the headspace in the can is an important factor. If the headspace is too great when the can is sealed, a relatively large amount of air will be trapped in the can with the result that a low vacuum will be formed on cooling. On the other hand, over-filling is to be avoided if the can ends are to remain flat and straining during retorting is not to occur. The filling level should be such that when cool, the surface of the contents is within 6 mm to 9 mm of the top of the can. An incidental advantage which accrues from heat-exhausting, and one which is often of great practical importance, is that it enables a shorter heat-process time to be applied due to the fact that the cans enter the retort partially heated.

Heat-exhausting can be applied in two ways; the contents may be heated immediately before filling into the can, which is then quickly sealed, or they may be filled cold into the can which is then passed through an exhauster (steam-heated chamber) before sealing. In the latter case the cans may be exhausted with clinched ends, that is, ends loosely attached so that escape of air is permitted, seaming being completed when the cans leave the exhauster.

Heat exhaust methods may be applied to flexible pouches, simple hot-filling procedures being suitable for many products. Exhaust box type methods or steam flushing are also feasible and the FMC Corporation has made proposals for a system in which the pouch is opened, filled and sealed in a steam atmosphere claiming this can be done without reducing seal performance.

Mechanical exhaust

In this exhausting procedure, which is often referred to as vacuumising, the cold material is filled into the can, which is then sealed in a vacuum-closing machine, the can being subjected to a vacuum during the seaming operation. The difficulty of completely removing pockets of air or gas from products such as fruit has led to the application of vacuum syruping as a preliminary to vacuum closing or steam injection. In this process, air or gas entrapped in or around the fruit is first removed by vacuum and the vacuum then dissipated by syrup.

The removal of air from flexible pouches may be achieved by a pressure plate system in which the pouch is squeezed slightly so that the total volume of the pack is reduced by squeezing the product into the headspace region whilst sealing. Another method is to seal the pouch in a partial vacuum; equipment for this purpose is widely used. Further methods have been proposed such as the use of two stage sealing stations, the first closing and sealing the pouch except for a small area in the seam through which a vacuum may be drawn before the final sealing operation is carried out. These latter methods necessitate the product being cool otherwise boiling may occur resulting in contamination of the seal. Whichever method is used the air content after processing should not exceed 10 ml per 500 g size pouch.

Steam injection

This method involves injecting a blast of steam into the headspace as the lid of the can is being positioned for sealing. The steam, which is injected through ports around the seaming head of the closing machine, sweeps the air from the headspace; a vacuum is formed when the steam condenses after the can is sealed. The method is most effective for products packed in liquids but may not be suitable for products containing much occluded air, or where the surface of the contents is very irregular and obstructs the free flow of steam. Methods and equipment for elimination of air from canned foods are reviewed by Boyd et al (1952).

Sealing of the container

The formation of the double seams for cans is described in Chapter 4.

Modern can closing machines operate at very high speeds and considerable progress has been made towards higher speeds for the automatic closing of glass jars. The sealing of flexible pouches relies upon the fusion of two thermoplastic materials through application of heat by means of heated pressure plates or jaws. The inner ply or fusion material is generally either polypropylene or a modified high density polyethylene. In either case the dwell time for the sealing operation is about 1 second.

Processing/sterilisation

After exhausting and closing, the containers must be heated for an accurately predetermined time and temperature in an atmosphere of saturated steam, in heated water, or, occasionally in an air-steam mixture. Other heating media such as fluidised beds of inert material of sand or hematite have been suggested and pilot studies of a suitable system have been made by Jowitt & Thorne (1971, 1972). The selection of the heating medium is largely determined by the type of package, those requiring an overpressure during processing, such as glass and pouch packs, generally being processed in superheated water with air overpressure or in steam-air mixtures.

The sterilising action of steam depends largely upon the transfer of its latent heat of vaporisation to the surface of the cans on which it condenses. Dry or superheated steam condenses less readily and is therefore less efficient than saturated steam in transferring heat. The complete elimination of air from the retort is a vitally important factor in steam processing and, unless special provision is made to maintain a uniform steam-air mixture, serious underprocessing may result. Retorts or cookers should be so constructed that removal of air is facilitated. This is brought about by a procedure known as 'venting', the purpose of which is to displace all air in the retort by steam before it is brought to operating temperature. Specifications for vent design and venting procedures are given by the National Canners' Association (1976). Steam enters either at the top or bottom of the retort through spreaders. Experiments on venting of vertical retorts with steam entry at the top are described by Hickman & Robinson (1973). In either case the vent should be remote from the steam inlet so that air is swept from the retort as the steam enters. Immersion of cans in the water at the bottom of the retort is to be avoided since the cans must be completely surrounded by steam if it is to exert its maximum heating efficiency.

One effect resulting from the presence of air in the retort is that for any given pressure the temperature is lower than that obtained with steam alone. The total pressure of a mixture of air and steam is equal to the sum of the partial pressures, but the temperature of the mixture will correspond to the pressure of the steam only. The effects of air in lowering the efficiency of steam sterilisation were investigated by Hoyt et al (1938). Their conclusions were:

1. Air reduces the retort temperature and this is an uneven phenomenon because air, being heavier than steam, tends to stratify below the steam
2. A mixture of air and steam at any temperature is not as efficient as saturated steam at the same temperature
3. Air in the retort cuts down the penetration of the steam.

From these considerations it will be clear that sole reliance in pressure processing should not be placed on pressure gauge readings; there must be agreement between the readings of both pressure gauges and thermometers, the accuracy of which is, of course, important and should be checked periodically. It should be noted, however, that while pressure and temperature agreement must be obtained when operating retorts, such agreement is not necessarily an indication that all air in the retort has been completely removed; air trapped between cans will not affect the readings of thermometers fitted in the shell of the retort.

The need for adequate venting of retorts cannot be over-emphasised and in this connection the design and operation of retorts is of paramount importance. Useful data on the installation of retorts and

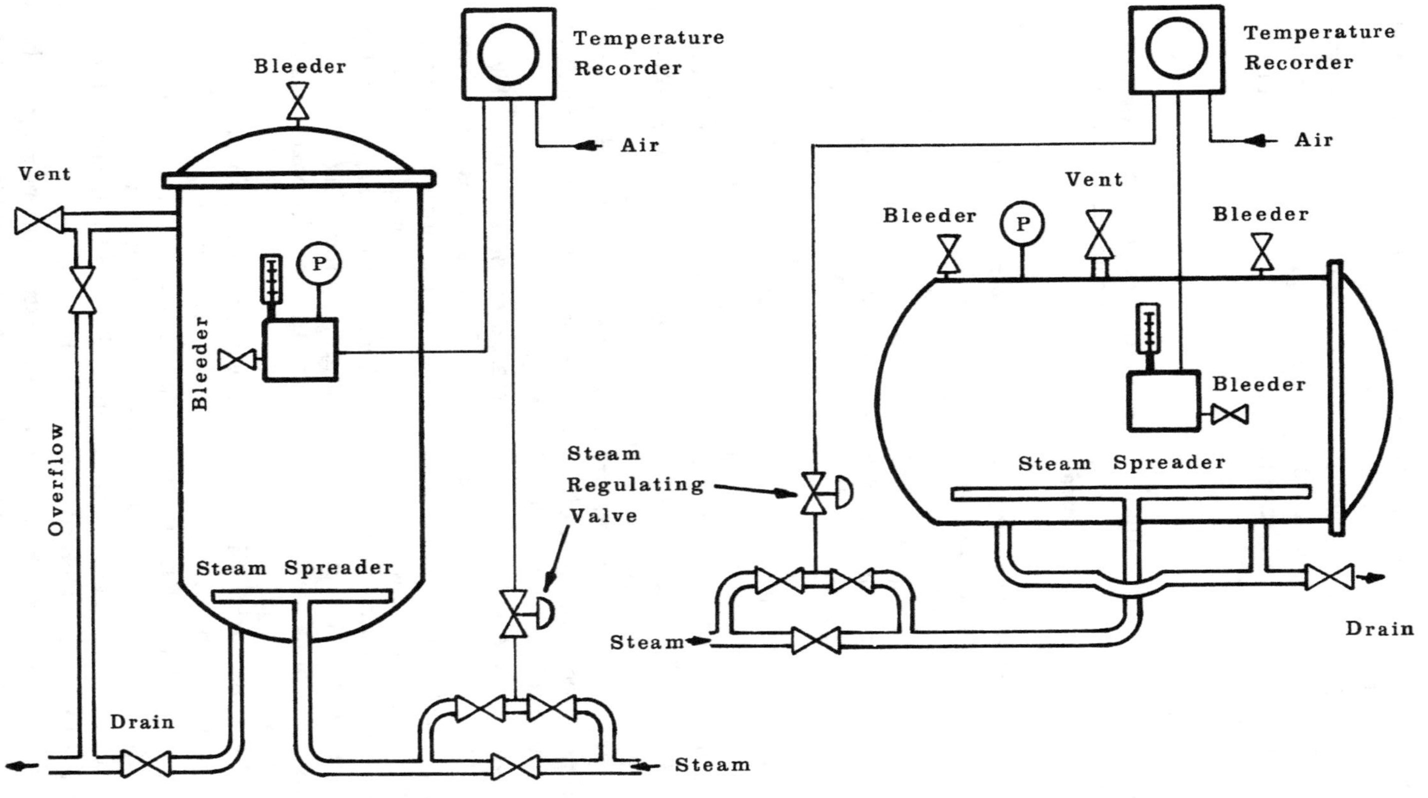

(a) VERTICAL RETORT (b) HORIZONTAL RETORT

Fig. 5 Batch retorts

operating procedures are provided in publications by Thorpe et al (1975) and the National Canners Association (1976).

Processing may be carried out in batch retorts or in continuous pressure cooker/coolers in which the steam pressure is maintained while the cans enter and leave through specially constructed ports or hydrostatic seals.

Batch retorts

A batch retort may be of the still or agitating type and may be designed to operate with saturated steam, with water and an overpressure of air or, less common, with air-steam mixtures.

1. Saturated steam retorts. These retorts are essentially pressure vessels into which the cans may be loaded, usually stacked or jumbled into crates, cars, baskets or trays. They may be of the vertical or horizontal type for top or front loading respectively (Fig. 5). Because of the need for adequate venting and good distribution of steam great care must be paid to the size of the steam headers, the size and positioning of steam inlets and vents and the provision of bleeders to remove air which enters with the steam. Details of retort design may be found in the bulletin 26L published by the National Canners Association (1976). Retorts must be provided with reference thermometers of the mercury in glass type and with pressure gauges to indicate steam pressures. Additionally, to provide a permanent record of the process, time/temperature recorders are fitted. Various types of instrumentation and controllers may be provided and for detail of these the reader is referred to Hughes (1971).

2. Overpressure retorts. These retorts are used for processing of large cans, glass packs and flexible pouches. Sterilisation is carried out under water with an overpressure of air which prevents excessive distension of cans and pouches or loss of closure on glass packs. In general the operating pressure ranges from 1.4 to 2.5 bar depending upon the pack type and the temperature of processing. Details of retorts and procedure may be found in the National Canners Association Bulletin (1971).

3. Air-steam retorts. Since the density of air is twice that of steam, there is a natural tendency for stratification to take place. Stratification of air, or inadequate mixing of air and steam, can lead to serious difficulties in a process designed to use pure steam. However, if a mixture of air and steam can be maintained at a uniform composition, generally by means of forced circulation within the retort using fans or blowers, then the overpressure required to restrain the package can be maintained by the air-steam mixture. Retorts of this type, such as the

Lagarde retort, are available and have been used, particularly for the processing of retort pouches.

Agitation of the can and its contents aids heat penetration which permits the use of higher process temperatures and shorter processing times. Some batch retorts are equipped with means of rotating the retort baskets during the heating and cooling process. This type of equipment is capable of providing an end-over-end type agitation (Fig. 6).

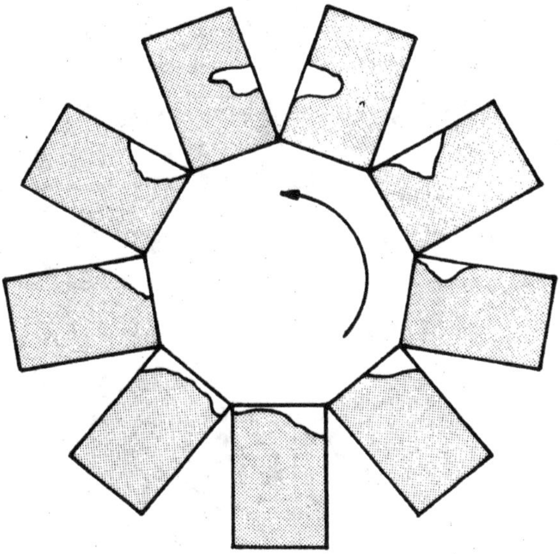

Fig. 6 End-over-end agitation

The Konservomat, which is widely used in Europe, is a fully automated batch retort of this type. The FMC Orbitort is a fully automatic batch steriliser-cooler in which the cans are loaded and discharged simultaneously from a spiral reel assembly. This assembly consists of an inner reel which carries flights that push the cans as they are fed into the stationary spiral on the inside of an outer reel. By this means the cans may be loaded and discharged from the retort. During processing, agitation of the cans is brought about by locking the two reels so that they move together, the flights carrying the cans around the reel. The type of agitation imparted to the cans is shown in Figure 7.

Continuous cooker-coolers

These are either of the rotary type in which the cans enter and leave the pressure shell by means of rotary self-sealing valves, or of the hydrostatic type in which the mechanical valves are replaced in whole or in part by hydrostatic legs of water which balance the internal pressure. The first type is often capable of applying agitation to the container while the hydrostatic type does not necessarily do so and may be considered as a pressurised steam retort through which the containers are conveyed continuously.

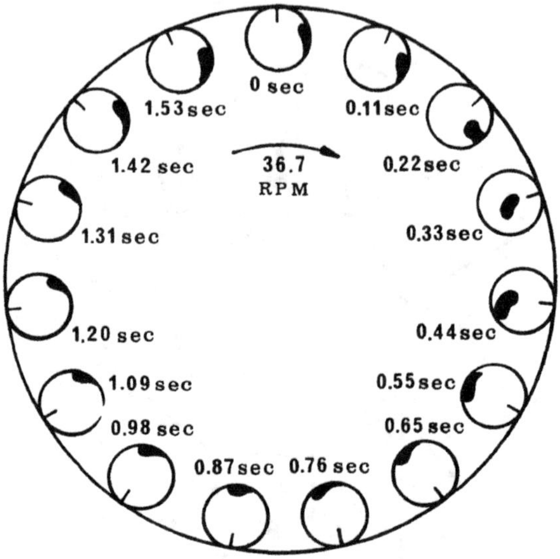

Fig. 7 Agitation in Orbitort steriliser

1. The rotary type. In the Sterilmatic cooker-cooler, made by FMC (Fig. 8), the cans enter the steam chamber through a self-sealing rotary valve and fall into compartments which are at the circumference of a rotating reel. The cans are pushed by the reel through a spiral permanently attached to the inside of the shell, the lead of the spiral progressing the cans through the chamber. A three-phase action is applied to the cans: (a) a spin type agitation at the bottom part of the shell where the cans roll on the inner surface; (b) fixed reel travel in the upper part of the shell where the cans sit in pockets and travel about a central axis and (c) sliding rotation in the transitional stage between the two cycles. When the cans reach the end of the steam chamber they are transferred to the cooler shell where they are cooled under pressure. The Sterilmatic cooker-cooler may be used for cans or for glass containers, the latter being first loaded automatically into cartridges before processing.

The Hydrolock (Fig. 9) is a continuous agitating cooker-cooler, the

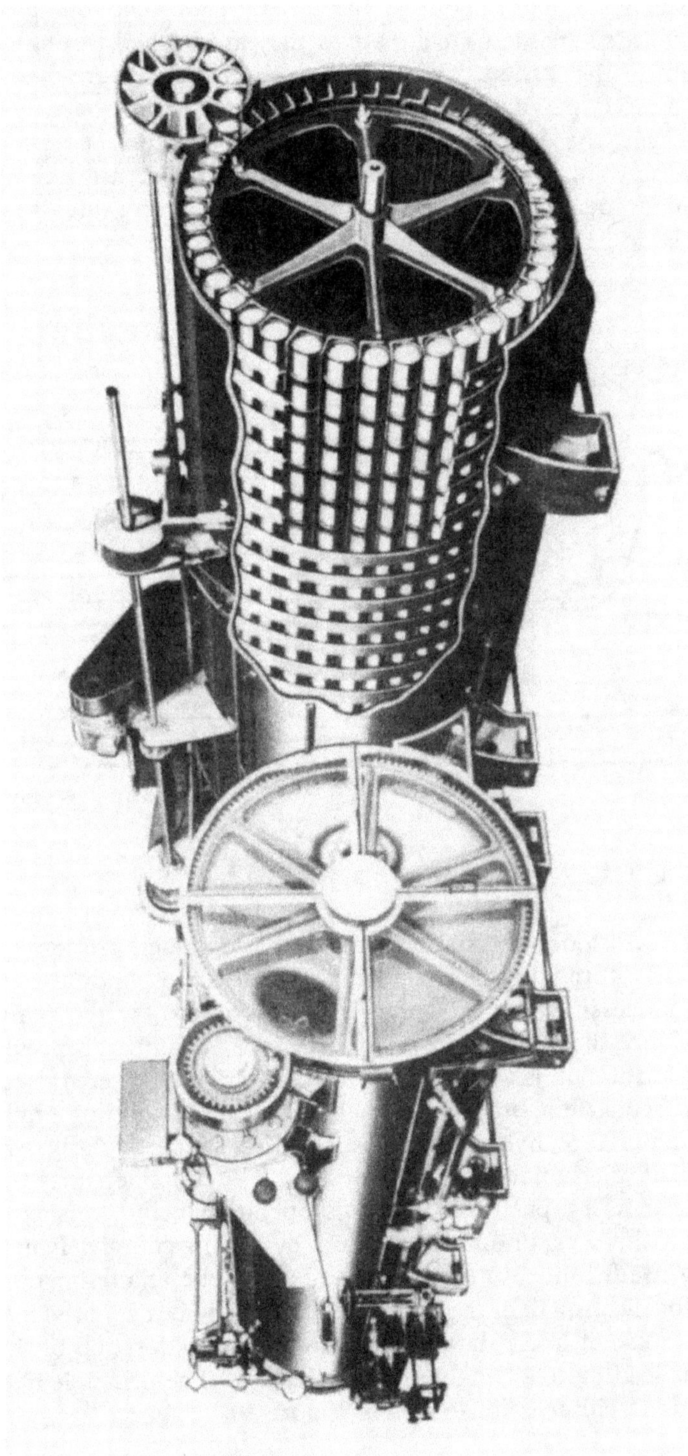

Fig. 8 Sterilmatic cooker-cooler (courtesy of FMC Corporation)

agitation being brought about by the action of the cans rolling on a fixed perforated stainless steel plate as they are pushed through the equipment by the carriers. The containers enter the pressure chamber through a rotary paddle wheel with spring-loaded sealing bars. Water which leaves the vessel through the chain opening enters the atmospheric water leg and flows over a weir to the return line. Any heating medium; saturated steam, water or steam-air mixtures may be used. It is described by Anon (1974).

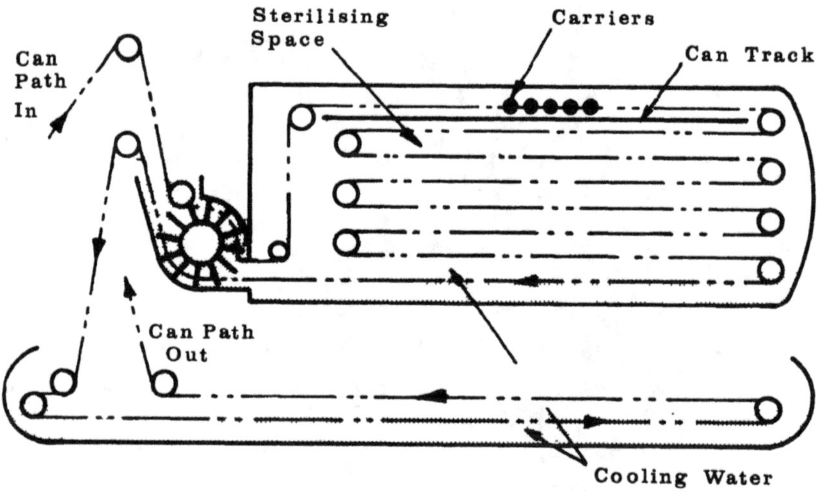

Fig. 9 Principle of Hydrolock steriliser

2. *The hydrostatic type.* In these retorts the containers are conveyed continuously through the chamber at a rate corresponding to the required processing time. The transfer valves consist of water columns which balance the internal pressure of the steam chamber. Since only gentle axial rotation is applied to the containers and they are raised to and from retort temperature gradually, with corresponding gradual changes in pressure, hydrostatic cooker-coolers are particularly gentle in treatment of product and container alike. The containers may leave the exit water leg partly cooled, this operation being finished by spraying with water, or, in some models, by cooling completely in the water leg itself. Since water in the exit leg may be at a temperature suitable for the growth of thermophiles, care has to be exercised in the design and use of hydrostatic cookers to avoid proliferation of these organisms leading to a danger of micro-leakage. The principle of the hydrostatic steriliser is illustrated in Figure 10.

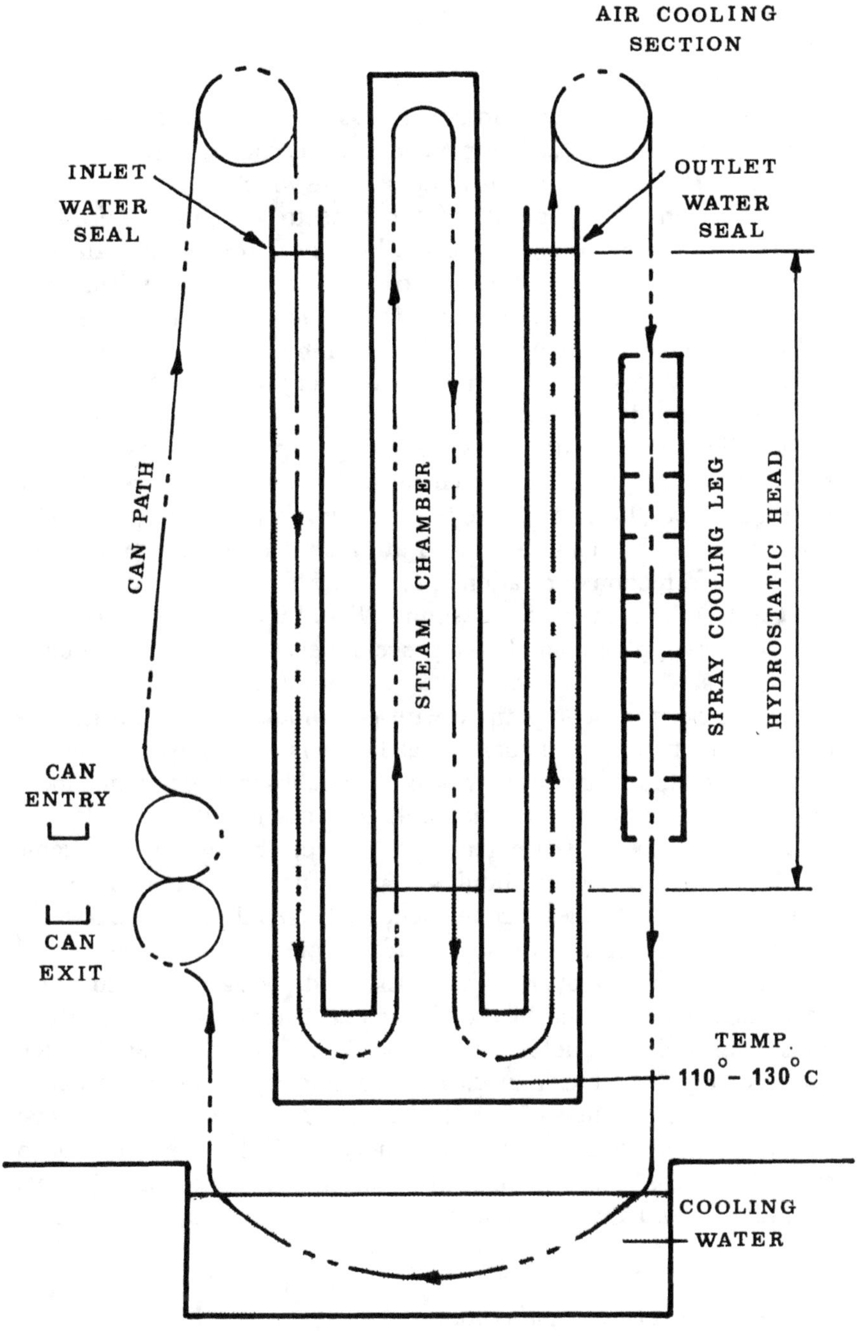

Fig. 10 Principle of hydrostatic steriliser

Flame sterilisation

This is a method of sterilising canned foods by the direct application of gas flames to the can which is rotated rapidly in order to induce convection. Invented in France by Cheftel and Beauvais (1957; 1958 a, b, c) it was called the Steriflamme process. It is described by Beauvais (1964). Commercial Steriflamme installations consist of four sections (Thomas, 1972): (1) a pre-heating section which uses atmospheric steam to bring the contents of the cans to a uniform minimum temperature; (2) a heating section in which the flames impinge directly on the spinning cans so that the contents are heated quickly to a predetermined temperature usually in the range 120° to 130°C; (3) a holding section which uses lower intensity flame burners, intermittently spaced, to maintain the correct processing temperature for the required sterilisation time, and (4) a water spray cooling section. The cans roll on tracks, the processing time being controlled by chain-driven push bars which cause the cans to rotate at 20 to 40 rpm. Hayward (1963) developed a steriliser in which the heat transfer was improved by using a reciprocating shuttle bar to periodically reverse the direction of rotation of the can. This was incorporated into the CSIRO flame steriliser (Casimir, 1975). Since the flame steriliser is capable of applying an HTST process the usual quality advantages accrue (*see* Ch. 10). According to Casimir (1972) the capital and operating costs are competitive with continuous steam retorts. It is claimed that there is no damage to can lacquers or seaming compounds and that lithographed cans may be used. The absence of pressure locks or transfer valves is a considerable advantage but the lack of a counterbalancing pressure tends to restrict the use of the flame steriliser to relatively thin liquids or particulates contained in brine, syrup or juice. With these products the heat is rapidly transferred from the inside of the can into the foodstuff thus preventing excessive build up of pressure due to overheating of the headspace gas. Even so it may be necessary to consider the use of heavier substance or redesigned can ends to minimise can deformation. The CSIRO flame steriliser, which uses the Hayward shuttle mechanism to improve heat penetration, is claimed to be suitable for thicker products than the original Steriflamme equipment. A series of papers by Leonard et al (1975a, b and c) discuss the principles of bacterial inactivation, thermal degradation and the theory of heat penetration as applied to flame sterilisation.

The processing of non-rigid containers

In recent years there has been considerable interest in the development of flexible and semi-rigid processable containers as an alternative to the tinplate can. In particular the flexible pouch which is made from a laminate of thermoplastic and aluminium has received wide acceptance in Japan where 400 million packs per year are produced. In Europe there are some commercial operations but generally the output is on a much smaller scale although the technology has aroused considerable interest both in Europe and in the USA. The pouches may be produced on horizontal form-fill equipment such as the Rexham Bartelt, Bosch Hoeler and Acma machines, or, pre-made pouches may be used. Machines of Japanese design such as that from Toyo Seikan Kaisha Ltd. are available in Europe through licensees such as the Metal Box Company, or a system is available from Bosch, who offer the Hoeler MU machine.

There are two methods in general use for processing pouches. They are heated under water with an air overpressure, or in a steam-air mixture as commonly used in Japan. Overpressure retorts of the water or steam-air type have been described earlier. For continuous operation specialist machines are offered by Stork, namely the pouch Hydromatic (Anon, 1975) and the retort car system from FMC. The FMC system is interesting in that only the retort car is filled with water and not the retort. Consequently cycle times are reduced and energy saved. This system is described by Robe (1978).

An important consideration in processing flexible pouches is pack orientation in the retort. Whilst it is generally advisable to adopt a loading pattern in a retort which encourages the flow of the heating medium, processing of pouches in a vertical position, particularly those containing thin products, would tend to bulge the bottom of the pouch resulting in an increase in cross-sectional area at the base. This changes the heat penetration characteristics. Whilst the pack geometry may be maintained by the use of racks which restrain the pouch thickness within predetermined limits, nevertheless, in Europe, the majority of pouches are processed horizontally. The processing of heat sterilisable flexible packs is reviewed by Rees (1974) and Lingelbach (1976).

The processing of glass containers

Because of the risk of fracture of the glass, lower temperatures and longer exposures are generally used for processing glass-packed foods than is customary for canned products. For the same reason, the heating and cooling of glasses must be carried out slowly. The normal method of processing glass containers entails heating in steam-heated water on which air pressure is superimposed. The headspace in the jar must be regulated to not less than 6 per cent of the container volume (measured at a sealing temperature of 55°C) to allow space for expansion of the contents. The containers are placed in the retort and covered with cold water. Air is admitted to give a pressure of about 7 bar and the water is then heated to the desired temperature by steam. Throughout the process the excess air pressure in the retort is maintained. At the conclusion of the heating period the steam is shut off and the cooling is carried out by slowly admitting cold water to the retort, the caps being prevented from 'blowing' off by the air pressure which is maintained during the whole of the cooling period. Details on the processing of products in glass containers are given by the National Canners' Association (1971).

Cooling

At the conclusion of the heating process the cans are cooled as rapidly as is consistent with avoidance of seam strain in order to prevent overcooking of the foods. This cooling operation is very important and if mismanaged may be the cause of considerable trouble and loss. During heat processing the pressure in the can increases due to the expansion of the residual air, the increase in the vapour pressure of the water in the product and the expansion of the contents. At the conclusion of the actual heating period, the pressure in the can is at its maximum, but is partially balanced by the steam pressure in the retort. Provided that the can is closed under conditions giving a minimum amount of residual air in the can, the difference between the internal pressure in the can and the external steam pressure is insufficient to cause serious straining during the heating period. If, however, the steam pressure in the retort is permitted to drop suddenly, the can may be permanently distorted or the ends may be forced out to such an extent that severe straining of the seam results with consequent risk of leakage. Such straining is especially liable to occur with the larger sized cans.

In order to obviate such straining, pressure cooling may be applied. Fundamentally, this operation consists of maintaining the retort pressure during the cooling period by replacing the steam with compressed air, the air pressure being maintained while the retort is flooded with cold water. Pressure cooling may also be carried out under steam pressure. For this operation a steam supply at the top of the retort is necessary to maintain the pressure as the cooling water enters at the bottom of the retort. Packs such as flexible pouches or glass containers which are processed with overpressure are invariably cooled under pressure usually by gradually replacing the hot water by cold whilst maintaining an air overpressure. Details of the operation of cooling procedures are given in the publication by Continental Can Co. Inc. (1950).

Where pressure cooling in the retort is not applied, the retort pressure is allowed to drop slowly to atmospheric pressure and the cans are then removed and sprayed with cold water or passed through a water tank. The cans should be cooled to an average temperature of 37°C; this will result in their retaining sufficient heat to ensure rapid drying and thus obviate rusting.

At all times cooling operations call for the use of bacteriologically clean water; this is especially true for pressure cooling.

Coding of cans

It is customary for canners to stamp the ends of cans with a code mark denoting the contents, date of production, etc. Each canner devises his own system of coding, the usual practice being to use a combination of letters or numbers. Codes are usually embossed on the canner's or closing end; careful adjustment of the embossing machine is necessary to avoid damage to lacquer or tin coating and a consequent tendency for corrosion to occur. Special marking inks applied by means of rubber stamps or by electrostatic deposition are also used for coding.

REFERENCES

Anon 1974 Food Production/Management 96 (June): 6

Anon 1975 Food Manuf. 50 (10): 27

Beauvais M 1964 Rev Conserve 19 (7): 209

Bomben J L, Dietrich W C, Farkas D F, Hudson J S, deMarchena E S, Sanshuck D W 1973 J Food Sci 38: 590

Boyd J M, Heinen J M, Parrin F W 1952 Engineering the air removal from canned foods. Res Dept Bull No 29, Cont Can Co Inc, Chicago

Casimir D J 1972 In: Flame sterilisation. Specialist Courses for the Food Ind No 2 AIFST-CSIRO, North Ryde, NSW, Australia, p 10

Casimir D J 1975 CSIRO Food Res Quart 35: 34

Cheftel H, Beauvais M 1957 Fr Pat 1: 154, 099

Cheftel H, Beauvais M 1958a Fr Pat 1: 164, 343

Cheftel H, Beauvais M 1958b Fr Pat 1: 180, 283

Cheftel H, Beauvais M 1958c Fr Pat 1: 189, 333

Continental Can Co Inc 1950 Retort installation, equipment and operating procedures, Res Dept Bull No 19, Chicago

Farrall A W 1976 Food Engineering Systems Vol 1. Avi Publishing Co, Westport

Hayward H M 1963 Australian Pat 268: 750

Hickman A, Robinson D J 1973 Process Biochem 8 (7): 21

Hoyt A, Chaney A L, Cavell K 1938 J Bact 36: 649

Hughes P 1971 Food Manuf 46 (11): 23

Jowitt R, Thorne S N 1971 Food Engng 43 (11): 60

Jowitt R, Thorne S N 1972 Food Trade Rev 42 (6): 9

Lazar M E, Lund D B, Dietrich W C 1971 Food Technol 25: 68

Leonard S, Merson R L, Marsh G L, York G K, Heil J R, Wolcott T 1975a J Food Sci 40: 246

Leonard S, Marsh G L, Merson R L, York G K, Heil J R, Fryer S, Wolcott T, Ansar A 1975b J Food Sci 40: 250

Leonard S, Marsh G L, Merson R L, York G K, Buhlert J E, Heil J R, Wolcott T 1975c J Food Sci 40: 254
Lingelbach H 1976 Fleischwirtschaft 56 (5): 615
Lock A 1969 Practical canning. Food Trade Press Ltd, London
Lopez A 1975 A Complete Course in Canning, 10th edn. The Canning Trade, Baltimore
Lund D B 1973 In: Proc 4th Nat Symp on Food Processing Wastes. Syracuse, New York (NTIS Rept PB-234606), p 167
National Canners Association 1971 Bull 30-L, 4th edn. Washington
National Canners Assocation 1976 Bull 26-L, 11th edn. Washington
Ralls J W, Maagdenberg H J, Yacoub N L, Zinnecker M E, Reiman J M, Karnath H D, Homnick D N, Mercer W A 1973 In: Proc 4th Nat Symp on Food Processing Wastes. Syracuse, NY (NTIS Rept PB-234606), p 178
Ray A 1975 Food Processing 36 (1): 64
Rees J A G 1974 FMF Rev 9 (1): 10
Robe K 1973 Food Processing 34 (1): 12
Robe K 1978 Food Processing 39 (1): 92
Slade F H 1967 Food processing plant, Vol 1. Leonard Hill Books Ltd, London
Thomas G 1972 In: Flame sterilization. Specialist Courses for the Food Industry No 2 AIFST-CSIRO, North Ryde, NSW, Australia, p 49
Thorpe R H, Atherton D, Steele D A 1975 Canning Retorts and their operation, Tech Manual 2, Campden Food Presn Assoc, Chipping Campden

6

Sources and control of contamination

Sources of contamination

Reference to the section dealing with the effects of heat on micro-organisms will show that a very important relationship exists between the number of spoilage organisms present in the food and the degree of heating necessary to ensure their destruction. Both practical and laboratory experience teach that greater heat treatment is required to destroy a large number of organisms than a small number. There are few canned foods which do not deteriorate when heated beyond a certain period which varies with the nature of the product. Since normal heat treatment is usually adequate to ensure sterilisation only when the product is not excessively contaminated by spoilage organisms, it follows that in the preparation of food for canning it is necessary to exercise control over the number of potential spoilage organisms entering the can before processing.

There remains the possibility that micro-organisms may gain access through the seams of the can after processing. Seam leakage is most likely to occur during the can cooling stage and the incidence of spoilage which results from such leakage is greatly influenced by the bacteriological quality of the cooling water.

For practical purposes the sources of microbial contamination can be considered under four main headings:

1. Raw materials
2. Plant, utensils, etc
3. Containers
4. Cooling water.

Raw materials

All raw materials for canning purposes must be sound and clean if the quality of the final product is not to suffer. The use of any material showing obvious signs of spoilage will inevitably result in some deterioration in the quality of the pack. Scrupulous cleanliness and strict adherence to the principles of sanitation are necessary in the preparation and packing of the food. Soil is frequently a prolific source of micro-organisms and many outbreaks of spoilage have been traced to contaminants derived from this source. It may be noted in this connection that infection with a particular organism may be confined to vegetables or fruit derived from a single locality. In our experience carrots supplied by a certain farm were found to be heavily infected with a gas-producing thermophile which caused an outbreak of spoilage; carrots from other localities, some not far removed from the infected area, were free from this organism. Cameron (1938) mentions a similar case in which the normal process was inadequate for asparagus grown in some sections of certain fields.

An essential preliminary to the canning of fruits and vegetables is the removal of adherent soil and dust from the products by a thorough washing process. Gillespy (1948) has published data showing the effect of initial infection on the sterility of canned peas and the reduction of contamination achieved by various washing methods. He concluded that improved washing of vegetables can be as effective in reducing the degree of non-sterility of cans as 5 or 10 minutes of process time at 116°C. Even after thorough washing, soil-contaminated products may still carry considerable numbers of spoilage organisms. If, because of inadequate cleansing, the product is packed with adherent soil, the number of organisms entering the can may be so great that normal processing conditions may be inadequate to deal with them. According to Mercer et al (1958), investigations carried out by the National Canners' Association in America have indicated that the major cause of spoilage in canned tomato products is the failure to remove heat-resistant micro-organisms of the butyric acid or flat-sour groups in the washing process. Analyses of soil samples taken from the tomato fields showed alarmingly high spore counts; furthermore the tomatoes, when received at the cannery, were usually heavily contaminated. Mercer et al (1958) state that the objective must be to remove soil from the fruit and to prevent soil build-up in the washers. They outline important factors affecting the efficiency of the washing process.

For the washing process, a clean, copious supply of running water is required. The use of a static or recirculated supply, even though it is discarded and renewed frequently is unsound practice since the build-up of contamination in the water under such circumstances may be so rapid that after a short time the material passing through the water becomes more heavily infected than it was originally. Ruyle et al (1946) found that spoilage developed in 32 per cent of the cans packed with blueberries washed in recirculated water, whereas less than 5 per cent of the cans spoiled when fresh water was used. In their paper

Mercer et al (1958) emphasise the importance of a final spray wash on tomatoes. In one system, without this spray, the use of 1100 gallons of water per ton of tomatoes gave a reduction of only 63 per cent in the bacterial load but with the total volume of water reduced to 500 gallons per ton, of which 200 gallons was in the form of a final spray, the reduction in bacterial load averaged 90 per cent.

The use of wetting agents in the water used for washing vegetables is of value in reducing infection. Knock (1955) reported that Deciquam 222 (didecyldimethylammonium bromide), used in a concentration of 1:1000 to 1:2000 with a contact period of 30 seconds, markedly reduced the number of flat-sour spores on peas. Washing of fruit has been shown to reduce contamination from the heat resistant moulds *Penicillium* and *Byssochlamys*.

In addition to the microbial contamination carried by extraneous soil or dust, practically all the ingredients commonly used in canned food products are potential sources of contamination. Because of their importance in canned food spoilage there have been numerous investigations into the sources of thermophilic organisms. These researches have mainly been carried out in America where higher prevailing temperatures cause these organisms to have a greater significance as spoilage agents in canned food than they normally possess for products packed in this country. As vehicles for infection with thermophilic organisms, sugar and starches are exceedingly important and it has been found necessary to set up standards limiting the thermophilic contamination of these ingredients for canning purposes.

Apart from sugar and starch, raw materials which may also contribute large numbers of thermophilic spores to the product are milk powder and tomato purée. Others are those, such as mushrooms, which are difficult to wash adequately. Allen et al (1946) have shown that crude beet sugar may be heavily contaminated with thermophilic organisms, many of which presumably survive in the finished product. Richmond & Fields (1966) surveyed a number of food ingredients for thermophilic spore-forming bacteria. They found *Bacillus coagulans* to be the most widely distributed followed by *B. subtilis* and *B. licheniformis*.

Miskovic & Petrovic (1971) investigating the thermophilic bacteria in raw peas found gas-producing anaerobic bacteria, flat-sour bacteria and sulphite-reducers to be present. Only the flat-sours remained after the canning process. Reddy et al (1975) determined the thermophilic counts of dried milk and found an average count of 1948 per g for the roller dried product and 1490 per g for the spray dried.

The incidence of mesophilic clostridial spores in fresh meat has been investigated by Steinkraus & Ayres (1964) and by Greenberg et al (1966). Contamination was found to be low, the latter workers reporting the higher figures averaging 2.8 putrefactive anaerobes per gram of meat. No spores of *C. botulinum* were found by Steinkraus and Ayres whilst Greenberg and colleagues obtained only one isolate from 2358 samples of meat. However, Roberts & Ingram (1976) quote a survey carried out in the UK which showed two cells of *C. botulinum* per kilogram of meat. Matsuda et al (1975) found low clostridial counts but reported *Bacillus* spores at levels of 10 to 100/g in frozen meat and 100 to 1000/g in thawed meat.

Apart from materials used in relatively large bulk, minor ingredients may be heavily contaminated. Jensen et al (1934) have reported spoilage of canned chopped ham caused by organisms carried by spices. Spoilage of canned frankfurter sausages due to *Clostridium* and *Bacillus* species contributed by the spices and the soya gluten has been reported by Fievez & Granville (1965). Thangamani et al (1975) found large variations in the contamination of spices but reported up to 51×10^5 and 5×10^3 thermophiles and spore-formers in turmeric and ginger respectively. Castell (1944) found aerobic and anaerobic thermophiles to be present in relatively large numbers in spices, unprocessed cereal products, food stabilisers and powdered milk products. Crossley (1938) found colouring materials and spices to be a prolific source of bacteria, including putrefactive anaerobes. Cipolla et al (1976) have reported heat resistant strains of *B. subtilis* and *B. stearothermophilus* in carob seed flour which is used as a thickener or stabiliser. Of 450 water samples from private wells and dairy water supplies, approximately 35 per cent contained aerobic types, usually in very small numbers; thermophilic anaerobes were present in about 25 per cent of the samples. Knock & Baumgartner (1947) have reported that meat extract may be a prolific source of heat-resistant 'flat-sour' spores.

Except in the case of thermophilic contaminants in sugar and starches, no general standards exist by which hygienic quality of canning materials can be measured. With regard to sugar and starches, the American standards for thermophilic spoilage organisms (*See* Ch. 16) have secured wide acceptance and undoubtedly the institution of these tests has resulted in an improvement in the bacteriological quality of sugar and starches for canning purposes.

Apart from the selection, based on bacteriological tests, of material relatively free from spoilage organisms, contamination control can sometimes be achieved by other methods.

Spices, in particular, are often subjected to some form of disinfection and methods of treatment have been reviewed by

Gottschalk (1977). Heat treatment, if carried out at a sufficiently high temperature, is effective in reducing the contamination but, in general, is inapplicable since it leads to some loss of volatile oil and to a reduction in quality (Thiessen, 1970, Thiessen, 1971). The use of gaseous fumigants such as ethylene oxide or propylene oxide may also drastically reduce the spore counts of raw materials (Pappas & Hall, 1952, Coretti, 1957). The use of ethylene oxide for sterilisation of foods and packing materials has been reviewed by Driessen & van Duin (1975). A disadvantage is that it reacts with inorganic chlorides to form chlorhydrins (Wesley et al, 1965) and if traces of salt or other chlorides are present undesirable residues remain in the food. In a study comparing the effects of ethylene oxide, microwave radiation and gamma radiation on the quality of six different ground spices, Vajdi & Pereira (1973) found that the use of ethylene oxide gave a reduction in the volatile and non-volatile oil content of black pepper and allspice. Coretti & Inal (1969) detected a definite change in the aroma of mustard treated with ethylene oxide. In view of the disadvantages of heat treatment and gaseous fumigation the use of ionising radiations has been suggested as a means of treating spices. Irradiation with gamma rays at doses between 1 and 2 Mrad, generally leads to complete sterilisation although the dose varies with the spice (Goto et al, 1971). Experiments by Inal et al (1975) on ground paprika, black pepper and caraway seed showed that with 1 Mrad all spices except one sample of paprika were sterile. A joint FAO/IAEA panel (1970) stated that radioappertisation, i.e. a dose of several Mrads employed to destroy all micro-organisms except viruses and some radiation resistant spores, is applicable for treating spices and condiments. Very often lower dose irradiation is adequate and Farkas (1973) considers 0.4 to 0.5 Mrad sufficient to reduce the cell count to the level required in the meat and canning industries.

Preliminary heat treatment of certain other canning ingredients may sometimes be employed as a practical means of contamination control. For example, in some circumstances, meat, fish and vegetables may be sterilised by retorting in bulk before incorporation in canned products and we have found that preliminary treatment of vegetables by boiling in 1 per cent citric acid solution is effective in reducing contamination by thermophilic spore-formers. Ultra-filtration is also a useful means of eliminating contamination from soluble ingredients such as meat extract and sugar.

Plant

Apart from aesthetic consideration, the necessity for conducting canning operations under sanitary conditions is clearly reflected in the literature dealing with the bacteriological spoilage of canned foods. In the bottling of fruit juices by the 'cold process', extreme precautions against contamination, even to the sterilisation of air in the bottling room, are employed. Although such measures are unnecessary in normal canning processes, it is desirable to minimise atmospheric contamination by regular cleaning of surfaces such as walls, window-sills, etc., on which dust may accumulate. An extensive outbreak of canned food spoilage investigated by Baumgartner & Wallace (1936) was caused by an organism present in air-borne dust derived from a neighbouring factory.

It is well established that some of the most important sources of contamination are those arising within the canning plant. This has particular reference to spoilage caused by the 'flat-sour' thermophilic organisms, the biological characteristics of these types being such that they are well adapted for growth in heated equipment such as fillers, blanchers, mixing tanks, etc. The role of such units in outbreaks of 'flat-sour' spoilage has been stressed by Cameron & Esty (1940), Bohrer (1949) and others. Similar observations have been made in England and the Fruit and Vegetable Canning and Quick Freezing Research Association directed attention to the possibility of flat-sour spoilage of canned peas arising from blancher contamination. Such contamination arises on the inner surfaces above the water level, and if measures are taken to maintain these surfaces above 74°C thermophilic development is greatly reduced or eliminated. Bashford (1945) carried out a bacteriological survey of plant and isolated heat-resistant spoilage organisms from a wooden framework inside a steam-heated continuous exhauster. Cans passing through the exhauster were open to receive condensed steam draining from the infected wooden surfaces. Rotary brushes in sieving machines and pipelines with blanked-out T junctions were also noted as contamination foci by Bashford.

We have noted an instance of extensive product contamination from compressed air and vacuum lines not included in the normal clean up. The air and vacuum lines were connected through a common manifold to a whisk so that the product could be sucked into the whisk and later expelled by compressed air. Small quantities of food sucked into the manifold provided a source of contamination which continually infected subsequent batches each time the compressed air was turned on.

The control of microbial contamination arising within the canning plant lies in the correct design and operation of equipment and efficient cleaning procedures properly carried out at the end of each production period.

Equipment design and operation

In the past canning equipment was frequently designed mainly with mechanical considerations in view and without sufficient attention to sanitation aspects. The risks of contamination foci arising in the plant can be minimised or eliminated by attention to certain principles of design and construction which may be summarised as follows:

1. The equipment should be constructed from non-porous materials which will not contaminate the food chemically and which will resist corrosion not only from foodstuffs but also from the chemicals to which it will be exposed during cleaning and sterilisation.

2. General layout and construction of the various units should be as simple as possible. A straight-through flow of material should be aimed at, thereby eliminating stagnant pockets in which spoilage organisms may develop and serve as a reservoir of contamination for the product passing through the plant.

3. Pipelines should be in short, internally flush-fitting lengths, with quick-release sanitary type couplings to facilitate easy dismantling and cleaning. Elbow joints should be round and 'dead-ends' eliminated. Where valves or taps are inserted in the line, the removable plug type with full-way opening should be employed; apart from being easily removed for cleaning, they afford no opportunity for small pockets of stagnant food to arise.

4. Equipment surfaces should be free from recesses, depressions and ledges in which food can lodge and decompose. Chutes, bucket conveyors, storage vessels, filling hoppers, etc., should have rounded corners and bottoms and be free from rough, pitted areas.

5. To obviate the accumulation of food debris, equipment should be raised from the floor in such a manner as will allow the floor to be cleaned; in the case of heavy machinery such as seamers, the base should be 'grouted in' on the floor in order to prevent seepage of water etc.

6. Wherever possible, wood must be eliminated from canning equipment. Wood is a sanitary menace in so far as its rough grained surface tends to become impregnated with food material in which bacteria may develop. Furthermore, contaminated wooden surfaces are extremely difficult to clean and sterilise. This has been clearly shown by Highlands & Williams (1944), who found that surface scrapings of hardwood packing-tables used in sardine canning contained as many as 700 000 organisms per gram in spite of the fact that the tables were thoroughly scrubbed with detergent solution at the end of each packing period. Similar findings are reported by Kelch & Palm (1958). The hygienic aspects of food plant design are discussed

in papers by Blois Johnson (1960), Brissenden (1966) and Salter (1969).

In connection with contamination derived from plant, difficulty may arise as the result of operating at temperatures favourable to the rapid growth of spoilage organisms. This applies especially to the thermophilic organisms which multiply rapidly in the range 49° to 71°C. As shown by Bohrer (1947), Reed & Bohrer (1951), and others, failure to maintain a temperature outside this range in filling hoppers, blending tanks, blanchers, etc., may lead to a rapid build-up of 'flat-sour' organisms. Where such control of temperature is impracticable, it may be advisable to empty the equipment and flush thoroughly with clean cold water every three or four hours in order to interrupt any possible build-up of contamination.

Cleaning

In considering the implementation of a cleaning programme various points must be taken into account, such as: (1) the complexity of the plant to be cleaned; (2) the type of soil to be removed, that is, whether it is fatty, proteinaceous, inorganic or a combination of these; (3) the material from which the plant is constructed, particularly in regard to its corrodability; (4) the possibility of carry-over of cleaning compounds to the food and (5) the cost of the operation. Methods range from fully automated clean in-place (CIP) systems to plant dismantling and hand-washing procedures. Most cleaning programmes are based on a combination of CIP and manual methods. Whatever the choice an important prerequisite is an adequate supply of water of suitable quality, that is, of potable quality and preferably not unduly hard. Cleaning may be achieved by purely physical and mechanical means but it is often supplemented by chemicals sometimes with the assistance of an added detergent. If detergents are used care must be exercised in their selection. Guidelines on the selection of detergents and biocides are given by Wildgoose (1972), Bensink (1974) and Anon (1975).

Detergents. In general, detergents are based on: (1) the alkalis and the alkaline salts such as sodium hydroxide, sodium metasilicate and sodium carbonate; (2) the phosphates such as sodium tripolyphosphate or tetra sodium pyrophosphate, used primarily for prevention of deposition of insoluble metallic salts; (3) surface active agents, generally either anionic or nonionic in type or (4) acids, such as phosphoric acid, which have a specific action against mineral soiling and may be used in addition to, and after the use of, the alkaline types which remove fat and protein. The selection of the detergent system depends on considerations as outlined previously but it should be noted that for CIP methods stronger detergents and higher temperatures may be used than in hand-washing procedures.

Biocides. The use of biocides should be regarded as an ancillary operation to the removal of food debris or soil deposit although they are often incorporated into detergent formulations. The biocides most widely used are based on: (1) the chlorine and chlorine releasing compounds; (2) the iodophors; (3) the quaternary ammonium compounds and (4) the amphoteric compounds.

Because of their low cost and wide spectrum of kill, chlorine and its compounds are widely used in the canning industry for factory and plant cleaning and for the disinfection of can cooling water. Since they are corrosive to most common metals care should be exercised in their use. Chlorination may be carried out with chlorine gas or with solutions of chlorine compounds such as sodium hypochlorite. For information concerning the merits and demerits of chlorinating materials and the use of chlorine for water disinfection the reader is referred to publications by Harris (1947) and by Malpas (1971).

In water containing ammonia, amino and other organic substances, the chlorine dosage may need to be considerably increased above that for clean water in order to give an effective residual. It should be noted also that chlorine has little power of penetrating into solid media particularly if they are of organic nature. That the efficiency of a given dosage of chlorine is dependent on purity of the water is shown in Table 3 given by Bashford (1945).

Table 3 Efficiency of chlorine dosage in waters of differing purities

	Organisms per ml	
Time of exposure to 0.8 to 1.0 p.p.m. Cl_2	Clean water	Water containing 1% sterile soil
0 minutes	1 850 000	1 500 000
20 minutes	Sterile	340 000
40 minutes	Sterile	12 000
60 minutes	Sterile	680

The reaction of chlorine with ammonia to form pure chloramines or with organic nitrogen compounds to form chloro derivatives is of importance in connection with the use of chlorine as a bactericide. The addition of chlorine to normal commercial waters as used in the food plant first produces such compounds, but, as the dosage is increased a point is reached, known as the 'break-point', at which they are oxidised. Increasing additions of chlorine beyond this point produce a residual of free chlorine present in the form of hypochlorous acid or the hypochlorite ion, the relative amounts of each being dependent upon the pH of the solution. Since the undissociated hypochlorous acid is the effective germicidal agent the efficiency increases as the pH becomes lower. The chlorine residual existing below the 'break-point' is known as the 'combined available residual', and that above as the 'free available residual'. In waters of normal pH 'free available residual' chlorine is a much more efficient bactericide than chlorine present in the combined form. Furthermore, 'combined available chlorine' is not able to oxidise to the same extent as is 'free available chlorine', many objectionable taste and odour producing substances sometimes being obtained with marginal chlorination. It has been reported that high levels of organic matter in the water, such as that found in recirculated supplies, make 'break-point' chlorination impracticable. In such cases, Welch & Folinazzo (1959) have found chlorine dioxide to be a more effective agent. The efficiency of chlorine and chloride dioxide as bactericidal agents has been compared by Benarde et al (1965).

A level of free chlorine of between 5 and 7 p.p.m. may be maintained continuously and applied during operation to conveyor belts, elevators and similar apparatus; for cleaning operations a level of 15 to 20 p.p.m. is often used. Reports on the use of 'in-plant' chlorination are numerous and Somers (1951) who investigated the effect of chlorination on bacterial populations and slime formation in several canneries, concluded that residual levels of 5 p.p.m. were adequate for normal operating conditions and produced lower populations of mesophilic aerobic bacteria and reduced slime formations at many points. No off flavours were detectable provided that water used for fruit syrup was not chlorinated; no increased corrosion was attributed to using a residual of 5 p.p.m. for normal operations and 20 p.p.m. for cleaning purposes. A detailed account of the use of chlorine in good plant sanitation is given by Mercer & Sommers (1957).

Iodophors are acidic solutions of iodine complexed with a nonionic surface-active agent. They release iodine when diluted with water providing a quick microbial action against a wide variety of micro-organisms, although not against bacterial spores. They are effective at relatively low temperatures, indeed they should not be used above 45°C, and if the concentration does not exceed 25 p.p.m. they may be used without an after rinse. They must be used at a pH below 4.0 which may be an advantage where hard water scale is a problem but they become ineffective under alkaline conditions and for this reason are often avoided in situations where alkaline cleaning has been carried out. Their use in the food industry has been reviewed by Davis (1971).

The cationic surface-active agents known as quaternary compounds (QAC's) are good disinfectants, although more effective against

Gram-positive than Gram-negative micro-organisms, non-corrosive and stable in solution. They are, however, inactivated by soaps and anionic detergents and all traces of these substances should be removed before QAC's are used. Their biocidal activity is affected by pH, being most effective under alkaline conditions. QAC's are particularly suitable for the sanitation of food equipment and may be left in contact with surfaces overnight, before rinsing, in order to prevent multiplication of micro-organisms in the plant during the shut down period.

The amphoteric compounds (alkyl or acyl amino acids), which combine detergent and disinfectant properties, are compatible with anionic, nonionic and cationic agents and are not affected by hard water. They are also non-toxic and non-corrosive. However, they are expensive and not very powerful as disinfectants although they have an action on both Gram-positive and Gram-negative bacteria.

In order to maintain the canning plant in a sanitary condition, the necessity for certain cleaning procedures during the operating period should not be overlooked. The accumulation of waste, spillage, etc., on floors and equipment should be prevented by cleaning at intervals throughout the day. Seaming machines and fillers, where spillage is apt to be especially heavy, should be hosed down during break periods and where necessary other equipment should be similarly treated taking care not to splash soil back on to clean equipment. Containers used for transporting and weighing perishable raw materials should be washed with a suitable detergent and rinsed before being used a second time.

Immediately after the plant is shut down, thorough cleaning of every item of equipment is essential since the development of spoilage organisms in residual food material left in pipes, fillers, hoppers, etc., will result in heavy contamination of the product subsequently passing through the plant. A high standard of cleaning is required and the procedure adopted should be based on a survey of the plant by a competent technician. Many items of equipment such as certain types of pumps, emulsifiers, fillers, blanchers, sorting and grading equipment etc., cannot be cleaned properly without dismantling and it is usually necessary to include a mechanic in the cleaning team which should comprise trained operators working under an efficient supervisor.

The following summary is offered as a guide for general cleaning procedure:

1. Flush through the normal route with warm water remembering to include any by-pass or return pipework. Where necessary, dismantle equipment and remove gross food residues with warm water. Very hot

water is not recommended for this initial cleaning phase as it tends to cause coagulation of food particles in the pipelines, nozzles, etc. Piper (1944) has pointed out that contrary to general practice, water at a temperature of 43° to 48.5°C is most suitable for the cleansing of dairy equipment.

2. After the initial wash, the plant is rendered physically clean (i.e. free from all traces of food debris, grease, etc.) by brushing and the use of a suitable detergent solution. A detailed discussion on the selection of the type of detergent is outside the scope of this book and the reader is referred to the publication prepared by The Association of Food Industry Sanitarians Inc (1952). However, the choice of detergents is determined by a number of factors, principally (a) the type of soil to be removed, that is, whether it is fatty, proteinaceous, inorganic or a combination of these, (b) the corrosive effect on the plant surface, (c) the temperature that can be attained in the cleaning operation (d) the possibility of taint and (e) the cost of the cleaning operation.

3. As a third stage in the cleaning procedure, the equipment should be flushed thoroughly and continuously with hot water until it becomes sufficiently heated to dry rapidly. By this treatment, residual detergent solution is removed and the risk of contamination build-up over night eliminated since there can be no development of spoilage organisms in clean, dry plant. In this connection it is important to ensure that steam valves are completely closed in order to prevent accumulation of condensate; where worn valves permit escape of steam, these should be renewed or the steam supply cut off at the main. The importance of drying plant after use has been noted by Hawley (1947) in connection with the control of bacterial contamination of condensed milk. Maxcy (1975) discussing the fate of bacteria exposed to washing and drying on stainless steel surfaces found this was destructive to many vegetative bacteria.

4. When all equipment has been cleaned and reassembled, the floor and gullies should be well hosed with warm water and then scrubbed with hot detergent solution. After rinsing with water, a solution of sodium hypochlorite (containing 0.02 per cent Cl_2) is applied over the whole area for at least 5 minutes. Finally, the area is rinsed with water.

Hydraulic-jet cleaning devices of various types are available. With these, detergent solutions may be drawn from a tank and mixed with the desired proportion of hot water in a venturi tube. For the interiors of tanks high pressure sprays may be used, the solutions being supplied and recirculated by pump. Special procedures designed to increase contact time between the detergent and the soiled surface, for example, foam cleaning and gel cleaning, are sometimes employed. In the former process a foaming agent is added to the detergent solution which is then sprayed through special equipment in order to produce a thick foam. In the latter process a gelling agent is added to the detergent solution enabling it to stick to the surface longer. These processes are described by Cox (1970).

The complexity of modern plant has necessitated the adoption of cleaning methods which do not involve extensive dismantling. Cleaning may be achieved by methods in which detergents and rinse solutions are circulated at high velocity through a closed circuit producing a turbulent scrubbing action in the pipelines. Such methods are referred to as clean in-place (CIP). In order for the cleaning to be effective due consideration must be given to the design of the valve, pump and pipe configurations. For further information on design, papers by Lloyd (1972) and Pardue (1973) may be consulted. In order to achieve the scrubbing action required a minimum velocity of about 1.5 to 2.0 m per second is maintained in the pipelines which should be pitched with a slope of at least 1 in 200 to allow drainage. Dead ends in pipework must be avoided since they will not receive an adequate treatment and may become a focal point for infection. On large plants completely automatic systems are often installed. Further information on CIP is given in papers by Lloyd (1969) and Elmore (1975). Whatever the cleaning procedure adopted the efficiency of the system must be established during the commissioning stage and then checked periodically by bacteriological tests as indicated in Chapter 16.

Containers

Apart from micro-organisms which may enter through seam or closure defects after packing, the container is the ultimate source from which the food may derive contamination. From the purely bacteriological aspect the importance of this depends on the type of pack concerned. For example, the bottling of fruit juices, certain types of sauce, milk and the canning of sweetened condensed milk demands a more vigorous attention to container contamination than is necessary in the case of pressure-processed foods. For any food product, however, there is no justification for the use of soiled, dusty containers which should be washed with hot detergent solution before filling. Normally, the bacterial content of cans, bottles, etc., is related to their dust content (air-borne bacteria being mainly associated with dust particles) and any container showing considerable dust contamination can be expected to give a high bacterial count.

That the bacterial contamination of containers may reach a high level has been shown by Fellers (1928). In a survey of cans and glasses from a number of canneries and container manufacturing plants, counts as high as 162 000 bacteria and 22 000 mould spores per can were found. The bacteria included a high proportion of spore-forming types, some of which were thermophilic. Lang (1935) also examined cans from a number of different factories and found 1 per cent yielded toxic type A strains of *Clostridium botulinum*. During a microbiological survey of apple juice processing, Forgacs (1942) found that cans, even after rinsing with warm water, gave counts of over 3 000 000 organisms per 25 sq cm of surface. It should be mentioned that most can manufacturers' pack and deliver cans in cartons or on quick turn-round pallets and this minimises contamination.

The normal method of washing cans is to subject them to a series of water sprays, but, for sensitive products in which all infection from the container must be avoided, more rigorous methods, such as gas flaming, have been employed. For some products, relying on regulation of water activity for preservation, the presence of residual water is undesirable and dry cleaning methods have been employed. Geddis et al (1974) describe the use of ionised air to charge the dust particles which are then blown out by compressed air and removed by a vacuum pump.

Cooling water

The bacteriological condition of the water used for cooling cans is important since the cans may become spoiled due to leakage of cooling water through the seams. Leaker spoilage is normally associated with faulty can construction, for example sub-standard seams, but it should be borne in mind that all cans are at risk during the immediate post-process operations. This is due to a number of factors such as: (1) the transition from an internal pressure to a partial vacuum within the can which tends to suck the cooling water through the seams; (2) the relatively soft condition of the lining compound due to the previous heat treatment (this is particularly important in the handling of cans filled aseptically and seamed at high temperature in Dole equipment) and (3) the motility of, or the mobility afforded to, micro-organisms when the outside surface of the can is wet.

Usually, water drawn from a municipal supply is satisfactory for can cooling purposes as it leaves the main but where, for economic reasons, such water is recirculated for use, its bacterial population may reach many millions per ml in a relatively short time; there are usually ample bacterial nutrients in the water in the form of food washed off the exterior of the cans. Raw river waters are especially liable to be excessively contaminated for use in cooling equipment and supplies drawn from wells should be checked bacteriologically, as these too may be found unsuitable. Gartland (1941) states that spoilage resulting from water contamination is negligible when the bacterial count is less than 100 per ml and it is generally agreed that the standard for cooling water should be similar to that for a potable water (about 100 organisms per ml).

The relationship between the degree of contamination of cooling water and the amount of leaker spoilage in canned foods is clearly illustrated in a number of published reports dealing with the use of chlorinated water for cooling purposes. For example, Scott (1937) found that as a result of reducing the bacterial contamination of cooling water by chlorination, spoilage could be reduced to one-fifth or less of that which occurred among cans cooled in highly contaminated water. Data from Scott's report are given in Table 4.

Table 4 The effect of chlorination of cooling water on spoilage rate (Scott, 1937)

		Swelled cans per 1000	
Can size	Product	Chlorinated cooling water	Non-chlorinated cooling water
306 x 302	Vacuum-packed corn	0.46	3.44
603 x 408	Vacuum-packed corn	0.39	5.17
303 x 406	Cream-style corn	0.29	1.17
603 x 700	Cream-style corn	1.37	7.80

Chlorination of can cooling water is the usual method of control. This may be carried out by injecting chlorine gas or by the addition of chlorine compounds such as sodium hypochlorite. Automatic dosing is widely employed usually in conjunction with recirculation and re-use of the water. The level of chlorination required will depend on the quality of the water but it is usual to aim at 4 to 5 parts total residual chlorine per million of water at the can cooling zone. Free chlorine should be present. In order to ensure a good bacteriological condition of the cooling water an adequate contact time between the chlorine and the micro-organisms must be maintained (Levine et al, 1942; Gillespy & Thorpe, 1961). Everton & Dutson (1964) recommend an exposure time of 20 minutes and point out that an adequate chlorine level in itself is insufficient to maintain satisfactory cooling water hygiene. The requirement of adequate contact time may be met by correct siting of inlet and outlet pipes and proper design of the cooling system. Information on the design of cooling water systems is given by Thorpe & Everton (1968).

There have been instances of botulism arising due to post-process leakage involving *Clostridium botulinum*, type E, e.g. in canned tuna (Johnston et al, 1963) and recently in canned salmon (Gilbert, 1978). Whilst canned salmon and canned tuna may not be water cooled, nevertheless, the chlorine resistance of spores in general and those of *C. botulinum* in particular is of interest to the canning microbiologist. The sporicidal properties of halogens have been investigated by Cousins & Allan (1967), Dye & Mead (1972) and others. Odlaug & Pflug (1976) have reviewed the work of various authors and in their paper they have recalculated data in terms of the time required for a 90 per cent reduction in spores as a function of the hypochlorous acid concentration. This analysis clearly showed that *Bacillus* spores are more resistant than *Clostridium* spores. The lethal effect of chlorine in solution was found to increase with (1) the free chlorine concentration, (2) a decrease in pH, and (3) an increase in temperature. They anticipated that the predominant spores in can cooling water would be resistant *Bacillus* spores if free chlorine levels were maintained at 2 to 5 p.p.m. and the pH in the range of 7 to 7.5. In practice the commonest types of micro-organisms found in can cooling water are chromogenic bacteria generally referred to as flavobacteria which are relatively resistant to the action of chlorine and tend to form clumps in cooling water systems. Bean & Everton (1969) have shown that some of these bacteria are capable of causing spoilage of canned milk products.

The effectiveness of chlorine on the spores of *C. botulinum*, types A, B and E has been investigated by Ito et al (1966). Exposure of spores to 4.5 p.p.m. of free available chlorine and phosphate buffer with a pH of 6.5 caused a 99.99 per cent reduction in 4 to 6 minutes for type E strain and in 3 to 8 minutes for types A and B. Thus it would appear that the survival and/or development of *C. botulinum* spores in properly chlorinated cooling water is unlikely.

Although chlorination of cooling water is a valuable means of reducing spoilage resulting from leakage it cannot be regarded as a safeguard for poor seaming. The cooling of poorly seamed cans in almost sterile water may prevent infection at the cooling stage but there will remain the risk of subsequent contamination or leakage of air through seams which may lead to internal corrosion.

Control of leaker spoilage

In order to maintain a satisfactory control of leaker spoilage, it is useful to understand the mechanism involved. This has been reviewed by Put et al (1972). The various investigations covered by them showed that a number of factors contribute to the bacterial reinfection of canned foods. These are: (1) the mechanical construction of the can as indicated by double seam characteristics and by side seam and lap construction; (2) the adequacy of soldering and the type, amount and disposition of the lining compound; (3) the diameter, profile, shape and length of the leakage path and the capillary action within this path and within the adjacent area of double seam; (4) the time the double seam remains wet and the number of bacteria in the water around the seams; (5) the viscosity of the contaminating liquid; (6) the number and severity of the reversible and non-reversible deformations to which the double seam is subjected whilst wet; (7) the vacuum within the can and (8) the occurrence of sudden fluctuations of pressure within the can during sterilisation and cooling. Therefore measures to minimise can leakage should take the above factors into account and will include steps such as ensuring sound can seams, maintaining the bacterial content of the can cooling water at a low level, careful manipulation of the retorting process in conjunction with adequate exhausting of the cans at closing and avoidance of abuse during handling.

Although micro-organisms normally enter with the water during the cooling process the danger of infection exists as long as the cans are still moist. Once the seams are externally dry this risk is considerably diminished. Thus can driers should be installed as near to the exits of continuous coolers as possible and cans cooled in retort baskets should be allowed to dry before they are handled by operatives. Runways and conveyors carrying wet cans soon become wet themselves and bacteria multiply on their can contact surfaces. Any method of reducing the length of wet runway is useful and the installation of heated runways prior to the can drier is to be recommended. Hollow section rails carrying steam at low pressure (0.5 bar) are effective in prevention of bacterial multiplication.

All post-process conveyors or handling equipment should be regularly cleaned at least once every 24 hours and regularly disinfected at more frequent intervals preferably by automatic means. Periodic swabbings should be carried out by the plant microbiologist to confirm the effectiveness of the treatment. Strasburger et al (1957) considered a level of infection higher than 125 bacteria per square inch as excessive. The effects of post-cooling handling equipment such as casers, dumpers and un-scramblers have been investigated by Demsey (1958) who concluded that spoilage rates are increased by (1) rough handling (2) contaminated cooling water and (3) handling cans while still wet and contaminated. Operatives who unload retort crates or handle wet cans in other ways may be able, in rare instances, to infect the contents via the film of water on the surface of the can. Thus a high standard of personal hygiene is required from these operatives and those with boils and septic sores or cuts should not be allowed to handle cans after processing. The subject of spoilage of canned foods due to leakage has been dealt with by Troy et al (1963) in a publication by the Continental Can Company, Inc. and in the paper by Put et al (1972). Practical recommendations for post-process sanitation in canneries are given by Thorpe & Everton (1968).

REFERENCES

Allen L A, Cooper A H, Cairns A, Maxwell M C C 1946 Proc Soc appl Bact No 1: 5
Anon 1975 Food Process 36 (2): 18
Bashford T E 1945 Food Manuf 20: 313
Baumgartner J G, Wallace M D 1936 Food Manuf 11: 10
Bean P G, Everton J R 1969 J Appl Bact 32: 51
Benarde M A, Snow W B, Olivieri V P, Davidson B 1965 Appl Microbiol 15: 257
Bensink J C 1974 CSIRO Food Res Q 34 (3): 49
Blois Johnson W 1960 Chem Ind, p 1203
Bohrer C W 1947 Canner 105 (22): 14
Bohrer C W 1949 Canner 109 (10): 12
Brissenden C H 1966 R Soc Hlth J 86 (6): 317
Cameron E J 1938 Food Res 3: 91
Cameron E J, Esty J R 1940 Food Res 5: 549
Castell C H 1944 Food Res 9: 410
Cipolla M, Cantoni C, d'Aubert S 1976 Industrie Alimentari 15 (7/8): 95
Coretti K 1957 Fleischwirtschaft 9: 183, 186
Coretti K, Inal T 1969 Fleischwirtschaft 49 (5): 599
Cousins C M, Allan C D 1967 J Appl Bact 30: 168
Cox D 1970 Food Manuf 45 (Oct): 37
Crossley E L 1938 J Hyg 38: 205
Davis J G 1971 Food Manuf 46 (Aug): 39
Demsey J N 1958 Nat Canners' Ass Inf Letter No 1666 (Convention Issue), p 61
Driessen F M, van Duin H 1975 Voedingsmiddelentechnologie 8 (32/33): 15
Dye M, Mead G C 1972 J Food Technol 7 (2): 173
Elmore D G 1975 Brewer 61: 305
Everton J R, Dutson J L 1964 Proc Int Food Ind Congr, London, p 71
FAO/IAEA Panel 1970 Microbiological specifications and testing methods for irradiated food. Technical Report Series 104, International Atomic Energy Agency, Vienna
Farkas J 1973 Aspects of the introduction of food irradiation in developing countries. Conference Proceedings, International Atomic Energy Agency, Vienna, p 43
Fellers C R 1928 Amer J Publ Hlth 18: 763
Fievez L, Granville A 1965 Annl Méd vét 109 (2): 143
Forgacs J 1942 Food Res 7: 442
Gartland B J 1941 The Crown 29: 38
Geddis J, MacLean R, Russell N, Robe K 1974 Food Process 35 (7): 83
Gilbert R J 1978 Director Food Hygiene Lab, London. Personal Communication
Gillespy T G 1948 Ann Rept Univ Bristol Fruit & Veg Pres Res Stn, p 25. Campden
Gillespy T G, Thorpe R H 1961 Cannery hygiene. Technical Bulletin (6) Fruit and Vegetables Pres Res Assoc, Chipping Campden
Goto A, Yamazaki K, Oka M 1971 Food Irradiation (Shokuhin-Shosha) 6 (1): 35
Gottschalk H M 1977 Fd Irradiat Inf (7): 7
Greenberg R A, Tompkin R B, Bladel B O, Kittara R S, Anellis A 1966 Appl Microbiol 14: 789
Harris J J 1947 Continental Can Co Res Dept Bull No 13
Hawley H B 1947 Proc Soc Appl Bact No 1: 1
Highlands M E, Williams O B 1944 Food Res 9: 34
Inal T, Keskin S, Tolgay Z, Tezcan I 1975 Fleischwirtschaft 55 (5): 675
Ito K A, Seslar D J, Mercer W A, Meyer K F 1966 The thermal and chlorine resistance of *C. Botulinum* types A, B and E spores. In: Botulism 1966. Proc of the 5th Int Sym

on Food Microbiol, Moscow, July 1966. Chapman and Hall, London
Jensen L B, Wood I J, Jansen C E 1934 Ind Eng Chem 26: 1118
Johnston R W, Feldman J, Sullivan R 1963 Public Health Report, Washington 78: 561
Kelch F, Palm A 1958 Fleischwirtschaft 10: 17
Knock G G 1955 Personal Communication
Knock G G, Baumgartner J G 1947 Food Manuf 22: 11
Lang O W 1935 Univ Calif Publs Publ Hlth 2 (1): 1
Levine M, Heller A, Bender R 1942 J Am Water Works Assoc 34 (12): 1787
Lloyd A K 1969 Food Process Ind 38 (Oct): 57
Lloyd A K 1972 Food Manuf 47 (Feb): 37
Malpas J F 1971 Chem Ind 111
Matsuda N, Matsumoto N, Ushizawa S, Kakegawa Y 1975 J Food Hygienic Soc of Japan (Shokuhin Eisegaku Zasshi) 16 (2): 99
Maxcy R B 1975 J Milk and Food Tech 38 (4): 192
Mercer W A, Somers I I 1957 Adv Food Res 7: 129
Mercer W A, Rose W W, Butler C E, Appleman M M 1958 Nat Canners' Ass Inf Letter No 1666 (Convention Issue), p 56
Miskovic P O, Petrovic K 1971 Hrana i Ishrana 12 (11/12): 489
Odlaug T E, Pflug I J 1976 J Milk Food Technol 39 (7): 493
Pappas H J, Hall L A 1952 Food Tech 6: 456
Pardue T A 1973 Food Engng 45 (9): 107
Piper P E 1944 Food Ind 16: 884
Put H M C, van Doren H, Warner W R, Kruiswijk J T 1972 J Appl Bact 35 (1): 7
Reddy P C, Atwal J S, Srinivasan R A 1975 Indian J Dairy Sci 28 (4): 289
Reed J M, Bohrer C W 1951 Canner 112 (17): 10
Richmond B, Fields M L 1966 Appl Microbiol 14: 623
Roberts T A, Ingram M 1976 Proc 2nd Int Symp on Nitrites in Meat Products. Tinbergen B J, Krok B (eds), Wageningen Centre for Agric Pub and Doc
Ruyle E H, Pearce W E, Hays G L 1946 Food Res 11: 274
Salter S C M 1969 Food Process Ind 38 (Oct): 43
Scott G C 1937 Canning Age 18: 190
Somers I I 1951 Food Tech 5: 46
Steinkraus K H, Ayres J C 1964 J Food Sci 29: 87
Strasburger L V, Gieseker L F, Jewell F 1957 Nat Can Corp Lab Tech Bull (102)
Thangamani I, Mattada R R, Sankaran R 1975 Indian Food Packer 29 (2): 11
The Association of Food Industry Sanitarians Inc 1952 Sanitation for the food preservation industries. McGraw Hill, New York
Thiessen F 1970 Fleischwirtschaft 50 (6): 813
Thiessen F 1971 Fleischerei 22 (11): 15
Thorpe R H, Everton J R 1968 Post-process sanitation in canneries. Technical Manual No 1 Fruit and Vegetable Preserv Res Assoc, Chipping Campden
Troy U S, Boyd J M, Folinazzo J F 1963 Spoilage of canned foods due to leakage. Research and Development Dept, Continental Can Company, Inc, Chicago
Vajdi M, Pereira R R 1973 J Food Sci 38 (5): 893
Welch J L, Folinazzo J F 1959 Food Tech 13: 179
Wesley F, Rourke B, Darbishire O 1965 J Food Sci 30: 1037
Wildgoose J G 1972 Process Biochem 7 (Nov): 13

7

Principal spoilage organisms in canned foods

Microbial spoilage in heat-preserved canned foods is due to the activity of micro-organisms which survive the heat-processing of the cans or those which gain access through leaks in the containers after processing. In under-processed canned foods, it is often possible from a knowledge of the processing conditions and the nature of the food to predict the type of organism which is responsible for the spoilage. Such prediction is possible as the result of accumulated knowledge concerning heat resistance, biological activities and other characteristics of the different types of organisms which are associated with various raw materials. In 'leaker' spoilage, however, the type of infection is usually unpredictable since the microbial condition of the cooling media (water and air), from which the spoilage organisms are derived, may be extremely varied.

Acidity classification of canned foods

American workers have shown that the type of microbial spoilage occurring in under-processed packs is related to the acidity of the food. Bigelow & Cameron (1932) proposed a classification of canned foods, based on acidity, as follows:

Non-acid foods: pH value above 6.0
Semi-acid foods: pH value between 4.5 and 6.0
Acid foods: pH value below 4.5.

Cameron & Esty (1940) have suggested modifications to this classification. They recognise four acidity groups, to each of which they have assigned special spoilage relationships:

Group 1. Low acid (pH 5.0 and higher). Meat products, marine products, milk and certain vegetables
Group 2. Medium acid (pH 5.0 to 4.5). Meat and vegetable mixtures, specialities such as spaghetti, soups and sauces
Group 3. Acid (pH 4.5 to 3.7). Tomatoes, pears, figs, pineapple and other fruits

Group 4. High acid (pH 3.7 and below). Pickles, grapefruit, citrus juices and rhubarb.

In so far as canned products in this country are concerned, the pH values given by Cameron and Esty appear to be low in certain instances, as, for example, meat and vegetable mixtures and most of the soup varieties which are usually above pH 5.0.

The principal demarcation in the acidity classification is between the medium acid and the acid groups, that is at a pH of 4.5. Below this point the growth of *Clostridium botulinum*, the most heat resistant of the food-poisoning organisms, is generally regarded as being inhibited, and for foods with pH values below 4.5 pressure processing is, therefore, considered to be unnecessary. With promulgation in the USA of 21 CFR 128b, now recoded as 21 CFR 113 (Thermally Processed Low-Acid Foods Packaged in Hermetically Sealed Containers) the FDA officially accepted a higher value of 4.6. The value of 4.5 continues to be used in Europe the difference representing the extent of the safety margin applied when considering the growth and toxin production of *C. botulinum*.

The lowest pH levels at which growth and toxin production occurs has formed the subject of various investigations. Townsend et al (1954) conducted an extensive study. They found inhibition occurred at a pH of 4.7 and noted that 'the maximum pH level for control purposes was agreed to be 4.6, although for commercial operating conditions canners were advised to keep the pH below 4.5'. Odlaug & Pflug (1978) have reviewed the subject matter and have stated that the minimum pH at which *Clostridium botulinum* will grow and produce toxin is dependent on the initial pH, incubation temperature, time and nature of the food substrate as well as the concentration, type and strain of *C. botulinum* spores involved. They found that in none of the studies was *C. botulinum* able to grow and produce toxin when the pH of the substrate was equal to or below the value of 4.7. However, in this survey no reference was made to reports from Russian workers who report growth at lower values (Rozanova, 1974; Prokhorovich et al, 1976). There is also a report by Hauschild et al (1975) of an experiment in which they challenged marinated mushrooms with spores of *C. botulinum* and showed toxinogenesis at a pH of 4.77 but not at 4.62.

In our view the minimum pH for growth and toxin production from *Clostridium botulinum* spores appears to be 4.7. It is possible that vegetative cells are able to grow and produce toxin at lower values but this is not significant in canned acid foods correctly processed. However, there are products containing solids where some temporary heterogeneity of pH may exist. In such products, particularly, we feel that the value of 4.5 should be kept so as to provide an adequate margin of safety. There are instances where the growth of *C. botulinum* has been reported in acid foods such as tomato juice. This is probably due to the concomitant growth of other micro-organisms. Huhtanen et al (1976) showed that tomato juice inoculated with *Cladosporium* sp. or *Penicillium* sp. developed pH gradients, with the upper portions near the mould mats having pH values near neutrality. *C. botulinum* spores grew out and produced toxin in these mouldy juices.

Cameron (1938) states: 'To the best of our knowledge all processes for non-acid products fulfil the basic requirement that they shall be adequate to destroy *C. botulinum* . . .' This generalisation is, except for certain special products, universally recognised throughout the canning industry. The group of cured or pickled meat products including ham, spiced ham, bacon, etc., usually receive a relatively low process since the curing salts present in the meat, in conjunction with the heat treatment, exert an inhibitory effect on the growth of *C. botulinum* and other putrefactive anaerobes (cf. Yesair & Cameron, 1942; Silliker et al, 1958). Otherwise they are stored under refrigeration. Rodel et al (1976) suggested a classification of meat products into highly perishable, perishable and shelf stable products; the classification being based on their pH and a_w. The highly perishable group (pH > 5.2 and $a\ \ > 0.95$) require refrigeration at $\leqslant 5°C$, the perishable group (pH 5.2 to 5.0 or a_w 0.95 to 0.91) require refrigeration at $\leqslant 10°C$ and the shelf stable group (pH $\leqslant 5.2$ and a_w $\leqslant 0.95$ or only pH < 5.0 or only $a_w < 0.91$) which do not require refrigeration.

In Cameron and Esty's classification the subdivision of the 'acid' group into 'acid' and 'high acid' was necessitated by the incrimination of acid-tolerant, spore-forming bacteria as the spoilage agents in certain canned fruits and tomato products with pH values of 4.5 to 3.7. At the lower end of the range spoilage of this type does not appear to have been experienced in this country. In our experience, the pH of 4.0, rather than 3.7, is a more realistic limiting value for survival, germination and growth of spores in properly heat-treated foods.

In the following brief review the principal spoilage organisms are treated under two main headings, according to their significance for low and medium acid foods and acid foods.

Low and medium acid canned foods

Aerobic spore-formers

The aerobic spore-formers of the genus *Bacillus* are widely distributed in nature. Originating in soil and water, they are almost always present in the raw materials used in canning. For most members of the group, the optimum growth temperature lies between 28° and 40°C, but many are also thermophilic, developing at 55°C and one species can grow at 70°C. With regard to their oxygen relationships, some of the organisms are strict aerobes but others are facultative anaerobes and their growth in canned foods is not inhibited by vacuum.

Many of the aerobic spore-formers, especially the thermophilic strains, may exhibit exceptional resistance to heat destruction and it is, therefore, not surprising that they are important spoilage agents in canned foods; indeed, Jansen & Aschehoug (1951) consider these organisms to be of greater significance than putrefactive anaerobes. The principal form of spoilage produced by the aerobic spore-formers is of the 'flat-sour' type, so called because the organisms attack carbohydrates with the production of acids but without gas formation; in canned foods so spoiled there is consequently no depreciation of vacuum, the ends of the cans remaining flat or concave. Some members of the group, however, can produce gaseous spoilage in which the ends of the cans become distended.

In connection with the role of aerobic spore-formers in canned food spoilage, the work of Smith et al (1946, 1952) on the identification and classification of this group, and the studies by Gordon & Smith (1949) and Allen (1953), must be considered. Smith et al found that many organisms previously described as separate species were merely variants of established species; they also found that certain characteristics, formerly used for identification purposes, were so variable as to be of very limited value. As a result, some former specific names of the *Bacillus* genus have been discarded by them and other species have been demoted to the rank of varieties. Similarly, the work of Gordon and Smith and Allen on thermophilic members of the *Bacillus* group has brought about certain changes in viewpoint regarding these organisms.

The effect of these developments is to make a review of the literature on spoilage of canned foods by aerobic spore-formers a difficult matter since many of the organisms referred to in the reports of spoilage are inadequately described or incorrectly identified. Especially in connection with the microbiology of canned foods, it has in the past been customary to consider the thermophilic strains of the aerobic spore-formers as a group distinct from the mesophilic strains.

The definition of thermophiles widely accepted in the canning industry has been that of Cameron & Esty (1926):

Obligate thermophiles — cultures growing at 55°C but not at 37°C
Facultative thermophiles — cultures growing at 55°C and 37°C.

The investigations of Gordon & Smith (1949), Smith et al (1952), Allen (1953) and others, however, show that the temperature range for growth is not a reliable means of separating aerobic spore-formers. Gordon and Smith examined 216 strains of reputed, and in some cases, named thermophiles (many from spoiled canned foods) and found that the majority could be identified as either *Bacillus stearothermophilus* or *B. coagulans*, the remainder proving to be mesophiles or strains of species normally recognised as mesophiles. A study of the growth temperature relationships of the organisms showed that the maximum and minimum growth temperatures may vary with the strain for any given species. Furthermore, Campbell (1954) and Long & Williams (1959) have demonstrated that the so-called obligate strains of thermophiles may be induced to grow at 37°C if special cultural conditions are provided. Table 5 taken from Gordon & Smith's (1949) paper, clearly indicates that the temperatures of 55°C and 37°C chosen by Cameron and Esty are of little value in the identification of species. Gordon and Smith suggested that the term 'thermophile' should never be used to describe a species; its use should be restricted to denote an organism capable of growth at 55°C. In the face of the evidence available, this recommendation can hardly be ignored. In addition the terms 'obligate' and 'facultative' should be used with caution having due regard to the cultural conditions employed.

Allen (1953), from a survey of the literature, concluded that most of the aerobic or facultative anaerobic thermophiles isolated from canned foods resembled *Bacillus stearothermophilus* or *B. coagulans*. Allen's view is that the former organism is a thermophilic variant of *B.*

Table 5 Thermophilic variants of mesophilic aerobes (Allen, 1953)

	Number of cultures	Number of cultures growing at various temperatures[1] °C								
		28°	33°	37°	45°	50°	55°	60°	65°	70°
B. stearothermophilus	87	0	10	73	81	87	87	87	87	45
B. coagulans	73	53	73	73	73	72	66	23	0	—
B. subtilis	154	154	154	154	150	105	17	0	—	—
B. brevis	57	57	57	57	38	16	7	0	—	—
B. circulans	55	55	55	51	18	6	1	0	—	—
B. pumilus	65	65	65	65	64	43	0	—	—	—
B. macerans	13	13	13	13	13	9	0	—	—	—
B. cereus	50	50	50	50	23	0	—	—	—	—
B. sphaericus	42	42	42	42	15	0	—	—	—	—

1 Water bath temperatures were used in all cases except for 28°C

circulans; she reported that thermophilic variants of a number of mesophilic aerobes can be obtained under natural conditions and in laboratory cultures and concluded that thermophilic strains of almost any mesophilic *Bacillus* may occur; she considers that there is little justification therefore for a separate classification of thermophiles.

With regard to the heat resistance of thermophiles, Bigelow & Cameron (1932) reported that 'obligate' thermophiles are generally more heat resistant than 'facultative' types. Lamanna (1942) and others have noted that the spores of organisms with high maximum growth temperatures generally have a high thermal resistance. Since Gordon and Smith found *B. stearothermophilus* and *B. coagulans* to have the highest growth temperature (*see* Table 5), these organisms can be expected to have a higher heat resistance than other aerobic thermophiles. In the light of these observations, Allen's conclusion that *B. stearothermophilus* and *B. coagulans* are the chief flat-sour thermophiles found in canned foods appears valid.

Knock (1954) found that numerous strains of flat-sour thermophiles isolated from spoiled canned peas in South Africa all conformed to the description of *B. stearothermophilus*; he later (1955) reported that *B. stearothermophilus* was the only flat-sour organism found in a variety of canned vegetables which had received fairly heavy processing. With these organisms, souring rarely occurred at 37°C and never at a temperature of 30°C or lower. Dormant spores of *B. coagulans*, which he finds to be considerably less heat resistant, are chiefly found in canned meats not so heavily processed as canned vegetables. Put & Wybinga (1963) reported the occurrence of a strain of *B. coagulans* with exceptionally high heat resistance, showing a $D_{110\,°C}$ value of 2.4 minutes in neutral buffer. Even so, this value is low compared with $D_{121,1\,°C}$ values of up to 5.0 minutes found for *B. stearothermophilus*.

Spoilage by *B. coagulans* is usually associated with acid products such as tomato juice which, because of its pH, receives a mild heat treatment. Furthermore *B. coagulans* is aciduric in nature, growing in media with a pH down to 4.02 (Pederson & Becker, 1949) whilst *B. stearothermophilus* does not grow below 5.0 (Smith et al, 1952). Jakobsen & Nielsen (1973) isolated 15 strains of thermophilic flat-sour organisms from vegetables at five different canneries and identified them as *B. stearothermophilus*. Inoculated into canned vegetables high sensitivities were established to pH (particularly in combination with temperature), salt and glycerol. They concluded that these characteristics offer possibilities for prevention of flat-sour spoilage. It is our experience that *B. stearothermophilus* is the principal flat-sour organism found in low-acid canned products which have received a fairly heavy processing and that such contamination does not result in souring when the products are rapidly cooled and stored under cool conditions.

Fields (1970) has reviewed the flat-sour bacteria and discusses classification, distribution, spore resistance, sources of contamination and methods of control.

In semi-preserves such as canned hams, which receive a light heat treatment, a more extensive bacterial flora survives. Therefore this type of product is normally stored under refrigeration. Some species of *Bacillus* survive the heat process and may give rise to spoilage if the product is held under unsatisfactory conditions of storage. According to Leistner (1960), the most important *Bacillus* species causing spoilage of canned hams are *B. subtilis, B. cereus, B. megatherium* and *B. polymyxa*.

The majority of the organisms in the *Bacillus* group attack carbohydrates with the production of acids only, but two species, *B. polymyxa* and *B. macerans*, form both acids and gas. Furthermore, gas production may also occur under certain conditions with strains not normally able to form gas from carbohydrates, and although economically the most important type of spoilage caused by the *Bacillus* group is flat-souring, a gaseous form of spoilage may also occur. For example, aerobic spore-formers have been incriminated in the gaseous spoilage of canned cured meats. Experiments on the mechanism of gas production by these organisms were made by Jensen & Hess (1941) who found that nitrate was essential for gas formation. According to Gibson (1943) some *Bacillus* species, grown in strongly buffered alkaline or neutral medium containing sugar and nitrate, produce copious gas as the result of denitrification. Eddy & Ingram (1956) described a denitrifying salt-tolerant *Bacillus* strain isolated from a blown can of bacon.

The production of gas by the *Bacillus* group, however, is not confined to canned cured meats. Michael & Tanner (1936) have observed that gaseous spoilage of canned corn may be caused by such organisms, and Normington (1919) reported gaseous spoilage of canned peas by an aerobic spore-former. Cameron et al (1936) isolated a *Bacillus* species, which they named *B. betanigrificans*, which was responsible for the blackening of canned beets. This organism, which has been found to be in synonymy with *B. macerans* by Smith et al (1952), produced gas in beet juice containing dextrose and in dextrose broth containing iron. Nichols (1939) found that *Bacillus* sp. caused swelling of canned evaporated milk and cream, whilst Aschehoug & Vesterhus (1941) and Jansen & Aschehoug (1951) reported that organisms of the *Bacillus* group were responsible for gaseous spoilage of Norwegian canned vegetables and meat and fish products.

Aerobic spore-formers have been noted as causal agents of a variety of defects in canned milk and cream; prominent among these is the defect characterised by thinning and a concomitant bitter flavour. Nichols (1940) noted a variety of defects produced by these organisms, ranging from definite bitterness and thinning to minor changes in flavour and consistency. Ruyle & Sognefest (1951) and other workers have noted the occurrence of gelling or curdling due to *Bacillus* species. The curdling may be 'sweet' (i.e. that produced by enzymic action), or that caused by acid produced by the organisms. Reports of various types of spoilage of canned milk products are reviewed by Nichols (1940), Morgan (1943), and Hiscox (1947). Grosskopf & Harper (1974) found spores of psychrotrophic organisms in sterilised aseptically packed milk.

The *Bacillus* group have also been incriminated in the spoilage of canned fish products. Fellers (1927) reported non-gaseous spoilage of salmon, crab and shrimp due to the activities of aerobic spore-formers such as *B. cereus*, *B. subtilis*, etc. Spoilage was chiefly characterised by production of unpleasant flavours or odours, alterations in colour and softening of the flesh. Norwegian examinations of incubated cans of fish and meat products, reported by Jansen & Aschehoug (1951), showed that underprocessed cans exhibited a type of flat-sour spoilage. *B. coagulans*, *B. circulans* and *B. alvet/B. circulans* intermediates were isolated.

Spore-forming anaerobes

The spore-forming anaerobes are mainly derived from soil, which fact accounts for their fairly wide distribution in milk, vegetables and other food materials. Some species are also commonly found in human and animal intestines and due to their presence in animal excreta they are frequently contaminants of meat, with the result that they have been prominently associated with the spoilage of canned meat products. These organisms, however, have also been responsible for losses in canned milk, cream, fish, vegetables and other products.

With reference to canned food spoilage, the spore-forming anaerobes of the genus *Clostridium* can be conveniently separated into a thermophilic and mesophilic group.

Of the thermophilic anaerobes, the most important are the saccharolytic organisms which do not produce hydrogen sulphide. The type species was named *Clostridium thermosaccharolyticum* by McClung (1935). The organisms of this group are markedly saccharolytic, producing large quantities of gas, chiefly carbon dioxide and hydrogen, from carbohydrates and consequently they cause spoilage of the 'swell' or gaseous type in which the ends of affected cans are distended. A butyric or 'cheesy' odour frequently accompanies spoilage by these gas-forming thermophiles. Proteins are not attacked. Cameron & Esty (1926) state that in products of moderate acidity (pH 5.0 to 4.5) these organisms are of greater significance as spoilage agents than the 'flat-sour' types.

The optimum temperature of the saccharolytic anaerobes is in the region of 55°C. Of the cultures examined by McClung (1935), only one showed slight growth after 30 days' incubation at 30°C. The members of this group are, therefore, not responsible for appreciable losses in this country since storage temperatures are not normally favourable for their growth. They are, however, a considerable source of trouble in products in hot countries where the storage temperatures may exceed 35°C.

The thermophilic anaerobes characterised by the production of hydrogen sulphide are responsible for the so-called 'sulphur stinker' spoilage in canned foods. The type species is *Clostridium nigrificans* (Werkman & Weaver, 1927) which is synonymous with *Desulfotomaculum nigrificans*. This organism is only feebly saccharolytic and infected cans usually remain flat due to the fact that the hydrogen sulphide is soluble in the contents which become black as the result of interaction between the sulphide and the iron of the container. Spoilage of this type is comparatively rare.

Next in importance to the flat-sour organisms as spoilage agents for canned foods are the mesophilic spore-forming anaerobes. Outstanding in this group is the food poisoning organism *Clostridium botulinum* which is discussed in the section on food-poisoning. Destruction of the spores of this organism is generally accepted as the minimum standard of processing for low-acid and medium-acid canned foods. This species is not, however, as heat resistant as some other spore-forming anaerobes.

The optimum growth temperature of most of the mesophilic anaerobes is about 37°C but many develop at 20°C or lower, and a few can grow at 50°C and above. Although classification of these organisms is primarily based on morphological and other characteristics, a rough grouping is possible on the basis of the relative ability of the various species to attack proteins and carbohydrates. Thus, there is a group which is predominantly proteolytic or putrefactive and which includes such species as *Clostridium hystolyticum*, *C. sporogenes* and *C. bifermentans*, and a predominantly saccharolytic group which is represented by *C. butyricum* and *C. pasteurianum*, *C. perfringens* and other species. The significance of the groups as spoilage agents is related to the type of food and general experience is that for low-acid or medium-acid products the putrefactive group is the most important except in the case of canned hams in which *C. perfringens* is sometimes found (Leistner, 1960).

Among the putrefactive anaerobes, *Clostridium sporogenes* and related strains appear to be most frequently involved in the spoilage of low and medium-acid canned foods. In a study of over 10 000 cans of a wide variety of Norwegian fish and meat products, Aschehoug & Jansen (1950) isolated 46 strains of putrefactive anaerobic spoilage organisms all of which proved to be or were closely related to *Clostridium sporogenes*. Crossley (1941) observed that *Clostridium sporogenes* types were the most serious cause of spoilage in canned meat and fish products, and the spoilage of canned vegetables such as beans, peas, potatoes and tomato products of low acidity was principally caused by organisms of the *Clostridium butyricum* type. In an earlier report, Crossley (1938) stated that spoilage in meat and fish pastes was mainly due to putrefactive anaerobes (*Clostridium sporogenes* and *Clostridium putrificum*) but noted that these organisms failed to develop in products with pH values below 6.0. According to Cameron & Esty (1940), putrefactive anaerobic species may show abnormal development resulting in non-gaseous spoilage in medium-acid products (pH 4.5 to 5).

Usually, the spoilage produced by the putrefactive anaerobes is of the gaseous type and is often accompanied by disintegration of the food and the production of characteristic foul-smelling compounds.

Yeasts, moulds and non-spore-forming bacteria

Apart from spoilage arising through leakage of the container, these organisms of relatively low heat-resistance have no significance in the spoilage of low-acid or medium-acid canned foods with the exception of sweetened condensed milk and canned cured meats such as ham and bacon.

Sweetened condensed milk is not heat processed, its keeping properties depending on its high sugar content. Nichols (1939) states that 'bloats' or gaseous spoilage of this product is usually due to sucrose-fermenting yeasts. Hiscox (1947) names two types of yeast, *Torula-lactis-condensi* (oval cells) and *T. globosa* (round cells), which cause blowing of sweetened condensed milk. *T. lactis-condensi* causes the most vigorous fermentation and infected cans may burst within a few days. *T. globosa* rarely gives rise to more than a distension of the can ends. These yeasts may occasionally be isolated from sound cans showing no trace of spoilage, and Hiscox suggests that in such cases the strains may be less sugar-tolerant than usual or that the initial infection was slight and that lack of oxygen in the can prevents extensive growth.

Mould species, notably *Aspergillus repens*, are associated with the formation of 'buttons' which are formed on the surface of sweetened condensed milk. The 'buttons', which assume a definite form after 8 to 10 weeks' storage, consist of a reddish brown curd mass 0.62 to 1.86 cm (¼ to ¾ in) in diameter. Nichols (1939) states that growth of the mould probably ceases after 2 to 3 weeks, the 'button' continuing to form as the result of the activity of a clotting enzyme. Lipase-secreting bacteria such as *Pseudomonas fluorescens* may cause rancidity as well as yeasts or moulds, and certain species of cocci may be responsible for thickening of sweetened condensed milk.

Although members of the *Clostridium* and *Bacillus* groups cause spoilage in canned cured ham, which is a lightly processed (pasteurised) pack, defects are also caused by non-spore-forming organisms. Buttiaux (1951) has reported that *Streptococcus liquefaciens* may produce marked liquefaction of jelly in canned hams. Coretti & Enders (1964) also attributed softening at the centre of pasteurised meat packs to organisms of this type. According to Buttiaux, faecal streptococci are fairly frequent contaminants, but these organisms do not produce any adverse changes in the pack. Ingram & Hobbs (1954) and others, however, report that off-flavours and odours in canned hams may be caused by faecal streptococci. According to Leistner (1960), streptococcal spoilage is common and is responsible for souring of the stored hams. Both *Strep. faecium* and *Strep. faecalis* are involved but *Strep. faecium* is the most important because of its greater heat resistance.

Organisms belonging to the *Enterobacteriacea* (coliform, *Aerobacter*, *Proteus* species, etc.) have also been found to cause swelling of canned hams (Buttiaux, 1951). The defect in canned hams known as 'soft core' is due to the development of *Streptococcus faecalis var. liquefaciens* (Gardner & Patton, 1975). The gaseous spoilage of canned bacon by micrococci has been reported by Hankins et al (1950). The microbiology of canned cured meats is further discussed in Chapter 13.

Acid products

In the spoilage of acid products a miscellaneous group of organisms is involved which includes non-sporing and sporing aciduric bacteria, yeasts and moulds. In most cases the organisms are readily controlled by relatively short processes at temperatures below 100°C.

Spore-forming bacteria

Townsend (1939) has described spore-forming saccharolytic

anaerobes of the *Clostridium pasteurianum* type which are responsible for gaseous spoilage in canned fruit and tomatoes. A strain studied by him grew under laboratory conditions in canned pear juice down to pH 3.55. Under commercial conditions, however, the development of these organisms in canned products has not been observed below pH 3.7 (cf. Cameron & Esty, 1940). Ikegami et al (1970) investigated gaseous spoilage caused by *C. pasteurianum* and concluded that none occurred at a pH of 3.67 to 3.75 but that a proportion of samples spoiled at pH in the range 3.9 to 3.97.

In phosphate buffer at pH 7.0, Townsend found that these organisms were destroyed in 40 minutes at 100°C. In tomato juice at pH 4.5 the destruction time was 20 minutes and in canned apricot juice at pH 4.15 it was 7 minutes. Spiegelberg (1940) has also reported spoilage in canned pineapple by saccharolytic anaerobes resembling *C. pasteurianum* which, however, were not as heat resistant or as acid tolerant as Townsend's strains. Bowen et al (1954) have investigated the growth of *C. pasteurianum* in tomato puree and whole tomatoes. Using an inoculum of 200 000 spores per can they found, at pH 4.3 to 4.4, that centre temperatures of 93.3° to 95.6°C were required to prevent spoilage, but, at pH 4.0 to 4.1, can centre temperatures as low as 83.9°C gave satisfactory results. Casolari & Giannone (1966) consider that the heat treatments employed by the Italian tomato industry may be inadequate in some cases, particularly if the pH rises above 4.4. They recommend that a centre temperature of 95°C should be reached for products with a pH above 4.2. York et al (1975) carried out research to define the conditions required to obtain commercial sterility in canned whole peeled tomatoes and concluded that an integrated sterilising value $F^{15}_{100\,°C}$ of 1.56 minutes should be safe.

Clark & Dehr (1947) isolated spore-forming anaerobes closely related to *Clostridium butyricum* from swelled canned tomatoes. The thermal death time of some of these organisms in tomato juice at pH 4.4 was between 10 and 15 minutes at 100°C. In Australia the CSIRO Division of Food Preservation have noted spoilage of canned fruits, such as pears, by organisms of the *C. butyricum* type. They suggest pH control and reduction of the number of contaminating organisms by thorough washing as a means of overcoming the difficulty. Hernandez & Feria (1971) investigated the flora of spoiled canned tomatoes and found that the butyric anaerobes *C. butyricum* and *C. pasteurianum* gave rise to the most important form of spoilage. They noted the high heat resistance of some *Clostridium* sp. such as *C. bifermentans* and suggest the existence of heat resistant mutants that give rise to spoilage in correctly processed products.

Another aciduric spore-forming organism which is an important spoilage agent for acid products in America is *Bacillus coagulans* (first described by Berry (1933) as *B. thermoacidurans*). This organism is a thermophile with the general characteristics of the flat-sour type, but is tolerant of an acid environment, growth occurring in the region of pH 4.2. It is of considerable importance in canned tomato juice in which it produces off flavours and souring. According to Anderson, et al (1949) the organism may survive heating in tomato juice (pH 4.4 to 4.5) for 22 minutes at 100°C and is therefore capable of surviving the thermal processing of canned juice. Knock et al (1959) reported the occurrence of flat-sour spoilage in South African tomato juice by *B. coagulans* and determined the heat resistance of the causative organism. The D values at 93.3°C, 96.1°C, 98.9°C and 106°C were 13.4 min, 6.3 min, 3.1 min, and 0.51 min respectively giving a z value of 16.1. Rice & Pederson (1954) investigated a number of factors influencing the growth of *B. coagulans* in tomato juice. They found that the minimum pH at which growth occurs rose as the spore concentration in the inoculum decreased. The acid tolerance varied with the strain of the organism; at a spore concentration of 20 or less per ml of inoculum only 2 out of 20 strains were able to germinate at pH 4.31 or lower.

Other instances of spoilage of acid products by spore-forming bacteria have been noted by American investigators. Woerz & Lenane (1943) have reported gaseous spoilage of canned tomatoes by an aciduric anaerobic thermophile. Vaughn & Stadtman (1946) have observed spoilage of canned fruits (pH 3.8 to 4.0) by a spore-forming aerobe — *B. macerans*. Vaughn et al (1952) have reported several outbreaks of gaseous spoilage of commercially canned fruits and vegetables by organisms of the *B. macerans* — *B. polymyxa* group. These organisms were able to grow at pH 3.8 to 4.0. Circumstantial evidence pointed to post-process leakage, the source of the organisms being cooling water. Spoilage of acid products due to spore-forming bacteria does not appear to have been reported in this country.

Non-sporing bacteria

The most significant group of organisms of this type in the spoilage of acid packs are the Gram-positive, lactic acid-producing bacteria. The group includes both coccus and rod forms, some of which are gas producing. These organisms, which usually develop best under conditions of reduced oxygen tension, are widely distributed and are characteristically associated with the fermentation of vegetable material. Some members of the group are important in several industrial processes, being involved in the desirable fermentations which take place in cheese ripening and in vegetable pickling.

The gas-forming *Lactobacillus brevis* (Orla Jensen), is well known as the cause of a vigorous fermentation in tomato ketchup, Worcester sauce and similar products. Pederson (1929) described a number of organisms of this type which are associated with the spoilage of tomato products and Juven (1976) isolated a strain of *Lactobacillus brevis* from a blown can of grapefruit segments which causes spoilage of citrus products with a pH 3.5. Savage & Hunwicke (1923) reported a coccus, *Leuconostoc pleofructi*, which also belongs to the lactic group, as a spoilage agent in fruit juices. This organism, which is characterised by the production of slime in sugar solutions, was also noted by Pederson as a cause of spoilage in tomato products. Spiegelberg (1940) has described the gaseous spoilage of canned pineapple by *Leuconostoc mesenteriodes* and this organism has also been reported as a cause of ropiness in canned peaches (Fabian & Henderson, 1950). These non-sporing organisms are controllable by moderate heat treatment at temperatures below 100°C.

Yeasts

Due to their low heat resistance, yeasts are rarely involved in the spoilage of heat-processed canned foods except in cases of gross under-processing or leakage. They are chiefly noted as causes of fermentation in acid sauces, jams and similar products, which rely on acids, sugar and salt, either alone or in combination, for their preservation.

Moulds

Noteworthy exceptions to the generalisation that moulds are insignificant spoilage agents in canned foods are the species *Byssochlamys fulva* and *B. nivea*. *B. fulva*, described by Olliver & Smith (1933), has been implicated in the spoilage of canned and bottled fruits. It causes complete disintegration of the fruit due to the breakdown of pectinous material; in addition, infected cans sometimes swell through the production of carbon dioxide. Its optimum temperature for growth lies between 30° and 37°C. In relation to other mould species it is unusually heat resistant. Gillespy & Thorpe (1962) state that it survives 30 minutes at 85°C and 10 minutes at 87.7°C. Eckardt & Ahrens (1977) found that treatment at 95°C for 15 minutes caused inactivation of the spores but not complete destruction.

Hull (1934-35) reported that canned fruit required to be heated to 87.8° to 90.6°C at the centre in order to control *Byssochlamys fulva*. Michener et al (1966), using thermal death-time cans of grape juice containing 10^5 asci per can, found survival times exceeding 62, 25 and 10 mins at temperatures of 87.8°C, 90.0°C and 92.2°C.

The degree of softening of the fruit caused by this species is unrelated to the amount of visible mycelium, complete disintegration taking place when to the naked eye mycelium is not apparent. The species can tolerate a low oxygen tension, growth occurring in a vacuum of 50.8 cm (20 in). According to Olliver & Rendle (1934) *Byssochlamys fulva* is present on fruit in the field and is probably derived from the soil. In a survey by Gillespy & Thorpe (1962) 15 ascospores were detected in 128.5 kg (283 lb) of strawberries; this was lower than in previous surveys which showed an average of a little over 1 ascospore per lb of fruit.

The related mould, *Byssochlamys nivea* has been reported in England (Gillespy & Thorpe, 1962) and Holland (Put, 1964). Gillespy and Thorpe reported that the ascospores of *B. nivea* are less heat-resistant than those of *B. fulva*, surviving 30 minutes at 76.6°C but failing to survive 10 minutes at 82.2°C. Put & Kruiswijk (1964) found the heat resistance of *B. nivea* to be similar to *B. fulva* and claim it to be more commonly the causal agent in spoiled strawberries. The properties of *B. nivea* are summarised by Yates (1964).

A heat-resistant species of *Penicillium* has been isolated from canned blueberries by Williams et al (1941). In blueberry juice, the thermal death time of the sclerotia of this mould was 300 minutes at 85°C although its ascospores were not unusually resistant, being destroyed in 15 minutes at 81°C. Growth occurred in a vacuum of more than 63.5 cm (25 in). Ruyle et al (1946), who found that spores of this mould were commonly present on the raw fruit, consider that control of infection and spoilage lies in thorough washing of the fruit, cleaning of plant and the complete elimination of oxygen from the cans.

Kavanagh et al (1963) isolated a heat-resistant species of *Aspergillus* from canned strawberries. The ascospores survived a temperature of 100°C for more than 60 minutes when suspended in distilled water. The mould was isolated from soil samples taken from farms where the strawberries had been grown. Thus thorough washing of fruit intended for canning is clearly indicated. *Aspergillus fischeri* var. glaber as a contaminant of canned strawberries was investigated by McEvoy & Stuart (1970). They state that whilst ascospores survived the canning process, growth of the fungus in the can was minimal and caused little damage to the fruit.

The survival of heat resistant enzymes in canned fruit has been reported from Australia. Harper (1971) identified a thermostable polygalacturonase produced by *Rhizopus nigricans* thought to be responsible for slow post-canning breakdown of fruits, particularly apricots and Harper et al (1972) noted a slow softening process in canned apricots following the growth of *Rhizopus stolonifer* on the fresh fruit.

REFERENCES

Allen M B 1953 Bact Rev 17: 125
Anderson E E, Esselen W B, Fellers C R 1949 Food Res 14: 499
Aschehoug V, Jansen E 1950 Food Res 15: 62
Aschehoug V, Vesterhus R 1941 Zbl Bakt, Section 2, 104: 169
Berry R N 1933 J Bact 25: 72
Bigelow W D, Cameron E J 1932 Ind Eng Chem 24: 655
Bowen J F, Strachan C C, Moyls A W 1954 Food Tech 8: 471
Buttiaux R 1951 2nd Int Congress Canned Foods (CIPC), Paris
Cameron E J 1938 Food Res 3: 91
Cameron E J, Esty J R 1926 J Inf Dis 39: 89
Cameron E J, Esty J R 1940 Food Res 5: 549
Cameron E J, Esty J R, Williams C C 1936 Food Res 1: 73
Campbell L L 1954 J Bact 68: 505
Campbell L L, Frank H A, Hall E R 1957 J Bact 73: 516
Casolari A, Giannone L 1966 Industria Conserve 41 (2): 95
Clark F M, Dehr A 1947 Food Res 12: 122
Coretti K, Enders P 1964 Fleischwirtschaft 16 (4): 304
Crossley E L 1938 J Hyg 38: 205
Crossley E L 1941 J Soc Chem Ind 60: 131T
Eckardt C, Ahrens E 1977 Chem Mikrobiol Technol Lebensm 5 (3): 76
Eddy B P, Ingram M 1956 J appl Bact 19: 62
Fabian F W, Henderson R H 1950 Food Res 15: 415
Fellers C R 1927 Univ Wash Publ Fisheries 1: 229
Fields M L 1970 Advances in Food Res 18: 163
Gardner G A, Patton J 1975 Proc of European Meeting of Meat Res Workers No 21, 52
Gibson T 1943 Proc Soc Agric Bact, p 13 (Abstract)
Gillespy T G, Thorpe R H 1962 Survey of *Byssochlamys fulva* ascospore infection on strawberries for canning. Tech Mem No 44 Fruit and Veg Canning and Quick Freezing Res Ass, Campden
Gordon R E, Smith N R 1949 J Bact 58: 327
Grosskopf J C, Harper W J 1974 Milchwissenschaft 29 (8): 467
Hankins O G, Sulzbacher W L, Kauffman W R, Mayo M E 1950 Food Tech 4: 33
Harper K A 1971 Chem and Ind 462
Harper K A, Beattie B B, Pitt J I, Best D J 1972 J Sci of Food and Agric 23: 311
Hauschild A H W, Aris B J, Hilsheimer R 1975 Can Inst Food Sci Technol J 8: 84
Hernandez E, Feria M A 1971 Revista de Agroquimica y Tecnologia de Alimentos 11 (1): 126
Hiscox E R 1947 Chem Ind p 365
Huhtanen C N, Naghski J, Custer C S, Russell R W 1976 Appl and Environmental Microbiol 32 (5): 711
Hull R 1934-35 Ann Rept Univ Bristol Fruit & Veg Pres Res Stn, Campden, p 65
Ikegami Y, Okaya C, Sawayama Z, Shimoda Y, Mori D, Oku M 1970 Canners J (Kanzume Jibo) 49 (11): 993
Ingram M, Hobbs B C 1954 J Roy San Inst 74: 1151
Jakobsen M, Nielsen J L 1973 Nordisk Veterinaermedicin 25 (11): 554
Jansen E, Aschehoug V 1951 Food Res 16: 457
Jensen L B, Hess W R 1941 Food Res 6: 75
Juven B J 1976 J Milk and Food Technol 39: 819
Kavanagh J, Larchet N, Stuart M 1963 Nature 198: 1322
Knock G G 1954 Private communication
Knock G G 1955 Private communication
Knock G G, Lambrechts M S J, Hunter R C, Riley F R 1959 J Sci Fd Agric 10: 337
Lamanna C 1942 J Bact 44: 29
Leistner L 1960 Proc of 12th Res Conf of Am Meat Inst Found, Univ of Chicago, Circular No 61

Long S K, Williams O B 1959 J Bact 77: 545
McClung L S 1935 J Bact 29: 189
McEvoy I J, Stuart M R 1970 Irish J Agric Res 9 (1): 59
Michael V M, Tanner F W 1936 Food Res 1: 99
Michener H D, King A D Jr, Ito K A 1966 Bact Proc A15, p 3
Morgan G F V 1943 Proc Soc agric Bact, p 31 (Abstract)
Nichols A A 1939 J Dairy Res 10: 231
Nichols A A 1940 J Dairy Res 11: 274
Normington R 1919 Mich Agric Exp Sta Tech Bull 47
Odlaug T E, Pflug I J 1978 J Food Protection 41: 566
Olliver M, Smith G 1933 J Botany 71: 196
Olliver M, Rendle T 1934 J Soc Chem Ind 53: 166T
Pederson C S 1929 NY State Agric Exp Stn Tech Bull No 150
Pederson C S, Becker M E 1949 NY State Agric Exp Stn Tech Bull No 287
Prokhorovich L F, Shenderovskaya T M, Saltykova L A, Klepa N F 1976 Konservnaya i Ovoshchesushil 'naya Promyshlennost' No 12: 33
Put H M C 1964 J Appl Bact 27: 59
Put H M C, Kruiswijk J T 1964 J Appl Bact 27: 53
Put H M C, Wybinga S J 1963 J Appl Bact 26: 428
Rice A C, Pederson C S 1954 Food Res 19: 115
Rodel W, Ponert H, Leistner L 1976 Fleischwirtschaft 56 (3): 417
Rozanova L I 1974 Trudy, Vsesoyuznyi Nauchno-issledovatel'skii Institut Konservnoi i Ovoshchesushil'noi Promyshlennosti 21: 54
Ruyle E H, Sognefest P 1951 J Milk Tech 14: 173
Ruyle E H, Pearce W E, Hays G L 1946 Food Res 11: 274
Savage W G, Hunwicke R F 1923 Canned fruitFd Inv Bd Spec Rept No 16 HMSO, London
Silliker J H, Greenberg R A, Schack W R 1958 Food Tech 12: 551
Smith N R, Gordon R E, Clark F E 1946 Aerobic mesophilic sporeforming bacteria. Misc Publ 559 USDA, Washington
Smith N R, Gordon R E, Clark F E 1952 Aerobic sporeforming bacteria. Agric Monog No 16 USDA, Washington
Spiegelberg C H 1940 Food Res 5: 115
Townsend C T 1939 Food Res 4: 231
Townsend C T, Yee L, Mercer W A 1954 Food Res 19: 536
Vaughn R H, Stadtman T C 1946 J Bact 51: 263
Vaughn R H, Kreulevitch I H, Mercer W A 1952 Food Res 17: 560
Werkman C H, Weaver H J 1927 Iowa Stat Coll J Sci 2: 57
Williams C C, Cameron E J, Williams O B 1941 Food Res 6: 69
Woerz J A, Lenane D J 1943 Canner 96 No 10: 11
Yates A R 1964 *Byssochlamys nivea*, a food spoilage mould of unusual nature. In: Report of the School of Agriculture, Univ of Nottingham
Yesair J, Cameron E J 1942 Canner 94 No 13: 89
York G K, Heil J R, Marsh G L, Ansar A, Merson R L, Wolcott T, Leonard S 1975 J Food Sci 40: 764

8

Effects of heat on micro-organisms

A knowledge of the fundamentals involved and the various factors which influence the heat-destruction of micro-organisms is essential for a proper understanding of the problems which arise in the heat-processing of canned foods. While some of the factors affecting the destruction of micro-organisms by heat are only imperfectly understood, much work has been expended on the subject in its various aspects and a great deal of information has been accumulated.

In any discussion on heat resistance it is necessary, at the outset, to define what is meant by the death of micro-organisms. From a practical point of view the criterion must be their failure to reproduce when, as far as is known, suitable conditions for reproduction are provided. The vegetative cells of bacteria, yeasts and moulds are destroyed almost instantly at 100°C and do not normally constitute a problem in the heat-processing of canned foods, but the spores of certain species of bacteria are extremely resistant to heat and prolonged exposures to high temperatures are necessary for their destruction. The rate at which destruction occurs is a function of time and temperature which vary with each other, and other factors being equal, the higher the temperature to which the cells are exposed the more rapid is their destruction. The lethal conditions for an organism cannot, therefore, be expressed by temperature alone, that is as a thermal death-point, but only as thermal death-time where both the temperature and the time of exposure are stated. In what follows it will be seen that the heat treatment required to destroy any given organism is markedly influenced by many factors such as the chemical and physical nature of the medium in which the organism is suspended, the number of the organisms, their age, etc., so that for a precise definition of lethal conditions a reference to time and temperature only is not sufficient.

Cause of death and mechanism of heat resistance

The mechanism by which heat causes death of bacteria is still in doubt. However numerous studies have been conducted into various aspects of thermal injury and recovery of vegetative cells. Reviews on this topic have been made by Alwood & Russell (1970), Brown & Melling (1971) and Tomlins & Ordal (1976). Brown and Melling conclude that evidence relating ribosomal heat stability with maximum growth temperature points to a key role for ribosomal RNA. Recent studies have demonstrated damage to the DNA of bacteria after thermal shock, possibly caused by the action of native nucleases which were released or activated by mild heat treatment. The involvement of nucleic acids in the thermal inactivation of vegetative micro-organisms has been reviewed by Gomez (1977) and Pierson et al (1978). Also, damage to the cytoplasmic membrane resulting in leakage of cell constituents has been suggested as leading to the inability of vegetative bacteria to recover from thermal injury. Brown and Melling point out that the ability of bacteria to repair thermal injury is probably related to their heat sensitivity.

During spore development there is an increase in the heat resistance of the organism. A number of theories have been advanced to account for the striking difference in resistance between vegetative cells and spores but two in particular merit special attention. These are the dehydrated protoplast and contractile cortex theory or modifications of it and one involving the role of dipicolinic acid (DPA).

Friedman & Henry (1938) first ascribed the heat resistance of the spore to the fact that a large proportion of its water content is in a form not available for participating in coagulation of the cell protein. There is abundant evidence that spores are more resistant when heated in dry materials. Murrell & Scott (1957, 1966) have confirmed that heat resistance increases in a gradual manner as the water concentration of the environment decreases. For the species tested, maximum heat resistance was found at a_w values between 0.2 and 0.4. It is interesting to note that when spores were equilibrated to these water activities large differences in heat resistance between species disappeared. The high heat resistance of some spores in aqueous media would necessitate the thin spore coat to be of extremely low water permeability and this is difficult to reconcile with the known permeability of organic materials. For this reason Lewis et al (1960) postulated the contractile cortex theory by which the spore cortex is thought to be maintained in a dry state by exertion of hydrostatic pressure following contraction. Murrell & Warth (1965) in a carefully designed experiment with various *Bacillus* sp. investigated the correlation between the heat resistance of spores and other factors such as their calcium and magnesium content and amount of mucopeptide constituents (α, ϵ -diaminopimelic acid (DAP) and hexosamine). The results were analysed in a manner designed to throw light on the mechanism which regulates heat resistance. They were interpreted by Murrell and Warth as supporting the theory that a contraction of spore cortex is connected with high heat resistance. It was suggested that the significant relation between Mg—Ca ratio and heat resistance may be involved in the contraction mechanism. Gould & Dring (1975) have postulated that the dehydrating pressure on the protoplast is maintained by an expanded rather than a contracted cortex.

Powell & Strange (1953) and Powell (1953) showed that the calcium salt of DPA constitutes up to 15 per cent of the dry weight of spores but not of vegetative cells. Upon germination this substance is released with loss of heat resistance (Powell, 1953; Woese & Morowitz, 1958). Church & Halvorson (1959) showed that the rate of heat inactivation of the spores of *Bacillus cereus* was related to their DPA content; Lechowich & Ordal (1960) reported that as the DPA content and cation content increased in the spores of *B. coagulans*, the thermal resistance also increased. Slepecky & Foster (1959) showed that heat sensitive spores were produced when sporulation occurs in a calcium deficient medium and that replacement of calcium by other cations produced spores which were less resistant than those possessing a normal calcium content. However, Byrne et al (1960) found no correlation between the heat resistance and DPA content of *Clostridium roseum*. Walker et al (1961) reported similar findings with the spores of several *Bacillus* species, although they noticed that there was generally a fall in heat resistance as the molar concentration of magnesium increased in relation to the DPA and calcium concentrations. Levinson et al (1961) demonstrated that as the molar ratio of Ca:DPA decreased the heat resistance also decreased. Thus whilst significant correlations of DPA with heat resistance applied only to certain species there is more evidence supporting a relationship between heat resistance and high Ca:DPA ratios. More recently, however, Zytkovicz & Halvorson (1972) and Hanson et al (1972) have found spore mutants of *B. subtilis* and *B. cereus* with low DPA content that are very heat resistant and this theory must be discounted. Holland & Taylor (1974) suggested that the chelation of cations in the spore by DPA may be responsible for dormancy.

There is less information available on the effect of dry heat although it has been known for over 90 years that it is less effective than moist heat for destruction of micro-organisms (Koch & Wolffhugel, 1881). As far as vegetative cells are concerned microbial destruction is thought to be connected with oxidative processes (Murrell, 1964; Ernst, 1968). With spores the D and *z* values (see later for an explanation of these terms) are generally found to be greater with dry heat than with moist heat. For example, Fox & Eder (1969) compared survivor curves of *Bacillus subtilis* spores when subjected to wet and dry heat and found this to be the case. They noted also differences in the shape of the survivor curve and suggested physiological differences between wet and dry heat destruction of bacteria. Undoubtedly the lower moisture content and water activity associated with dry heat sterilisation are significant factors in protecting the spore from the effect of heat. Also it has been suggested that at the temperature of dry heat sterilisation, chemical injury to the genetic material (DNA) occurs (Zamenhof, 1960; Northrop & Slepecky, 1967).

During canning, wet heat is usually involved in the sterilisation process since bacteria are normally suspended in an aqueous food (but see section on fat protection). However in aseptic canning the containers and ends may be sterilised by hot air or by superheated steam, which behaves more like a hot gas, giving rise to dehydration during the thermal treatment. Thus dry heat sterilisation is of interest to the canning microbiologist and there is a need for more data on thermal death rates of spores treated under dry heat conditions.

The mechanism of heat deactivation and the reasons for the special resistance of the spore is still not completely explained. DPA was thought to play a major part in conferring heat stability to the spore but recent work seems to discount this. DPA probably has other roles such as the maintenance of dormancy and/or the initiation of germination. Current theories favour an explanation based on osmotic dehydration of the protoplast by the surrounding cortex which itself is expandable and impermeable to multivalent cations. In dry heat destruction of spores, the more rigorous conditions required are probably accounted for by supplementary dehydration effects, and, at the higher temperature then required, injury to spore DNA may result.

Order of destruction

If bacterial spores are exposed to heat at a constant temperature and the surviving fraction plotted against time the resulting curve is generally considered to follow a logarithmic course with equal percentages of surviving cells dying in each successive unit of time. The graph obtained by plotting the logarithm of the number of viable cells against the time of heating at a constant temperature is known as a survivor curve or thermal death-rate curve (Fig. 11). The slope of the survivor curve determines the Decimal Reduction Time (D) which is defined as the time of heating, in minutes, to reduce the numbers of the survivors to one tenth of the original, i.e. the time for the curve to span one log cycle. The temperature at which the D value was determined is often denoted by a subscript, e.g. D_x. If the D values equivalent to a number of temperatures are plotted on a logarithmic scale against their corresponding temperatures a straight line is normally obtained, the slope of which is designated *z*. The value of *z* varies with the organism and the medium in which the heating and recovery is carried out.

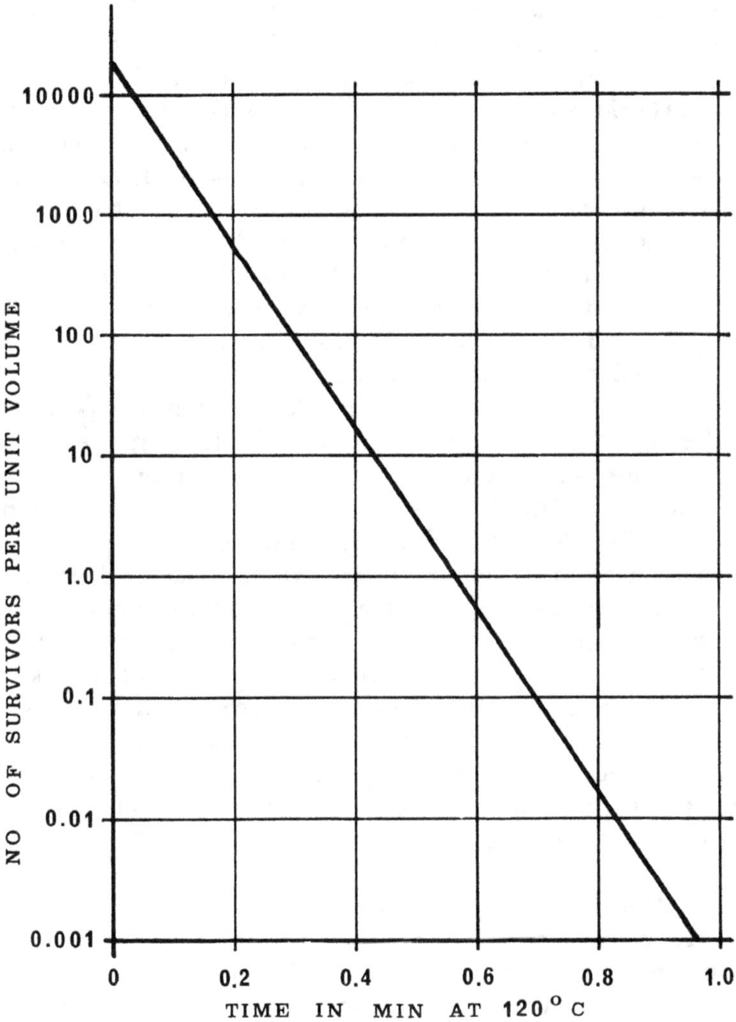

Fig. 11 Thermal death-rate curve

There is evidence of deviation from the logarithmic order of death (Anderson et al, 1949; Reed et al, 1951; Reynolds & Lichtenstein, 1952; Pflug & Esselen, 1954; Frank & Campbell, 1957; Humphrey & Nickerson, 1961; Moats et al, 1971; Casolari, 1974). The most commonly observed deviations are an initial shoulder and a residual tail so that a sigmoidal survivor curve is obtained (Fig. 12). The initial shoulder is sometimes explained as being caused by clumping of cells or by inadequate heat activation of the spores. Undoubtedly the latter explanation is valid in some instances. As pointed out by Vas (1970) the residual tail, sometimes reported on survivor curves, is of greater practical significance. It has been attributed to two or more types of cells with different heat resistances being present in the population. Hayakawa (1977) states that a close examination of the survival curves presented by Moats et al (1971) could be interpreted by one initial curvilinear segment and two or more straight line segments. A reason for conflicting reports is undoubtedly due to the fact that the same strain of organism is able, under different conditions, to give different survivor curves. Roberts & Ingram (1965) found a resistant tail on the curve for *Clostridium botulinum* type E when open tubes were used for heating the spores whereas this was not apparent when totally immersed sealed ampoules were used. Walker & Matches (1966) found that the shape of the curve depended upon the composition of the buffer in which the spores were heated. The temperature at which the spore crops were produced also influenced the shape of the curve (Lechowich & Ordal, 1962) as did extent of initial contamination (Casolari, 1974). Pflug & Smith (1977) also noted the effect of the suspension medium and the test temperature and concluded that the shape of the survivor curve appeared to be a unique property of the spores being evaluated, the solution in which they are suspended and the test temperature. Pflug & Schmidt (1968) regret the lack of nomenclature to realistically interpret non-linear survivor curves and suggested, for example, that the shoulder could be defined by reporting D values as usual and supplementing them by a value of the intercept on the vertical axis obtained by extrapolation of the straight line portion of the curve.

Factors affecting heat resistance

The heat resistance of bacterial spores is greatly affected by environmental factors during: (1) sporulation; (2) heating and (3) recovery. In addition, because of the logarithmic nature of the thermal death-rate curve, the higher the initial number of spores the longer the time necessary to effect their complete destruction. It should be noted, however, that this holds only for cells from identical suspensions.

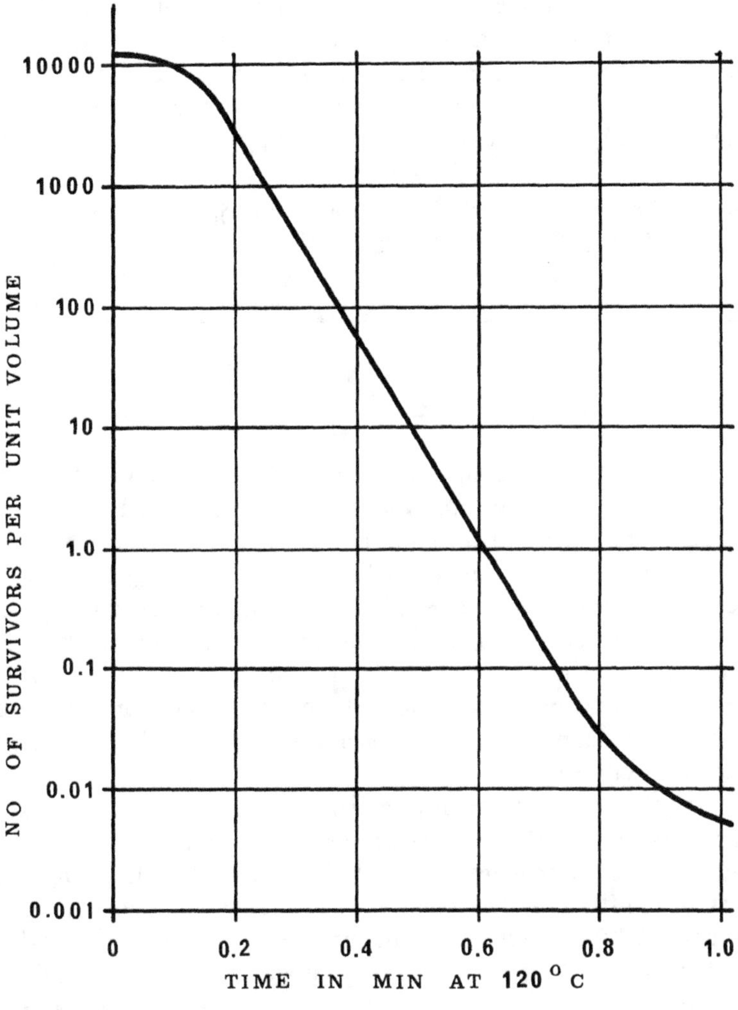

Fig. 12 Sigmoidal survivor curve

Lang (1935) has stated that the heat-resistance of an individual suspension is a definite property which is rarely comparable with other suspensions, even when they are produced under the same conditions. There is, however, no doubt that for the same suspension, heated under the same conditions, the thermal death time increases with increasing concentration of cells. In view of this relation of heat resistance to the number of cells, it is to be expected that the greater the initial contamination in a given food the longer will be the period of heating required to sterilise it. Commercial experience has amply confirmed this view. As an example, Cameron & Yesair's (1931) work with sugar contaminated with thermophilic 'flat-sour' spores may be quoted. The results of their experiments clearly demonstrated that spoilage was related to the degree of contamination as shown in Table 6. In this experiment sugar A1 contained approximately 60 'flat-sour' spores per 10 g whilst sugar A2 contained approximately 2500 per g.

Table 6 Effect of contaminated sugars on spoilage rates. Spoilage in cans of corn (No. 2) due to the presence of 'flat-sour' bacteria in sugar

	Percentage of spoilage in corn with: No Sugar	Sugar A1	Sugar A2
70 minutes	0	0	95.8
80 minutes	0	0	75.0
90 minutes	0	0	54.2

In the heat processing of canned foods, the importance of reducing contamination to a minimum cannot be overemphasised. Any marked increase in the degree of contamination of raw materials or plant may have the result that a process which has been adequate for years suddenly becomes inadequate. Also the presence of clumps of spores in a suspension is known to affect heat resistance and Amaha (1961) pointed out that this cannot be ignored in determining the thermal processing conditions for canned foods, since, under natural conditions, the spores of aerobic bacilli may frequently be present in the form of clumps.

Apart from the number of organisms present their age may affect heat resistance and this has been investigated by many workers with varying results. According to Esty & Meyer (1922), young moist spores of *Clostridium botulinum* are more resistant than old spores. Sommer (1930) suggested that maximum resistance of *C. botulinum* spores was reached after 4 to 8 days' incubation. Curran's (1934) work with aerobic spore-forming species led him to conclude that ageing up to 1 year increased the heat resistance of spores; there was no appreciable change in the first 3 months. Williams (1936), using various aerobic and anaerobic sporing organisms, failed to find any correlation between age of the culture and heat resistance over a period of 5 to 30 days. Maximum resistance of the spores of a strain of *C. sporogenes* was reached in 7 to 14 days in Bashford's (1942) experiments. Non-spore-forming organisms have also been studied in this connection. Anderson & Meanwell (1936) observed an increased resistance of thermoduric streptococci during the early logarithmic growth phase. Using *Escherichia coli*, Ellicker & Frazier (1938) found greater heat resistance during the initial stationary phase, the resistance decreasing when reproduction commenced until it reached a minimum during the period of most rapid reproduction. The divergent results obtained in the above and other investigations do not allow of a general statement on the subject, but it is evident that for some organisms at least, heat resistance varies with age.

Effect of environmental conditions during spore formation

The environment in which the spores of bacteria are produced may have a considerable effect on their heat resistance. Williams (1929) and others have reported that changes in the composition of the nutritive medium may produce deviations from the normal resistance. Vinton et al (1947) found that spores produced in pasteurised or sterilised meats were, on the average, about three times more resistant than spores grown in raw meat. Mayou (1970) showed that spores of *Bacillus stearothermophilus* produced in milk were more heat resistant than those produced on nutrient agar supplemented with 40 p.p.m. of manganese sulphate. Sugiyama (1951), using *Clostridium botulinum*, showed that a reduction in the Fe^{++} and Ca^{++} concentration in the sporulating medium below a certain value, decreased the thermostability of the spores. The presence of fatty acids, particularly the longer chain members, tended to increase the heat tolerance. Amaha & Ordal (1957) showed that manganese and calcium increased the heat resistance of *Bacillus coagulans* and that for development of maximum heat resistance there must be adequate amounts of both cations in the medium. The availability of calcium may be affected by the phosphate concentration of the medium since El-Bisi & Ordal (1956a) have observed that a high phosphate level depressed the thermal resistance of the spores. Levinson & Hyatt (1964), using spores of *B. megaterium*, found that supplementation with calcium chloride increased heat resistance but that addition of L-glutamate, L-proline or phosphate yielded spores with reduced heat resistance. The composition of the sporulation medium influenced also the germination characteristics and heat shock requirements.

The temperature at which the spores are produced may affect their heat resistance. Spores formed at high temperatures are often more resistant than those formed at lower temperatures (Williams & Robertson, 1954; El-Bisi & Ordal, 1956b; Cook & Gilbert, 1968a). Sugiyama (1951) said that the spores of *Clostridium botulinum* produced at 37°C are of higher heat resistance than those developed at 41°C, 28°C or 24°C. Lamanna (1942) found that with aerobic spore-forming bacteria, the heat resistance of the spores is related to the maximum temperature for growth, those with high maxima possessing highest resistance. Lechowich (1959) found similar results with the spores of *Bacillus subtilis* and *B. coagulans*; the heat resistance increased with the sporulation temperature. Lechowich & Ordal (1962) confirmed and extended this work to show that the DPA and cation content (Ca^{++}, Mg^{++} and Mn^{++}) of the spores also increased with the sporulation temperature.

In connection with the influence of environment, it has been suggested that spores as they occur naturally, for example in soil, may exhibit a different resistance to heat than those artificially produced. Curran (1935) found that spores produced and aged in soil and oats were more resistant than those developed on artificial medium. Gillespy (1947) found an appreciable increase in the resistance of spores of thermophilic anaerobes when soil was included in the sporulating medium. No significant differences could be found by Gortzen (1937) in the resistance of spores of anaerobic organisms obtained either directly from the soil or from cultures. *Clostridium sphenoides* (found once) was the only exception, its native spores having a considerably higher heat resistance than those from cultures. Sobernheim & Mundel (1938) observed that soil with its natural spore content, required six to eight times longer heating than did a similar number of spores cultured from the same soil. They concluded that the higher resistance of spores in natural soil was due to some physico-chemical effect of the soil and not to any difference between soil and culture spores.

In view of all the evidence available it would appear that laboratory findings on the heat resistance of artificially produced spores should not be too readily interpreted as representing conditions existing in factory practice. Cameron et al (1936) carried out an investigation on a spore-forming organism which caused spoilage of canned beets. The process given the beets was 20 minutes at 116°C which, from heat penetration data, should have been sufficient to destroy spores with a resistance of approximately 10 minutes at this temperature. Under laboratory conditions, however, there was no survival after 5 minutes' heating at 110°C. The investigators believed that the original resistance must have been materially higher than that indicated by the laboratory tests and the loss of resistance was presumed to be due to laboratory manipulation of the organism.

Effect of environmental conditions during heating

The heat resistance of micro-organisms is greatly influenced by the nature and composition of the suspending medium. There is evidence to suggest that the physical nature has an effect. Reed et al (1951) reported that the thermal resistance of spores varied depending upon whether the food was homogenous or not. The most striking influence, however, is due to compositional influences such as a_w, pH and the presence or absence of protective substances on the one hand and antimicrobial substances on the other.

A_w effects and protection by solutes. It is widely observed that moist heat is more effective than dry heat as a sterilising agent. However, the effect of moisture, defined in terms of the water activity of the medium in which the organisms are heated, is difficult to quantify, since the extent of the response seems to depend upon the organism involved and the manner in which the limitation of moisture is achieved. Murrell & Scott (1966) found that the heat resistance of spores was at a maximum when the a_w of the substrate, in this case controlled by equilibration with the vapour phase, was between 0.2 and 0.4. The response of different spores varied, *Clostridium botulinum*, type E, showing the greatest effect giving a 100 000 increase in heat resistance, whilst *Bacillus stearothermophilus* showed only a 20 fold increase. Similar results were obtained by Harnulv & Snygg (1972) when control of a_w was achieved by addition of glycerol. With the other solutes tested (LiCl, NaCl and glucose) the effect varied widely depending on the solute used. Similar results using *B. stearothermophilus* were obtained by Harnulv et al (1977). When heated in egg powder, fish protein concentrate and wheat flour the spores showed maximum resistance at an a_w in accordance with the results obtained in water vapour and glycerol solutions.

Corry (1974, 1976a) using vegetative cells of salmonellae and osmophilic yeasts heated in solutions of sugars and polyols, found that as the a_w decreased the heat resistances increased. More significant than the a_w was the type and total amount of solute present which afforded protection to the organisms. The extent of the protection from each solute correlated with the degree of plasmolysis they produced. For example, sucrose caused the greatest shrinkage of cells both in salmonellae and yeasts and also gave the greatest increase in heat resistance; glycerol caused little shrinkage and increased the heat resistances the least (Corry, 1976b). Hsieh (1976) investigated the death kinetics of *Staphylococcus aureus* and *Salmonella anatum* in brain–heart infusion to which was added glycerol to control the a_w within the range 0.11 to 1.0. Death rates of both organisms were at a minimum when the a_w was 0.7 to 0.8. Experiments with a solid medium (semolina plus whole egg solids) gave the same result. The heat resistance of mould spores, equilibrated to different a_w values, showed the heat resistance to increase as the a_w value was lowered (Lubieniecki-von-Schelhorn & Heiss, 1975). Using solutes to lower the a_w of the medium in which the spores of *Aspergillus flavus* and *A. parasiticus* were heated other effects were apparent, sodium chloride being more protective than sucrose and sucrose more protective than glucose, at the same a_w (Doyle & Marth, 1975).

The protective effect of sugar, glycerol and other non-toxic substances has been noted by several workers. When such materials are present in adequate concentration the protection is in some cases sufficiently marked to be significant in canning practice. Peterson et al (1927) found that yeasts were more rapidly destroyed at 100°C in distilled water than in syrup. In syrup at 24°Be the thermal death-time was 6 minutes and in a syrup of 36°Be it was 28 minutes. Wallace & Tanner (1931) stated that in the presence of 10, 25 and 50 per cent sugar, moulds showed greater resistance to heat. Baumgartner & Wallace (1934) found that the survival of *Escherichia coli* at 70°C was increased from 4 to 6 minutes by 10 per cent sucrose; in 30 per cent sucrose the organism survived for 30 minutes. Baumgart & Stocksmeyer (1976) determined the heat resistance of the ascospores of *Byssochlamys fulva* and *B. nivea* in apple juice with varying levels of added sucrose from 17 to 47 g per litre and found increasing heat resistance with increasing concentrations of sugar. With spore concentrations of 10^6 per ml resistance times of 10 minutes at 101°C for *B. fulva* and 15 minutes for *B. nivea* were recorded for the 47 g per litre concentration.

The effect of various sugar concentrations on the thermal resistance of some organisms commonly associated with canned food spoilage is given in a report by Braun et al (1941).

These authors' investigation of the spoilage of canned sweet potatoes has clearly demonstrated the practical importance of sugar-protection. They found that the spoilage arose through the addition of dry sugar to the cans in place of the usual sugar syrup. The sugar failed to dissolve and become distributed evenly throughout the pack as expected, but formed a heavy syrup layer which protected the spoilage organisms against the effects of the normal heat-processing.

Effect of fat. A number of investigators have found that organisms suspended in fat or oil are more difficult to destroy by heat than when suspended in aqueous media, and Lang (1935) suggested that this may be accounted for by the poor conductivity of oil media. Lang observed that the spores of *Clostridium botulinum* survived beyond all reasonable expectation when heated in oil suspensions. He stated that the eventual germination of oil trapped spores constitutes a definite problem in the sterilisation of canned fish.

Jensen (1954) reported the results of experiments with streptococci which indicated great differences in heat resistance depending on whether the organisms were heated in fatty or aqueous media. Thus, when the organisms were heated in moist melted butter at 100°C they were destroyed in 15 minutes. In dry butter the killing time at 115°C was 50 minutes and at 120°C it was 20 minutes. In milk, the organisms were destroyed in 30 minutes at 61.7°C. Zaleski et al (1971) found the heat resistance of *Staphylococcus aureus* greater in soya bean oil than in broth but the oil had to be dehydrated to obtain maximum effect.

Zuccaro et al (1951) showed that a yeast, *Saccharomyces* sp., had a higher heat resistance in oil than in buffer solution of the same pH. Pasteurisation of a French salad dressing in excess of that required to kill yeasts in the aqueous phase was inadequate to prevent spoilage.

Molin & Snygg (1967) compared the apparent heat resistance of spores of *Bacillus megaterium*, *B. subtilis*, *B. cereus*, *B. stearothermophilus* and *Clostridium botulinum*, type E, in lipids and in phosphate buffer. The most pronounced increase in heat resistance was found for *B. subtilis* and *C. botulinum*, type E, the extent varying with the type and water content of the lipid used. Fuchs & Clausen (1976) investigated the thermal inactivation of *B. cereus*, *B. subtilis* and *B. coagulans* in phosphate buffer, butter water emulsions and dry butter fat. When heated in dry butter fat they noted that 100 to 1000 fold longer exposure times were required to achieve the same degree of destruction than in buffer or butter water emulsions.

The investigation of Yesair et al (1946) on the fat protection of cocci is of particular interest in relation to canned foods. Cultures of micrococci isolated from heat-processed fatty products were used in their experiments. Moist cocci heated in broth were killed in 30 to 45 minutes at 55°C whereas cocci which had been dried and subsequently heated in broth were slightly more resistant, being killed in 45 to 60 minutes at 55°C. When dried cocci were heated in moist fat, an exposure of 15 to 20 minutes at 100°C was required for their destruction. In dry fat, the organisms showed a very pronouced increase in resistance, surviving 135 minutes heating at 100°C. The authors remark that the available evidence indicates that micrococci which survive heating in canned products by reason of fat protection are incapable of subsequent growth because they remain in the fat. They state that the mechanism of fat protection in foods appears to rest upon the localised absence of moisture.

Senhaji & Loncin (1977) and Senhaji (1977) attributed the protective effect of fats to the availability of water during the heat treatment. This varies according to the temperature of the treatment. For example, if the temperature of a system in which the fat is saturated with water ($a_w = 1$) is raised, the solubility of the water in the fat increases and as a result the a_w falls. If ample water is available, the value will return to unity, the time required depending upon the rate of diffusion of water into the lipid phase which, in turn, is related to the globule size of the fat. Thus the effectiveness of heat treatments on micro-organisms embedded in fat will depend upon whether a reservoir of water is available, the dispersion of the fat and the time and temperature of the heat treatment. It may be anticipated that low temperature–long time heat treatments would be more effective than high temperature-short time processes in such cases.

The subject of fat protection is further discussed in the chapter dealing with the microbiology of sound canned foods.

Dallyn & Everton (1970) carried out a survey of heat processes applied to packs of canned tuna fish in oil manufactured in Italy and found that the applied processes were theoretically inadequate to destroy the spores of *Clostridium botulinum*. F_0 values were generally less than 2.45 although the products had a good public health record. They were able to show the presence of a water soluble constituent in some olive and groundnut oils which significantly reduced the heat resistance of the spores of P.A. 3679 and suggested that the presence of this substance explained the commercial stability of such products.

Effect of pH. The hydrogen ion concentration of the heating medium may have a profound effect on heat resistance. For most spore-bearing bacteria maximum resistance usually occurs in the region of neutrality. Esty & Meyer (1922) found that *Clostridium botulinum* showed maximum resistance between pH 6.3 and 6.9. Williams (1929) observed that *Bacillus subtilis* spores were most resistant between pH 6.8 and 7.6. Tsuji et al (1960) measured the thermal death-rates of P.A. 3679 in aconitate buffer at pH values varying between 4.5 and 5.3 and at temperatures of 104.4°C, 110°C, 115.5°C and 121.1°C. They found the pH to have a highly significant effect on the destruction rate at the lower temperatures but not at 121.1°C. Similar results have been obtained by Clausen (1978) for strains of *B. coagulans*, *B. circulans*, *B. cereus* and *B. subtilis* for which the z value increased with increasing pH so that the main effect of pH on D value was seen at low temperatures (105°C rather than 118°C). Mazokhina et al (1973) determined the heat resistance of *B. stearothermophilus* in canned foods of various pH values and claimed that even a 0.1 drop in pH causes almost a 20 per cent reduction in the $D_{121°C}$ value which could be expressed by the equation $D_{121°C} = 1.7 \text{ pH} - 7.5$ for the spores used. Using spores of *C. sporogenes* a straight line relationship between pH and D value was again found $D_{121°C}$ being expressed as 1.30 pH - 3.69. Amaha (1961) observed that the spores of *C. sporogenes* in M/15 phosphate buffer at 115°C showed maximum survival near neutrality with their resistance falling off as the pH decreased or increased. Lowik & Anema (1972) investigated the heat resistance of *C. sporogenes* in minced meat. They found that a drop in pH from 6.0 to 4.8 decreased the D value by 40 per cent irrespective of the temperature (110° to 120°C) i.e. the z value was independent of pH.

Aref & Cruess (1934) found that yeasts were most resistant at pH 3.8 to 4.0. With one exception, however, the strains of yeasts studied by Beamer & Tanner (1939) showed greater resistance at pH 6.8 than at pH 3.8. Zuccaro et al (1951), using *Saccharomyces* sp. found a greater heat resistance in buffer at pH 5.6 than at pH 4.0. Gillespy (1936—37) found that the maximum heat resistance of the ascospores of *Byssochlamys fulva* occurred at about pH 5.0; the resistance at pH 3.0 was considerably greater than at pH 7.0. Splittstoesser (1973) showed that at a particular pH the heat resistance was greatly affected by the acid used, fumaric acid being the most effective followed by lactic and acetic acids. Doyle & Marth (1975) used various buffer solutions to adjust the medium in which they heated strains of *Aspergillus flavus* and *A. parasiticus*. When they used sodium acetate plus acetic acid as a buffer the conidia were more rapidly inactivated as the pH was reduced down the range from 6.0 to 3.5. However, when citric acid plus disodium hydrogen phosphate was used heat resistance increased as the pH was reduced. Baumgartner & Knock's (1940) experiments showed that non-sporing bacteria gave maximum heat resistance at a pH at which the resistance to disinfection by alcohol and stability in suspension was also at a maximum. It was concluded that at this pH value the hydration of the bacterial protein was at a maximum and the organisms were, therefore, less susceptible to harmful agencies.

Increase in the hydrogen ion concentration usually causes a corresponding decrease in heat resistance. Where canned foods are concerned, the effect of moderate changes in pH on the heat resistance of spores appears to be variable. Esty & Meyer (1922) stated that above pH 5.0 some factor other than hydrogen ion concentration was more important in its effect on the time required to sterilise foods. With canned marine products, Lang (1935) found no correlation between pH and the heat resistance of *Clostridium botulinum* spores within the range of 5.2 to 6.8. At pH 4.9, however, a marked reduction in resistance occurred. Nichols (1940) found that over a narrow range of pH values which covered that found in commercial canning (pH 5.95 to 7.25), changes in pH value of cream in which *Bacillus subtilis* spores were heated gave inconsistent results. Sognefest et al (1948) state that the lower the pH within the range 4.5 to 9, the lower the heating required to prevent spoilage of a variety of vegetable products. In the region of pH 5.5 there appeared to be a decided increase in the heating necessary to ensure a commercially sterile product. Between pH 6 and 9 only a slight increase in the heating required occurred with rise in pH. The effect of pH and heat on the thermal resistance of *C.*

botulinum spores in specific food products was investigated by Xezones & Hutchings (1965), who showed that the decimal reduction time increased with increasing pH particularly at low sterilisation temperatures.

Schmidt (1957) pointed out that in food substrates pH changes not only may have an effect directly on the organism, but may alter the degree of dissociation of substances or produce a shift in the oxidation–reduction potential and these may indirectly exert effects either upon survival or upon the ability of surviving organisms to develop. Alderton & Snell (1963) suggested that spores exhibited a base exchange behaviour, the pH influencing the type of ions adsorbed on the spore surface so reducing the heat resistance at low pH.

Anderson et al (1949), investigating the survival of *Bacillus coagulans* in tomato juice, reported reduced thermal death times after the addition of varying amounts of acetic, lactic and citric acids. Based on the percentage of acid used, the relative efficiency of the acids in lowering the destruction time was: lactic > citric > acetic; if based on the pH, the order was acetic > lactic > citric. The adjuvant effect of acidity in the heat destruction of micro-organisms is utilised in the processing of certain vegetables and other non-acid packs which cannot adequately be sterilised without impairment of quality. The liquors in which foods are packed are acidified with the result that the resistance of contaminating organisms is lowered. This procedure has been used extensively, for example in the canning of artichokes in America.

Effect of salt. The presence of salt in the heating medium may affect the apparent heat resistance of micro-organisms. However, there appears to be no concensus of opinion on the extent of this effect and this is probably due to complications arising due to: (1) the sensitisation to salt by heat treatment and (2) protection against the heat treatment by dissolved salt. Since not all micro-organisms react in the same way, or indeed at all, to these influences, results are bound to show some divergences.

Numerous investigations have been made concerning the effect of salt on the heat resistance of micro-organisms but, no general agreement has been reached. Weiss (1921) found that the resistance of *Clostridium botulinum* spores was considerably reduced by the presence of 3 per cent salt; with higher concentrations, the resistance decreased rapidly. Working with spores of the same species Esty & Meyer (1922) found that the resistance was increased by 1 to 2 per cent salt. Above 8 per cent, the resistance decreased. *C. welchii* exhibited greater resistance in 3 per cent salt when tested by Headlee (1931); at 10 per cent concentration a definite decrease was noted. In phosphate buffer at pH 7.0, 3.5 per cent salt was found by Yesair & Cameron (1942) to cause a reduction in the heat resistance of *C. botulinum* spores at temperatures below 110°C but no appreciable effect was noted in the range 110° to 112.7°C. Roberts et al (1966) found that the heat resistance of *C. sporogenes* spores was the same in 3 per cent and 6 per cent salt solutions as in water, but that heating to 0.1 per cent survival rendered the surviving spores sensitive to inhibition by sodium chloride in the recovery medium. Duncan & Foster (1968) showed, in micro-culture, that concentrations of sodium chloride above 6 per cent hindered germination of heat treated spores of the putrefactive anaerobe 3679 and, although at concentrations of 3 to 6 per cent germination occurred, cell division was often blocked. Pivnick & Thacker (1970) found that survivors from heat-treated spores of *C. botulinum* were sensitive to salt and the more severe the heat treatment the greater the inhibition of survivors by salt in the recovery medium. Jarvis et al (1976) also noted a relationship between the severity of the heat treatment and the salt sensitivity. Spores heated at 70° or 80°C were not significantly sensitive to 2.5 per cent and 3.5 per cent concentrations but were sensitive to 4.5 and 5.5 per cent. If the spores were heated to 85°, 90° or 95°C they were sensitive to the lower salt concentrations. Thus it seems clear that the addition of salt to the heating medium at the levels tested does not directly reduce the heat resistance of clostridial spores but the presence of salt at levels of 3 to 6 per cent and above in the recovery media may interfere with the ability of the spores to germinate and develop, the extent depending on the severity of the heat treatment. This constitutes a major factor in the inhibition of the growth of clostridial spores in canned cured meats.

Salt in amounts up to 4 per cent was found by Viljoen (1926) to afford protection for the spores of several organisms heated in canned pea liquor. Anderson et al (1949) found that amounts of 1, 2, 4 and 8 per cent in tomato juice caused a progressive decrease in the destruction times for *Bacillus coagulans*, although part of this effect may have been due to the slight reduction in pH which resulted from the addition of salt. Briggs & Yazdany (1970) studied the effect of sodium chloride, in concentrations of 2, 4 and 8 per cent of the heating medium, on the heat resistance of *B. stearothermophilus*, *B. pumilis*, *B. subtilis* and *B. subtilis* var. niger. They found that whilst increasing concentrations of sodium chloride reduced the heat resistance of *B. stearothermophilus* no effect was observed with the other spores. In all cases increasing heat damage rendered the spores sensitive to salt in the recovery medium.

Anderson et al (1951), working with a yeast in an acidified liquor, found that salt in concentrations of 7.5 and 10 per cent had a slight protective action but at a level of 15 per cent thermal death-times were slightly reduced. A general statment to the effect that salt exerts a protective action for micrococci against heat is made by Jensen (1954). Support for this was forthcoming from Bean & Roberts (1975) who found that the heat resistance of *Staphylococcus aureus* NCTC 10652 in neutral buffer was increased by the addition of 4 or 8 per cent salt. At a slightly lower pH of 6.5 the $D_{60\,°C}$ value, both in buffer and in meat macerate, was increased by the addition of 8 per cent salt. A strain of *Staphylococcus aureus*, investigated by Stiles & Witter (1965), showed an apparent reduction in resistance when heated in sodium chloride solutions of increasing concentrations within the range 1 to 15 per cent. This was shown to be due to increased sensitivity of the organism to sodium chloride following heat injury.

It appears that the effect of salt is variable, depending upon the concentration, the suspension medium and the organisms involved. Where it appears to reduce heat resistance the effect can probably be largely attributed to salt sensitisation leading to inhibition of growth in the recovery medium. In other cases, particularly with salt tolerant micro-organisms such as the micrococci, sodium chloride may exert a protective action thus enhancing the heat resistance.

Inhibitory substances. The presence in the heating medium of any lethal or inhibitory substance can be expected to cause a reduction in heat resistance. Berry et al (1938) and Coulthard (1939) have shown the value of small amounts of germicides in increasing the efficiency of low temperature sterilisation. Pederson & Beavens (1940) state that 22 parts per million of sulphur dioxide may lower the pasteurisation temperature of fruit juices by 5.5°C. Gillespy (1946), reviewing his earlier work, records that very small concentrations of sulphur dioxide, such as might be introduced into canned fruit with sugar, markedly reduce the heat resistance of ascospores of *Byssochlamys fulva*. For example, the time required to bring about 99 per cent destruction at 85°C at pH 3.0 was reduced from 40 to 14 minutes by 2 parts per million of sulphur dioxide. In later experiments Gillespy showed that the concentration of sulphur dioxide required for a given degree of lethal action is dependent on the pH of the heating medium, the amount necessary increasing as the pH rises.

Michener et al (1959) carried out an extensive survey into the effect of over 650 substances on the heat resistance of *Clostridium sporogenes*. Apart from the antibiotics, among the more active substances were the following: ethylene oxide, diepoxybutane, hydrogen peroxide, formaldehyde, dodecylguanidine hydrobromide, cetyltrimethyl-ammonium bromide, 2 methyl-1, 4-naphthoquinone (vitamin K_4) and garlic oil.

The effect of certain spice oils on the thermal resistance of micro-organisms has been investigated. Kosker et al (1951) demonstrated that 10 p.p.m. of allyl isothiocyanate (volatile oil of mustard) added to buffer solutions and to apple and grape juice caused a marked reduction in the thermal resistance of *Aspergillus niger* and *Saccharomyces ellipsoideus*. The effect was less marked against *Bacillus coagulans*. Oil of garlic and oil of onion had less effect on this organism than had allyl isothiocyanate. Within the limits of practical use, clove, cinnamon and garlic dill flavour oils were shown by Anderson et al (1953) to have little effect on the spores of *Bacillus coagulans* suspended in tomato juice. Cinnamon and garlic oils were shown by these workers to be effective in lowering the thermal resistance of a pickle spoilage yeast, the tests being carried out in acidified brine. Anderson et al (1951) have shown that the additions of a dill spice oil composed of both garlic and dill weed oils at an acceptable flavouring level of 258 p.p.m. and mustard oil at a level of 20 p.p.m. considerably reduced the thermal death-time of a test yeast suspended in an acidified brine. The effect of a number of spices on the lowering of heat resistance of *Bacillus subtilis* was investigated by Bose & Roy (1960); chillies and mustard oil were found to be the most effective. Michener et al (1959) found that garlic oil at a concentration of 300 p.p.m. and allyl isothiocyanate at a concentration of 500 p.p.m. brought about a 50 per cent reduction in the heat resistance of the spores of a putrefactive anaerobe (P.A. 3679) suspended in Anderson's pork pea medium. It appears from results of these experiments that certain spice oils may have a considerable effect on the thermal stability of yeast cells but their action is less marked against spores of bacteria.

La Baw & Desrosier (1953) and others have demonstrated that the crude extracts from many edible plants exert an inhibitory action on the growth of micro-organisms. In addition La Baw and Desrosier showed that some of the extracts reduced the thermal stability of a flat-sour spoilage organism at mild levels of heat treatment. Both of these effects will tend to reduce the apparent survival times of micro-organisms in processed foods and may account to some extent for variations of heat resistance in different food substrates.

Certain peptides have antimicrobial properties and the effect of Nα-acyldipeptides on the apparent heat resistance of *Bacillus subtilis* and *B. megaterium* spores was investigated by Lundgren et al (1965). They found these substances, particularly those with long fatty acid chains, inhibited spore outgrowth and the apparent heat resistance varied with the pH of the medium, maximal effect being obtained near neutrality.

Heated spores of *Clostridium* and *Bacillus* species are sensitive to curing salts. The effect of sodium chloride, one of the main constituents, has already been considered. Heated spores of P.A. 3679, *C. perfringens*, *B: subtilis* and others are also sensitive to nitrite (Duncan & Foster, 1968; Labbe & Duncan, 1970).

Antibiotics. Anderson & Michener (1950) first reported that an antibiotic (subtilin) coupled with a mild heat treatment prevented food spoilage and the growth of toxic clostridia. This stimulated numerous studies concerned with the possibility of using subtilin or other antibiotics to reduce the severe heat processes necessary to sterilise low acid canned foods. The antibiotics which have received most attention are subtilin, nisin and tylosin. O'Brien et al (1956) and Michener et al (1959) found that subtilin and nisin were the most active of a large number of antibiotics tested at that time for effectiveness against the putrefactive anaerobe P.A. 3679; Denny et al (1961b) carried out a screening of the antibiotic tylosin against a number of spoilage types.

Le Blanc et al (1953) studied the heat resistance of the spores of *Clostridium botulinum* and the putrefactive anaerobe P.A. 3679 in pea puree. In the presence of 14 p.p.m. of subtilin, P.A. 3679 had heat resistance equal to 47 per cent of its resistance in subtilin-free purée. In the same suspending medium the spores of *C. botulinum* exhibited 63 per cent of normal resistance. Similar results were obtained by O'Brien et al (1956) who demonstrated a reduction of the D value for a number of strains of *C. botulinum* when heated with subtilin or nisin in pea puree. In the experiments made by Le Blanc et al (1953) a limited number of tests were included to determine whether the carry-over of the antibiotic to the sub-culture medium would inhibit spore germination or growth of the organisms studied by them. Their results indicated that the small quantities of subtilin involved did not influence spore germination. More extensive studies by O'Brien & Titus (1955) have demonstrated quite clearly, however, that 0.03 p.p.m. of subtilin in the recovery medium would markedly reduce the survivor count of drastically heat treated P.A. 3679 by preventing the outgrowth of about 85 per cent of the spores. Anderson (1952) and Michener (1955) formed the opinion that subtilin could be adsorbed by heated spores which may germinate and subsequently be killed by the antibiotic. Campbell & Sniff (1959) and Thorpe (1960), working with *Bacillus coagulans* and *B. stearothermophilus*, demonstrated that the apparent reduction in the resistance of spores when heated with nisin appears to be due to the nisin being bound to the spore resulting in inhibition of outgrowth. There seems to be general agreement that antibiotics such as nisin are sporostatic and not sporicidal. It is thought by most workers that germination of the spore is not affected, merely that outgrowth or active vegetative growth is prevented. For example, with nisin, Hitchins et al (1963) concluded that rupture or lysis of the spore coat during emergence was prevented but germination was not inhibited. A similar mode of action is attributed to subtilin (Gould, 1964). It appears, however, that tylosin acts by inhibition of vegetative cell growth. Denny et al (1961a) demonstrated that tylosin has no effect on the heat resistance of spores of *Bacillus stearothermophilus* or *Clostridium botulinum*, but it is inhibitory to growth of the organisms and could be effectively used to control spoilage. Greenberg & Silliker (1962) confirmed that tylosin has no effect on the spore and did not prevent germination but could prevent active growth after germination. It does not seem clearly resolved whether antibiotics such as nisin must be heated with the spores to be effective or whether apparent lowering of heat resistance may be attributed to carry-over in the recovery media as indicated by the work of Tramer (1966).

In addition to these considerations the possibility of antibiotic resistant strains must be borne in mind and, in fact, Campbell & Winiarski (1959) have isolated a subtilin-resistant strain of *Clostridium botulinum*. These studies add weight to an early statement by Cameron & Bohrer (1951) that antibiotics have failed to meet the basic test, namely, the destruction of the spores of *C. botulinum*. For this reason they should not be used to reduce the processes for non-acid canned foods below the basic minimum required to render them safe against *C. botulinum*. There are, however, circumstances when risk from botulism does not arise, for instance, with acid canned foods having a pH below 4.5 or for non-acid canned foods which have received the accepted minimum sterilising dose for the destruction of the spores of *C. botulinum*. In such circumstances the use of combined antibiotic/heat treatments, if proved effective, might be considered as a means of reducing heat processes without increasing the risk of spoilage. Numerous studies have been made with this objective in mind, often with disappointing results.

Williams & Campbell (1951) found that subtilin and mild heat treatment could not destroy the spores of a number of thermophilic organisms isolated from spoiled canned foods. Adams et al (1952), working with comminuted beef and the spores of a putrefactive anaerobe, found that high levels of subtilin with mild thermal processes did not permanently inhibit anaerobic spores in beef; a combination of relatively drastic heat treatment with a high level of subtilin was required. Spilde (1951) noted that canned fish balls could not always be successfully preserved with subtilin. Denny & Bohrer (1959) compared the thermal death rates of spores of P.A. 3679 and

Bacillus stearothermophilus in the presence and absence of subtilin, tyrothricin, methylol gramicidin and rhatany root extract. None significantly affected the death-rate of P.A. 3679 and subtilin was not effective against *B. stearothermophilus* 1518 although the other antibiotics accelerated the thermal death-rate.

On the other hand, subtilin and nisin have been reported to reduce the thermal resistance of the spores of a number of spoilage bacteria or to reduce the incidence of spoilage in practical canning trials or runs in thermal death time cans. The work of Le Blanc et al (1953) and O'Brien et al (1956) has already been mentioned. The latter workers included *Clostridium thermosaccharolyticum* 3814, *Bacillus coagulans* 43P and *B. stearothermophilus* 1518 in their experiments with nisin and subtilin and found that all the bacteria tested except *Clostridium thermosaccharolyticum* 3814 showed an appreciable reduction in the D value, particularly *Bacillus coagulans* with nisin. Campbell et al (1959) using thermal death time cans presented data to show that subtilin and nisin effectively reduced the process requirements necessary for the control of food spoilage bacteria such as *B. stearothermophilus* 1518 and *B. coagulans* 43P. Wheaton et al (1957) demonstrated that 5 p.p.m. of subtilin combined with a mild heat treatment prevented flat-sour spoilage of tomato juice and Gillespy (1957) reported that the hard swell spoilage of canned beans with tomato sauce, caused by *Clostridium thermosaccharolyticum*, and the flat-sour spoilage of canned peas could be controlled by nisin. Wajid & Kalra (1976) reported that 100 Reading Units (now designated International Units) of nisin per ml in sterilised, homogenised milk caused a 45 per cent reduction in the D value of *B. subtilis* 9144 at 109°C and a 71 per cent reduction in that of *B. stearothermophilus* 1-63 at 115°C. The effect on *B. stearothermophilus* at 109°C was not significant.

Wheaton & Hays (1964) demonstrated that flat-sour thermophilic spoilage in canned foods could be controlled by 1 p.p.m. of nisin and suggested its use in hot vended foods for this purpose. More details on the use of nisin in processed food products may be found in papers by Heinemann et al (1965), Hall (1966) and Fowler & McCann (1972). The use of tylosin for the control of spoilage has been studied by Denny et al (1961a, 1961b) and by Wheaton & Hays (1964).

The conclusion must be reached that these antibiotics neither affect the heat resistance of the spore nor prevent germination, but act as inhibitors of outgrowth after germination. Therefore it is important that inhibitory levels are maintained at all times and careful studies of the loss during processing and on storage must be carried out in order to decide whether antibiotics may be used to advantage in supplementing the heat process. For this reason and others, such as the existence or development of resistant strains, antibiotics should not be used to reduce thermal processes of non-acid canned foods below the accepted minimum sterilising dose for the destruction of the spores of *Clostridium botulinum*. With non-acid foods they may be useful in supplementing normal processing schedules as a means of controlling heat-resistant spoilage organisms such as thermophiles or in acid products with a pH below 4.5 as a means of controlling *Bacillus coagulans*. In the UK provision has been made in the preservation regulations to allow the use of nisin in canned foods which have received at least an $F_0 = 3$ process or those with a pH below 4.5.

Effect of the recovery medium

The constitution of the recovery medium in which the organisms are placed after being heated may exert a considerable effect on the apparent thermal destruction time. Nelson (1943, 1944) and others have shown that bacteria which have been exposed to sub-lethal heat treatments are more exacting in their growth requirements than unheated bacteria. In addition it has been demonstrated that the sensitivity to inhibitors of spore germination increases with the severity of the previous heat treatment of the spores (Olsen & Scott, 1950: Murrell et al 1950). For maximum recovery the composition of the medium should be (1) suitable for the initiation of the spore germination process and (2) capable of stimulating and supporting vegetative growth. The importance of simple specific substances such as amino acids or glucose in the initiation of germination has been demonstrated by many workers (Hills, 1949a, 1949b; O'Brien & Campbell, 1957; Hitzman et al, 1957). On the other hand there are many substances chemically very similar which are antagonistic to these initiators and therefore act as inhibitors (Caracò, et al 1958). After germination the spore must be provided with a medium which is nutritionally adequate. Hyatt & Levinson (1957, 1959) have shown, for instance, that sulphate and phosphate are necessary for the growth of *Bacillus megaterium*. O'Brien & Campbell (1957) demonstrated that *B. stearothermophilus* spores would germinate in a glucose-mineral salts medium, but required amino acids and growth factors to support further growth.

The effect of the recovery medium on the apparent thermal resistance of various spores has been adequately demonstrated; for example, with thermophiles and mesophiles of the genus *Bacillus* by Schmidt (1955) and with P.A. 3679 by Frank & Campbell (1955) and Augustin (1965). Amaha (1961), summarising earlier experiments on the survival of *B. natto* spores, points out the importance of the various constituents of the recovery media. In these experiments, the spore suspension became progressively more selective in its nutritional requirements for recovery as the severity of the heating process was increased. This resulted in a progressive increase in the apparent heat resistance as the recovery media was made more complex.

The degree of anaerobiosis is obviously an important factor in the recovery of clostridial spores. Amaha (1953) demonstrated that the use of a thioglycollate agar for stratification of recovery cultures gave greater apparent survival times than when plain agar was used. Barnes & Ingram (1956) working with vegetative cells of *Clostridium perfringens* studied the effect of different redox potentials and found that at Eh — 45 Mv there was hardly any lag in growth but that the lag phase increased with the redox potential.

Recovery temperature. The work of Williams & Reed (1942), confirmed by Sugiyama (1951), showed that the incubation temperature of recovery cultures of heat treated *Clostridium botulinum* spores had an effect on their apparent heat resistance. Cultures incubated at 24°C, 27°C and 31°C showed greater recovery than parallel cultures held at 37°C. This is of practical significance in the testing of canned foods and several laboratories concerned with the canning industry have adopted incubation temperatures of 30° to 32°C for work with mesophilic clostridia.

Similar results have been obtained with aerobic spores. Cook & Gilbert (1968b) found that the optimum recovery temperature for spores of *Bacillus stearothermophilus* dropped from 50°—65°C to 45°—50°C after heating. Edwards, et al (1965) noted that *B. subtilis* spores surviving UHT treatments were counted more effectively at 32°C than at 45°C which was the reverse of that apertaining to recovery of unheated or mildly heated spores. Prentice & Clegg (1974) reported similar results for *B. subtilis* spores.

Adams & Busta (1972) suggested that the change in optimum recovery temperature for *Bacillus subtilis* spores from 45° to 32°C was the result of inactivation of one germination system and simultaneous substitution by another. Prentice & Clegg (1974) thought that out-growth, rather than germination was the mechanism controlling the optimum incubation temperature. Adams (1978), surveying heat injury of bacterial spores, suggested that both may be involved, depending upon the heating temperature, since earlier work (Adams, 1973a) had shown that for *Clostridium perfringens* spores, damage to the germination and out-growth systems had different temperature coefficients.

Effects of sub-lethal heat treatment

Dormancy. The delayed germination or dormancy of heated spores has been observed and studied by many workers and is a problem of particular significance in the microbiology of canned foods. Spores which have received sub-lethal exposures to heat sometimes remain dormant in favourable media for long periods. Dickson (1928) has recorded the delayed germination of heated *Clostridium botulinum* spores for more than 6 years. Esty & Meyer (1922) reported that heated spore suspensions (*C. botulinum*) remained dormant for 378 days at 36° to 37°C. Mossel & Mol (1956) reported an incident of delayed spoilage in a sucrose-coffee milk product by *Bacillus coagulans*. These examples may represent extreme cases, but dormancy periods of weeks are commonly met in the routine testing of canned foods.

The cause of dormancy is still a matter of conjecture. According to Murrell (1961) dormancy is believed to be due to previously unsuspected inhibitors of germination or to unfavourable conditions of the medium. Studies into the kinetics of heat activation (Busta & Ordal, 1964; Keynan et al, 1964a) have shown the energy requirements for activation to be high and similar to that for the denaturation of protein macromolecules. It has also been shown that several reducing agents, or exposure to low pH, may activate spores (Keynan et al, 1964a; Keynan et al, 1965) and this has led Keynan and colleagues to suggest that spore activation changes the tertiary structure of the protein coat responsible for dormancy. Keynan et al (1961) found a correlation between dormancy and calcium dipicolinic acid content, and Ca^{++} might be a factor in maintaining dormancy. The observation that calcium and DPA deficient mutants of *Bacillus cereus* do not readily germinate led Holland & Taylor (1974) to suggest that a minimum threshold amount of DPA is required to break dormancy. Aoki & Slepecky (1974) postulated that manganese is associated with the DPA and plays a role in inducing a requirement for heat shock before germination can occur.

It is difficult to dissociate the factor of cultural environment from dormancy and some effects of the recovery medium have been discussed earlier (*see* p. 161). Morrison & Rettger (1930a, 1930b), Curran & Evans (1937), Nelson (1943, 1944) and others have shown that dormancy of heat-treated organisms can be eliminated or reduced by the provision of suitable recovery media. In a study of the germination of *Clostridium botulinum* spores, Wynne & Foster (1948a) have observed that the addition of 0.1 per cent soluble starch to the medium has a striking effect on the germination process. They state that the starch acts mainly by absorbing, and thus rendering inactive, small amounts of substances, present in all media, that repress spore germination. Wynne & Foster (1948b) also found that small quantities of carbon dioxide were necessary for the germination of spores of *C. botulinum* in a synthetic medium; the absence of carbon dioxide delayed germination considerably in a complex organic medium. Other anaerobic organisms were not affected, but Reynolds et al (1952) have reported higher spore counts of P.A. 3679 when sodium bicarbonate was added to the medium. Anderson (1951) and others have confirmed the necessity of incorporating bicarbonate into media used for counting spores of *C. botulinum*. Cassier & Sebald (1969) showed that plating medium containing egg yolk or lysozyme initiated germination of spores of heat damaged *C. perfringens* type A presumed no longer viable. Similar results were obtained by Duncan et al (1972). Using *C. botulinum* spores Alderton et al (1974) demonstrated a dramatic increase in the apparent heat resistance of type E and a more moderate increase for type A by adding lysozyme to the recovery medium. However, it has been shown that not all spores are sensitive to lysozyme (Adams, 1974) without treatment with substances such as EDTA (Adams, 1973b) and sodium hydroxide (Barach et al, 1975). Nevertheless the use of lysozyme in the recovery medium offers interesting possibilities for recovery of small numbers of heat damaged clostridial spores from canned foods.

Observations on the germination rate of spores suggest that metabolic changes may occur in resting spores. Powell (1950) found that young fresh spores of *Bacillus subtilis* germinated slowly, but if stored in water at 20°C the rate of germination increased to a maximum after 20 days' storage. Spores aged in the growth medium behaved as young spores. Murty & Halvorson (1957) observed that the spores of *B. cereus* which had been stored in frozen distilled water lost the need for L-alanine for the activation of oxidative enzymes, an indication of germination.

The above observations are important in connection with the interpretation of sterility in canned foods. For example, heat injury rather than inactivation probably plays a major role in conferring stability to canned cured meats. In these circumstances, particularly, there is danger (1) in changing process and product conditions which have been established by long experience or (2) of extrapolating adequate thermal processes from one situation to another without complete and adequate testing. Finally the implications in connection with sterility testing of canned foods are obvious.

Heat activation. The development of a bacterial endospore into a vegetative cell has been shown to consist of three sequential processes: activation, germination and out-growth. Activation is the process necessary to condition spores for germination and is reversible. Many of the characteristic features of spores are retained, such as heat resistance, non-stainability and refractility but the spores are no longer dormant. The main change is probably the release of enzymes which are dormant in the resting spore (Gould et al, 1966). Curran & Evans (1945), who earlier (Evans & Curran, 1943), had shown that relatively mild heating brings about more rapid germination of spores of mesophilic aerobes, examined a group of heat-tolerant aerobes isolated from spoiled canned evaporated milk and demonstrated that sub-lethal heating of spores lowers the minimal temperature at which germination can occur. A similar phenomenon was noted by Edwards et al (1965), who showed that heat-injured spores of *Bacillus subtilis*, after ultra-high temperature treatment, showed survival when incubated at 32°C rather than 45°C and when maintained at 3°C before plating out at 45°C. This conditioning to development at low temperatures, which Evans and Curran term 'heat activation', applies when the heating is mild and non-lethal (e.g. 95°C for 10 to 30 minutes) and when it is so drastic that relatively few spores survive. The phenomenon of induced germination by sub-lethal heating is shown by many, but not all, types of spores. Heat activation of the spores of P.A. 3679 has been reported by Stumbo et al (1950) and of *Clostridium botulinum* by Treadwell et al (1958). Riemann (1957) observed that heat activation increased the rate of germination of P.A. 3679 and *C. botulinum* rather than the outgrowth from the spore.

Environmental factors influence heat activation. For example, water must be present (Levinson & Hyatt, 1969). Heat activation is pH dependent. Keynan et al (1964b) have shown that *Bacillus cereus* spores exposed to a pH lower than 2 at 65°C for 30 minutes do not germinate even in the presence of potent initiators such as L-alanine or adenosine. Gibbs (1967) has reported that activation of *Clostridium bifermentans* is optimal at high pH and is inhibited at low pH. Other environmental factors such as the presence of certain cations, e.g. potassium, may interfere with activation (Splittstoesser & Farkas, 1966). For further details the reader is referred to the review by Berg & Sandine (1970) on the activation of bacterial spores.

Methods of estimating heat resistance

In view of the many factors influencing the heat resistance of micro-organisms it is evident that an accurately standardised technique is required in securing reliable data. Under the most carefully controlled conditions, however, erratic results may be obtained and unless substantiated by repeated tests, heat resistance data should be cautiously applied to food processing problems. Some of the important factors which were found by Townsend et al (1938) to influence thermal death-time estimations, using the same organisms in the same medium, were:

1. Variations in the resistance of individual spores in a given suspension
2. Change in the resistance of spores in a stock suspension on storage
3. Loss of a portion of the suspension in transferring from the heated container to the enrichment medium
4. Variation in different lots of the same enrichment medium affecting its suitability for the germination of spores
5. Possible lack of germination of a few surviving spores within the period of incubation. With many organisms, a certain population is necessary for demonstrable growth, or even germination, to occur.

In the case where heating is carried out in phosphate buffer or other non-nutrient reference substrate, it is necessary to sub-culture into nutrient medium. The modern trend is to obtain data on spores suspended and heated in a food product when survival may be determined by sub-culture or product incubation. The indicated survival time may vary considerably depending upon whether sub-culture or product incubation is used. This, no doubt, is due to many factors, the most important of which appear to be the influence of pH, the presence of inhibitory substances and the nutritional requirements of the surviving organisms.

Methods of estimating heat resistance which have been mainly used in the solution of canned food processing problems are briefly outlined below:

Method of Bigelow & Esty (1920). This is known as the single tube method and involves heating a definite amount of a suspension of the organism or spores in small glass tubes (7 mm bore, wall 1 mm thick and 250 mm long). After sealing in a flame, the tubes are immersed in an oil bath of constant temperature and at short intervals a single tube is removed, rapidly cooled and cultured in a suitable medium. With this procedure, the death-time is taken as lying between the time of heating the last tube to show growth on sub-culture and the time of heating the first tube showing no growth. One disadvantage of the method is the occurrence of 'skips', this being the term given to tubes which give no growth on sub-culture although heated for a shorter time than tubes containing viable organisms.

Method of Esty & Williams (1924). This 'multiple tube method' was proposed by Esty & Williams in order to overcome difficulty with 'skips'. Instead of a single tube being removed from the heating bath at short intervals, a large number of tubes (25 to 30) are removed at each of four widely-spaced times. The tubes are cooled rapidly and cultured into suitable media. For each heating period, the percentage of tubes giving growth is plotted on semi-log paper and from the resulting straight line the thermal death-time of the organism at the temperature used can be found by extrapolation.

Esty (1928), discussing the relative merits of the single and multiple tube methods, quotes a typical result in which a test using single tubes indicated complete destruction of spores in 20 minutes, yet 39.4 per cent of tubes in a multiple test gave growth under the same conditions. He points out that single tube tests are useful in locating the range within which all but the occasional spore is destroyed. The multiple tube method is the one which has been most frequently used in making process calculations.

Method of Williams et al (1937). These authors designed equipment which enables a suspension to be heated under pressure and to be sampled without disturbing the conditions of the heating test. The apparatus, which is essentially a miniature retort fitted with a propeller for agitating the suspension and special valves through which the samples are drawn, was primarily designed for use in determining spore destruction rates. For this purpose, the originators claim that the apparatus is more convenient and accurate than Esty and Williams' procedure.

Method of Townsend et al (1938). For the determination of thermal death-times in raw food-products, Townsend, Esty and Baselt used special cans two and one-half inches in diameter and three-eighths inch deep. These small cans containing the raw material and test organisms were sealed in vacuum, heated in specially designed small retorts, cooled rapidly and finally incubated. Growth of the test organism was determined by bulging of the cans caused by gas formation. The construction of a small retort, suitable for use with thermal death-time cans or plugged tubes, is described by Schmidt et al (1955).

Reynolds et al (1952) have introduced concentrated media into the cans after heating. The cans were punctured, and the media introduced by a syringe and needle; after sealing with a tip of solder the cans were incubated in the normal way. In most instances longer survival times were indicated when using supplemented food substrates than with straight product incubation.

Method of Stumbo et al (1945). In their studies on the thermal processing of canned meats, Stumbo and his co-workers heated the inoculated ground meat in 10 × 75 mm glass tubes. A grease gun fitted with an artery pumping needle was used for filling the tubes which were then sealed in a flame and heated in a lard bath. The authors stated that the advantages of their method are: (1) it is possible to include wide variations in processes employed, wide variations in treatment of meats used, and replications of each study of variant combinations involved, without the time and space required becoming prohibitive; (2) meat can be readily introduced mechanically into the glass tubes; (3) processed tubes of meat can be readily cultured into a medium favourable to growth of the test bacteria with very little danger of contamination; (4) unopened processed tubes of meat can be introduced conveniently and (5) progressive spoilage of the processed meat can be readily observed through the incubated glass tubes.

Method of Stumbo (1948). The thermoresistometer described by Stumbo was designed for heat destruction studies in the temperature range of 115.6° to 132.2°C. Secrist & Stumbo (1958) pointed out that the instrument should not be used for studies in the temperature range beyond 138°C since the short exposure times involved demand accuracy requirements beyond its capability. The apparatus consists of three chambers connected to a steam reservoir and arranged so that each chamber can be pressurised or exhausted independently of the other two. The inoculated samples are spread in small metal cups approximately 7 mm in diameter and 1.5 mm deep which rest in aluminium boats. These are moved through the three previously sterilised chambers in turn. The first chamber is used for loading the cups, the second, which is the contact chamber, is pressurised by steam at the working temperature and in the third the boats, cups and samples, which have been cooled by evaporation, fall into tubes of culture medium. The process time in the centre chamber is accurately measured by an electric clock controlled by micro switches which are actuated when the sample enters the chamber and when the steam pressure is released. Advantages claimed for this method are: (1) greater accuracy, particularly in the higher temperature range, because of rapid heating and cooling of the sample; (2) comparative simplicity of operation; (3) the entire sample is cultured and (4) the chances of contamination are greatly reduced. A disadvantage is that evaporation may occur if there is delay between loading and introduction into the exposure chamber. Secrist & Stumbo (1958) have demonstrated that with phosphate suspensions this may lead to erratic results.

Method of Schmidt (1950). In this method, the test suspension was heated in small glass tubes fitted with cotton-wool plugs which permitted the addition of sub-culture medium direct into the tubes. Spattering of the liquid and blowing of the plugs was prevented by a pressure cooling system fitted to the miniature retort used for the heating and cooling operations. A thermo-couple inserted into one of the tubes allowed more accurate measurement of temperature lag, enabling the equipment to be used in the higher temperature range — 121.1° to 132.2°C.

Method of Pflug & Esselen (1953). An apparatus of the same principle as the thermoresistometer has been described by these authors. The sample is contained in metal cups and is introduced into and removed from a steam chamber by means of a sample-carrying piston. Loading and sub-culture of the cups is carried out in a second chamber. Cooling is achieved by quenching in the sub-culture medium and not by evaporation as in the thermoresistometer. Sterile compressed air is used to prevent boiling of the sample after it leaves the pressure chamber and before it is automatically sub-cultured.

Method of Stern & Proctor (1954). In this method the suspension of spores is contained in sealed capillary tubes each with an inner diameter of 0.8 mm, an outer diameter of 1.5 mm and a length of 7.6 cm (3 in). The tubes, held in place in a carrier tray, are immersed automatically into a hot bath, this action starting an electric timer. At the end of a set immersion time a 2.25 kg (5 lb) lead weight falling free rotates a cable drum which drives the carrier assembly in a semi-circular path from the hot bath to a cold bath. This action takes between 0.37 and 0.40 seconds. The apparatus has been used for heat resistance studies in the temperature range between 100°C and 140°C, correction factors based on the heat penetration characteristics of an infinite cylinder being applied. One advantage is that ten or more samples can be treated simultaneously, permitting the data to be subjected to statistical analysis.

The screw-cap tube technique. This method, described by Kooiman (1974), relies upon the addition of 0.1 ml of concentrated spore suspension to 10 ml of a temperature equilibrated medium contained in tubes with a screw cap and neoprene septum. The spores are added through the septum by means of a syringe and mixed with the hot medium so that temperature equilibration is achieved. After exposure of the spores to heat for the required time the tubes are cooled to sublethal temperatures by immersion in ice-water.

Bulbed capillary tubes. In our laboratory we use small bulbs of 5 mm outside diameter in which are placed, by means of an accurate syringe, 0.05 ml of spore suspension. The bulbs are blown from capillary tubing leaving a 20 mm stem which is sealed in a gas flame after addition of the suspension. Before use the bulbs are screened by means of a gauge to ensure that they conform to the size specification. We find the bulbs easier to fill than capillary tubes; furthermore because of their shape they may be used also for placement at a specified position in food products, e.g. at the slowest heating point in a can of food. (This use is described in Ch. 9.) The time required for the bulb and contents to reach bath temperature (up to 121°C) is about 15 seconds. The contribution of the come-up time to the total process is equivalent to 6 seconds at the operating temperature, therefore, 9 seconds is subtracted from the immersion time to give the time of exposure.

Methods for high temperature work. The determination of the thermal resistance of micro-organisms at temperatures in excess of 130°C demands short exposure times and, as a result, the come-up period during the heating of capillary tubes becomes significant. Perkin et al (1977) developed a computer programme to correct for errors due to this effect. A more direct approach is to employ a method which reduces come-up and cooling delays. Various devices have been used for this purpose. Wang et al (1964) used gas under pressure to force hot liquid medium and spore suspension into a mixing chamber where temperature equilibration was accomplished in 0.6 milliseconds. The required holding time was achieved by allowing the mixture to flow through a constant temperature reactor tube. Cooling was achieved by allowing it to expand instantaneously into a flash chamber provided with a cold water condenser. Perkin (1974) describes the equipment used in the United Kingdom a modified version of which is now known as MISTRESS (Milk Into Steam Treatment Research Equipment Small Scale). This is essentially a small-scale, product-into-steam, UHT steriliser in which the product is sprayed into a vessel filled with saturated steam at the required temperature. After holding the product for the desired time it is cooled by expansion through an orifice plate into a vacuum chamber. Oquendo et al (1975), describe another installation in which softened water is pneumatically conveyed through a heat exchanger which heats it, under pressure, to a few degrees above the test temperature. The heated water and suspension of micro-organisms are allowed to mix in a ratio which reduces the temperature to the required value. Once again cooling is achieved by expansion into a flash chamber.

Treatment of thermal resistance data

The thermal resistance data obtained by one of the above methods may be used to calculate the D value. Stumbo (1973) describes two methods: (1) by construction of the survivor curve from which it may be obtained directly and (2) by assuming the order of death is logarithmic and calculating the D value from the equation:

$$D = \frac{t}{\log a - \log b}$$

where t is the time of heating, a the 'initial' number of spores and b the number of survivors.

The first method is the most straightforward and involves establishing the survivor curve by counting colonies that result following the heat treatment. The number of replicates is usually small (less than five) and the time of heat treatment is regulated to give about 300 survivors. Needless to say this method is not suitable if direct incubation of the heat treated sample is intended. The difficulty is in establishing the true shape of the curve due to the problems in recovering heat damaged organisms in a reproducible and accurate manner. If a straight line survivor curve is obtained there is no difficulty in reading off the D value but if a straight line is not obtained a judgement has to be made as to which portion of the curve should be used. In making this decision due regard should be paid to the use to which the data will be applied. If interest is centred in applications where only one or two decimal reductions are involved then that portion of the survivor curve is the part of interest and the direct plating method is applicable. If, however, the interest is in applications where the probability of survival is low, as is usually the case when considering the sterility of canned foods, then a method using some sort of 'end-point' technique is preferred. The second of the two methods is suitable in these circumstances provided the heat resistance is gauged to give a very low number of survivors. In practice this involves preparing replicates and subjecting batches of these to different heating times all of which will reduce the number of survivors to a very low level, but, regulated in such a manner that the shortest heating time will give positives in all the replicates, the next longest time will give some positives and some negatives and the longest time will give all negatives. Methods of analysis of the resulting data are given by Stumbo et al (1950) and Schmidt (1957). It is desirable that some knowledge of the general shape of the survivor curve is available even when using this method. Provided this is the case the number of replicates may be reduced to 10 or even 5. Finally, it should be borne in mind that when using this method it is essential that the spores are fully activated before determining the 'initial' number otherwise the shoulder on the survivor curve will introduce an error.

Methods of carrying out heat resistance tests and ways of treating the data obtained are comprehensively covered in a review by Pflug & Holcomb (1977).

REFERENCES

Adams A T, Ayres J C, Tischer R G, Ostle B 1952 Food Tech 6: 421
Adams D M 1973a Appl Microbiol 26: 282
Adams D M 1973b J Bact 116: 500

Adams D M 1974 Appl Microbiol 27: 797
Adams D M 1978 Adv in appl Microbiol 23: 245
Adams D M, Busta F F 1972 Appl Microbiol 24: 418
Alderton G, Snell N S 1963 Biochem biophys Res Commun 10: 139
Alderton G, Chen J K, Ito K A 1974 Appl Microbiol 27: 613
Alwood M C, Russell A D 1970 Adv in appl Microbiol 12: 89
Amaha M 1953 Food Res 18: 411
Amaha M 1961 Proc 4th Int Congress on Canned Foods (CIPC), Berlin
Amaha M, Ordal Z J 1957 Bact 74: 596
Anderson A A 1951 J Bact 62: 425
Anderson A A 1952 J Bact 64: 145
Anderson A A, Michener H D 1950 Food Tech 4: 188
Anderson E B, Meanwell L J 1936 J Dairy Res 7: 182
Anderson E E, Esselen Jr W B, Fellers C R 1949 Food Res 14: 499
Anderson E E, Esselen Jr W B, Handleman A R 1953 Food Res 18: 40
Anderson E E, Ruder L F, Esselen Jr W B, Nebesky E A, Labbee M 1951 Food Tech 5: 364
Aoki H, Slepecky R 1974 In: Barker A N, Gould G W, Wolf J (ed) Spore research 1973. Academic Press, London, p 93
Aref H, Cruess W V 1934 J Bact 27: 443
Augustin J A L 1965 Dissert Abst 26: 302
Barach J T, Flowers R S, Adams D M 1975 Appl Microbiol 30: 873
Barnes E M, Ingram M 1956 J appl Bact 19: 117
Bashford T E 1942 Proc Soc agric Bact (Abs), 37
Baumgart J, Stocksmeyer G 1976 Alimenta 15 (3): 67
Baumgartner J G, Knock G G 1940 J Soc Chem Ind 59: 53T
Baumgartner J G, Wallace M D 1934 J Soc Chem Ind 53: 294T
Beamer P R, Tanner F W 1939 Zbl Bakt Section 2 100: Nos 9/13, 20
Bean P G, Roberts T A 1975 J Food Technol 10: 327
Berg R W, Sandine W E 1970 J Milk and Food Technol 33: 435
Berry H, Jensen E, Siller F K 1938 Quart J Pharm 11: 729
Bigelow W D, Esty J R 1920 J Infect Dis 27: 602
Bose A N, Roy A K 1960 J Sci Ind Res 19c: 277
Braun O G, Hays G L, Benjamin H A 1941 Food Ind 13: No 2, 47. No 3, 64
Briggs A, Yazdany S 1970 J appl Bact 33: 621
Brown M R W, Melling J 1971 In: Hugo W B (ed) Inhibition and destruction of the microbial cell. Academic Press, London, p 1
Busta F F, Ordal Z J 1964 J Food Sci 29: 345
Byrne A F, Burton T H, Koch R B 1960 J Bact 80: 139
Cameron E J, Bohrer C W 1951 Food Tech 5: 340
Cameron E J, Yesair J 1931 Canning Age 12: 239
Cameron E J, Esty J R, Williams C C 1936 Food Res 1: 73
Campbell Jr L L, Sniff E E 1959 J Bact 77: 766
Campbell Jr L L, Winiarski W 1959 Appl Microbiol 7: 285
Campbell Jr L L, Sniff E E, O'Brien R T 1959 Food Tech 13: 462
Caraco A, Falcone G, Salvatore G 1958 Giorn Microbiol 5: 127
Casolari A 1974 Industria Conserve 49: 235
Cassier M, Sebald M 1969 Ann Inst Pasteur, Paris 117: 312
Church B D, Halvorson H 1959 Nature 183: 124
Clausen M 1978 Personal communication
Cook A N, Gilbert R J 1968a J Food Technol 3: 295
Cook A N, Gilbert R J 1968b J Food Technol 3: 285
Corry J E L 1974 J appl Bact 37: 31
Corry J E L 1976a J appl Bact 40: 269
Corry J E L 1976b J appl Bact 40: 277
Coulthard C E 1939 Pharm J 142: 79
Curran H R 1934 J Bact 27: 26 (Abs)

Curran H R 1935 J Infect Dis 56: 196
Curran H R, Evans F R 1937 J Bact 34: 179
Curran H R, Evans F R 1945 J Bact 49: 335
Dallyn H, Everton J R 1970 J appl Bact 33: 603
Denny C B, Bohrer C W 1959 Food Res 24: 247
Denny C B, Sharpe L E, Bohrer C W 1961a Appl Microbiol 9: 108
Denny C B, Reed J M, Bohrer C W 1961b Food Tech 15: 338
Dickson E C 1928 Proc Soc Exp Biol and Med 25: 426
Doyle M P, Marth E H 1975 J Milk Food Technol 38: 750
Duncan C L, Foster E M 1968 Appl Microbiol 16: 406
Duncan C L, Labbe R G, Reich R R 1972 J Bact 109: 550
Edwards J L Jr, Busta F F, Speck M L 1965 Appl Microbiol 13: 858
El-Bisi H M, Ordal Z J 1956a J Bact 71: 1
El-Bisi H M, Ordal Z J 1956b J Bact 71: 10
Ellicker P R, Frazier W C 1938 J Bact 36: 83
Ernst R R 1968 In: Lawrence C A, Block SS (eds) Disinfection, sterilization and preservation. Henry Kimpton, London, p 703
Esty J R 1928 Newer knowledge of bacteriology and immunity. Jordan and Falk, Chicago
Esty J R, Meyer K F 1922 J Infect Dis 31: 650
Esty J R, Williams C C 1924 J Infect Dis 34: 516
Evans F R, Curran H R 1943 J Bact 46: 513
Fowler G G, McCann B 1972 Food Ind of S Africa 25 (6): 49, 51 and 55
Fox E, Eder B D 1969 J Food Sci 34: 518
Frank H A, Campbell Jr L L 1955 Appl Microbiol 3: 300
Frank H A, Campbell Jr L L 1957 Appl Microbiol 5: 243
Friedman C A, Henry B S 1938 J Bact 36: 99
Fuchs A, Clausen M 1976 Schweizerische Milchwirtschaftliche Forschung 5 (July): 91
Gibbs P A 1967 J Gen Microbiol 46: 285
Gillespy T G 1936-37 Ann Rept Univ Bristol Fruit & Veg Pres Res Stn Campden, p 68
Gillespy T G 1946 Ann Rept Univ Bristol Fruit and Veg Pres Res Stn Campden, p 31
Gillespy T G 1947 Ann Rept Univ Bristol Fruit and Veg Pres Res Stn Campden, p 34
Gillespy T G 1957 Fruit & Veg Canning and Quick Freezing Res Ass Res Leaflet, No 3, Campden
Gomez R F 1977 Adv in Biochem Engng 5: 49
Gortzen J 1937 Zbl Bakt Section 1 Originale, 138: No 3/4, 227
Gould G W 1964 Effect of food preservatives on the growth of bacteria from spores, in Microbial Inhibitors in Food. Proc of the 4th Int Symp on Food Microbiology, Göteborg. Almqvist and Wiksell, Uppsala
Gould G W, Dring G J 1975 In: Gerhardt P, Costilow R N, Sadoff H L (eds) Spores VI. American Society for Microbiology, p 541
Gould G W, Hitchins A D, King W L 1966 J Gen Microbiol 44: 293
Greenberg R A, Silliker J H 1962 J Food Sci 27: 64
Hall R H 1966 Process Biochem 1: 461
Hanson R S, Curry M V, Garner J V, Halvorson H O 1972 Canadian J Microbiol 18: 1139
Harnulv B G, Snygg B G 1972 J appl Bact 35: 615
Harnulv B G, Johannson M, Snygg B G 1977 J Food Sci 42: 91
Hayakawa K-I 1977 Adv in Food Res 23: 75
Headlee M R 1931 J Infect Dis 48: 468
Heinemann B, Voris L, Stumbo C R 1965 Food Tech 19: 592
Hills G M 1949a Biochem J 45: 353
Hills G M 1949b Biochem J 45: 363
Hitchins A D, Gould G W, Hurst A 1963 J Gen Microbiol 30: 445
Hitzman D O, Halvorson H O, Ukita T 1957 J Bact 74: 1
Holland K T, Taylor C E 1974 In: Barker A N, Gould G W, Wolf J (eds) Spore Research 1973. Academic Press, London, p 185

Hsieh F H 1976 Dissertation Abs Int B36 (8): 3850
Humphrey A E, Nickerson J T R 1961 Appl Microbiol 9: 282
Hyatt M T, Levinson H S 1957 J Bact 74: 87
Hyatt M T, Levinson H S 1959 J Bact 77: 487
Jarvis B, Rhodes A C, King S E, Patel M 1976 J Food Technol 11: 41
Jensen L B 1954 Microbiology of Meats. Garrard Press, Champaign
Keynan A, Evenchik A, Halvorson H O, Hastings J W 1964a J Bact 88: 313
Keynan A, Issahary-Brand G, Evenchik Z 1965 Activation of bacterial spores. In: Spores III. A Symposium held at Allerton Park, Illinois. American Society for Microbiology
Keynan A, Issahary-Brand G, Evenchik Z 1964b In: Campbell L L, Halvorson H O (eds) Spores III, American Society for Microbiology, p 180
Keynan A, Murrell W G, Halvorson H O 1961 Nature 192: 1211
Koch R, Wolffhugel G 1881 Mitt Gesundh Amte 1: 301
Kooiman W J 1974 In: Barker A N, Gould G W, Wolf J (eds) Spore research 1973. Academic Press, London, p 87
Kosker O, Esselen Jr W B, Fellers C R 1951 Food Res 16: 510
La Baw G B, Desrosier N W 1953 Food Res 18: 186
Labbe R G, Duncan C L 1970 Appl Microbiol 19: 353
Lamanna C 1942 J Bact 44: 29
Lang O W 1935 Univ Calf Publs publ Hlth 2: No 1, 1-182
Le Blanc F R, Devlin K A, Stumbo C R 1953 Food Tech 7: 181
Lechowich R V 1959 Thesis for Ph.D in Food Tech, University of Illinois, quoted by Murrell W G (1961)
Lechowich R V, Ordal Z J 1960 Bact Proc p 44
Lechowich R V, Ordal Z J 1962 Can Journ Microbiol 8: 287
Levinson H S, Hyatt M T 1964 J Bacteriol 87: 876
Levinson H S, Hyatt M T 1969 In: Campbell L L (ed) Spores IV. American Society for Microbiology, p 262
Levinson H S, Hyatt M T, Moore F E 1961 Biochem and Biophys Res Communication 5: 417
Lewis J C, Snell N S, Burr H K 1960 Science 132: 544
Lowik J A M, Anema P J 1972 J appl Bact 35: 119
Lubieniecki-von-Schelhorn M, Heiss R 1975 In: Duckworth R B (ed) Water relations of foods. Academic Press, London, p 339
Lundgren L, Molin N, Snygg B G 1965 Physoil Plant 18: 921
Mayou J L 1970 Dissertation Abs Int B31 (3): 1423
Mazokhina N N, Naidenova L P, Rozenova L, Dashevskaya T V 1973 Acta Alimentaria Academiae Scientiarum Hungaricae 2: 385
Michener H D 1955 J Bact 70: 192
Michener H D, Thompson P A, Lewis J C 1959 Appl Microbiol 7: 166
Moats W A, Dabbah R, Edwards V M 1971 J Food Sci 36: 523
Molin N, Snygg B G 1967 Appl Microbiol 15: 1422
Morrison E W, Rettger L F 1930a J Bact 20: 299
Morrison E W, Rettger L F 1930b J Bact 20: 313
Mossel D A A, Mol J H H 1956 Appl Microbiol 4: 69
Murrell W G 1961 Spore formation and germination as a microbial reaction to environment. In: Symposium No 10 Soc Gen Microbiol, published for the society, CUP, London
Murrell W G 1964 Aust J Pharm 540
Murrell W G, Scott W J 1957 Nature 179: 481
Murrell W G, Scott W J 1966 J Gen Microbiol 43: 411
Murrell W G, Warth A D 1965 Composition and heat resistance of bacterial spores. In: Spores III. A symposium held at Allerton Park, Illinois. Published by American Soc Microbiol
Murrell W G, Olsen A M, Scott W J 1950 Austral J Sci Res B3: 234
Murty G G K, Halvorson H O 1957 J Bact 73: 235

Nelson F E 1943 J Bact 45: 395
Nelson F E 1944 J Bact 48: 473
Nichols A A 1940 J Dairy Res 11: 274
Northrop J, Slepecky R A 1967 Science 155: 838
O'Brien R T, Campbell Jr L L 1957 J Bact 73: 522
O'Brien R T, Titus D S 1955 J Bact 70: 487
O'Brien R T, Titus D S, Devlin K A, Stumbo C R, Lewis J C 1956 Food Technol 10: 352
Olsen A M, Scott W J 1950 Austral J Sci Res B3: 219
Oquendo R, Valdivieso L, Stahl R, Loncin M 1975 Lebensmittel-Wissenschaft & Technologie 8 (4): 181
Pederson C S, Beavens E A 1940 Food Ind 12: 61
Perkin A G 1974 J Dairy Res 41: 55
Perkin A G, Burton H, Underwood H M, Davies F L 1977 J Food Technol 12: 131
Peterson E E, Levine M, Buchanan J H 1927 Iowa State Coll J Sci 2: 31
Pflug I J, Esselen W B 1953 Food Tech 7: 237
Pflug I J, Esselen W B 1954 Food Res 19: 92
Pflug I J, Holcomb M S 1977 In: Block S (ed) Disinfection, sterilization and preservation, 2nd edn. Lea and Febiger, Philadelphia p 933
Pflug I J, Schmidt C F 1968 In: Lawrence C A, Block S S (eds) Disinfection, sterilization and preservation. Kimpton, London, p 63-105
Pflug I J, Smith G M 1977 In: Spore Research 1976 Academic Press, London, vol II, p 501
Pierson M D, Gomez R F, Martin S E 1978 Adv in Appl Microbiol 23: 263
Pivnick H, Thacker C 1970 Canad Inst Food Technol J 3 (2): 70
Powell J F 1950 J Gen Microbiol 4: 330
Powell J F 1953 Biochem J 54: 210
Powell J F, Strange R E 1953 Biochem J 54: 205
Prentice G A, Clegg L F L 1974 J appl Bact 37: 501
Reed J M, Bohrer C W, Cameron E J 1951 Food Res 16: 383
Reynolds H, Lichtenstein H 1952 Bact Rev 16: 126
Reynolds H, Kaplan A M, Spencer F B, Lichtenstein H 1952 Food Res 17, 153
Riemann H 1957 J appl Bact 20: 404
Roberts T A, Gilbert R J, Ingram M 1966 J appl Bact 29: 549
Roberts T A, Ingram M 1965 J appl Bact 28: 125
Schmidt C F 1950 J Bact 59: 433
Schmidt C F 1955 Bact Proc Soc Amer Bact 40
Schmidt C F 1957 In: Reddish G F (ed) Antiseptics, disinfectants, fungicides and sterilisation. Henry Kimpton, London, p 843
Schmidt C F, Bock J H, Moberg J A 1955 Food Res 20: 606
Secrist J L, Stumbo C R 1958 Food Res 23: 51
Senhaji A F 1977 J Food Technol 12 (3): 217
Senhaji A F, Loncin M 1977 J Food Technol 12 (3): 203
Slepecky R, Foster J W 1959 J Bact 78: 117
Sobernheim G, Mundel O 1938 Z Hyg 121: 90
Sognefest P, Hays G L, Wheaton E, Benjamin H A 1948 Food Res 13: 400
Sommer E W 1930 J Infect Dis 46: 85
Spilde O 1951 Tids Hermetikind 37: 297. Chem Abstr 1952 46: 7672
Splittstoesser D F 1973 Abs of Ann Meeting Amer Soc Microbiology 73: 17
Splittstoesser D F, Farkas D F 1966 J Bact 92: 995
Stern J A, Proctor B E 1954 Food Tech 8: 139
Stiles M E, Witter L D 1965 J Dairy Sci 48: 677
Stumbo C R 1948 Food Tech 2: 228
Stumbo C R 1973 Thermobacteriology in food processing, 2nd edn. Academic Press, London
Stumbo C R, Gross C E, Vinton C 1945 Food Res 10: 260
Stumbo C R, Murphy J R, Cochrane J 1950 Food Tech 4: 321

Sugiyama H 1951 J Bact 62: 81
Thorpe R H 1960 J appl Bact 23: 136
Tomlins R I, Ordal Z J 1976 In: Skinner F A, Hugo W B (eds) Inhibition and inactivation of vegetative microbes. Academic Press, London, p 153
Townsend C T, Esty J R, Baselt F C 1938 Food Res 3: 323
Tramer J 1966 Chem and Ind p 446
Treadwell P E, Jann G J, Salle A J 1958 J Bact 76: 549
Tsuji K, El-Bisi H M, Esselen W B 1960 Food Tech 14: abst 200
Vas K 1970 J appl Bact 33: 157
Viljoen J A 1926 J Infect Dis 39: 286
Vinton C, Martin S, Gross C E 1947 Food Res 12: 173
Wajid H R A, Kalra M S 1976 J Food Science and Technol, India 13 (1):6
Walker H W, Matches J R 1966 J Food Science 30: 1029
Walker H W, Matches J R, Ayres J C 1961 J Bact 82: 960
Wallace G I, Tanner F W 1931 J Bact 21:32
Wang D I, Scharer J, Humphrey A E 1964 Appl Microbiol 12:451
Weiss H 1921 J Infect Dis 28:70
Wheaton E, Burroughs J D, Hays G L 1957 Food Tech 11:286
Wheaton E, Hays G L 1964 Food Tech 18: 549
Williams C C, Merrill C M, Cameron E J 1937 Food Res 2: 369
Williams O B 1929 J Infect Dis 44: 421
Williams O B, Campbell Jr L L 1951 Food Res 16: 347
Williams O B, Reed J M 1942 J Infect Dis 71: 225
Williams O B, Robertson W J 1954 J Bact 67: 377
Williams T F 1936 J Bact 32: 589
Woese C, Morowitz H J 1958 J Bact 76: 81
Wynne E S, Foster J W 1948a J Bact 55: 61
Wynne E S, Foster J W 1948b J Bact 55: 331
Xezones H, Hutchings I J 1965 Food Tech 19: 1003
Yesair J, Cameron E J 1936 J Bact 31: 1 (Abs)
Yesair J, Cameron E J 1942 Canner 94: No 13, 89
Yesair J, Bohrer C W, Cameron E J 1946 Food Res 11: 327
Zaleski S, Sobolewska-Ceronik K, Ceronik E 1971 Annales de l'Institut Pasteur de Lille 22: 263
Zamenhof S 1960 Proc Nat Acad Sci US 46: 101
Zuccaro J B, Powers J J, Morse R E, Mills W C 1951 Food Res 16: 30
Zytkovicz T H, Halvorson H O 1972 In: Halvorson H O, Hanson R, Campbell L L (eds) Spores V. American Society for Microbiology, p 49

9

Principles of thermal processing

The primary object of the thermal processing of canned foods is to destroy living organisms capable of causing deterioration of the food or endangering the health of the consumer. Coincidental with this is the necessity of retaining the organoleptic and nutritive properties to the greatest possible extent and the need for scientific adjustment of heat processes arises because a process which is adequate from the culinary point of view may be insufficient to eliminate spoilage organisms. The extent to which canned foods may be heated to achieve a compromise between these needs, that is, the processing standard, must be known and defined. This knowledge in conjunction with information on the heat-resistance of the contaminating micro-organisms, the chemical and physical nature of the food and the rate at which heat penetrates the contents of the can form the basis of process evaluation.

Thermal death-time data

In the preceding chapter methods of measuring the heat resistance of spore suspensions were briefly described. The data available from determinations of this type is normally presented in the form of survivor curves which relate to specific temperatures at which studies were carried out. For process evaluation work thermal destruction data relating to more than one temperature is required and this is usually presented in the form of a thermal death-time curve. This curve is obtained by plotting the thermal death-time on a log scale against temperature of heating. Thermal death-time (TDT) may be defined as the time required at any specified temperature to inactivate an arbitrarily chosen proportion of the spores, the higher the proportion the greater the margin of safety. The extent of inactivation required in canned foods is discussed later in the section dealing with standards of processing. The slope of the TDT curve is defined as z which is the number of degrees for the line to traverse one log cycle of thermal death-time. The value for *Clostridium botulinum* is normally taken as 10 degrees Celsius, the value obtained by Esty & Meyer (1922) in their classical studies on heat resistance of *C. botulinum* spores in phosphate buffer. It is used in process calculations to account for the relative resistance of this organism to different temperatures in a particular medium. The value of z will be dependent upon the organism concerned and the medium in which it is suspended during heating and recovery. Perkins et al (1975) and Stumbo et al (1975) consider that a z value of 7.8°C is more realistic for most foods. Thus the spores are more resistant at temperatures below 121.1°C and less resistant at temperatures above 121.1°C than predicted using the value $z = 10$°C. The paper by Perkins et al (1975) does not necessarily imply that processes obtained using the classical data are unsafe since they obtained lower D values than did Esty and Meyer in their work. However, they point out, that apart from convection heating packs processed at 121.1°C or above, in most canned foods the slowest heating point achieves the scheduled F_0 value by the cumulative heating effect at lower temperatures. In such cases, particularly where the slowest heating point does not reach temperatures in excess of 110°C, for example large cans of conduction heating product, then the processes indicated from the classical data may be too low.

The arbitrary proportion of spores to be inactivated will depend on the degree of assurance against spoilage required. A process with a duration of D minutes, for example, will inactivate 90 per cent of the spores and a plot of D against the corresponding temperature will give a TDT curve any part of which will determine the time and temperature conditions required to attain this objective. For reasons discussed later, when plotting TDT curves for use in process evaluation work, it is usual to aim for a degree of inactivation for *Clostridium botulinum* equivalent to a 12D process. The curve resulting is shown in Figure 13. Points taken from this curve provide data for use in thermal process evaluations as applied to canned foods.

For sterilisation processes using temperatures in excess of 125°C it is important to know whether the z value remains constant with temperature since it is common practice to extrapolate thermal death-time (TDT) curves to higher temperatures than the experimentally determined values. Edwards et al (1965) investigated the response of *Bacillus subtilis* to UHT processing and observed some unusual responses. Busta (1967) following up this work could not demonstrate non-linearity for spores of P.A. 3679 or for *B. stearothermophilus* heated in skim milk although unusually high z values were obtained for P.A. 3679. A series of papers from the National Institute for Research in Dairying (Davies et al, 1977; Perkin et al, 1977; Burton et al, 1977) on the thermal death of *B. stearothermophilus* at ultra high temperatures demonstrated a behaviour in accordance with Arrhenius

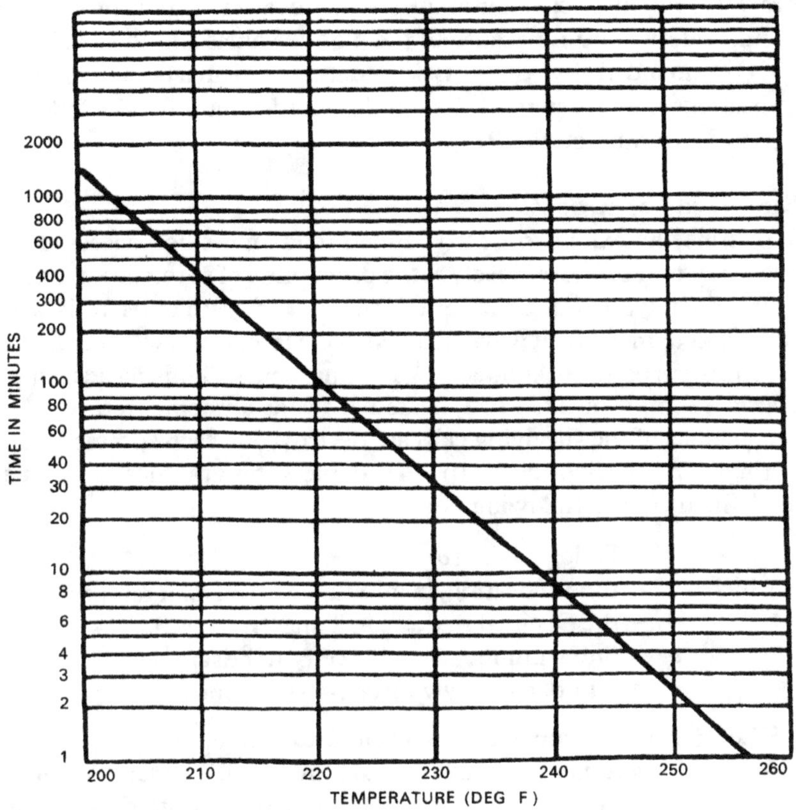

Fig. 13 Thermal death-time curve of *C. botulinum*

kinetics at temperatures up to 150°C provided that appropriate corrections for come-up time were applied. Thus a curvilinear thermal death-time curve is inferred with the *z* value increasing with temperature. They found also that the method of determining the heat resistance influenced the *z* value, the capillary tube method giving higher values than a steam into product UHT plant. The divergences of D value obtained by the two methods were greatest at the lowest temperatures and the values become coincident at 145°C. Currently there is no reliable UHT data for the spores of *Clostridium botulinum* and this information is badly needed to ensure an adequate safety margin on foods processed at temperatures in the region of 140°C as is the case in the continuous flow sterilisers used in UHT plants. At the time of going to press the results of an investigation being carried out by the British Food Manufacturing Industries Research Association were not available but should the results show the same pattern as that found for *B. stearothermophilus* at NIRD then the safety of UHT processes might be called into question. However there is probably an inbuilt safety margin because of the high D values and the z values of 10°C demanded by the classical data, particularly as there are indications that lower z values should be used in most foods (Perkins et al, 1975; Stumbo et al, 1975).

Standards of processing

To define the extent to which canned foods must be processed it is necessary to look more closely at the probability of survival of bacterial spores during the heat treatment. The order of death of bacteria when subjected to moist heat was discussed in Chapter 8 and was seen, for all practical purposes, to follow a logarithmic course. This means simply that in a sample of fixed size, heated at a constant temperature, equal proportions of the surviving cells are destroyed in each successive unit of time. The survivor curve (p. 143) depicts this graphically and may be defined by the following equation:

$$t = \mathrm{D}(\log a - \log b)$$

in which t = time of heating, in minutes, at a constant temperature,
D = the Decimal Reduction Time (*see* p. 142),
a = initial number of viable cells in the sample,
b = number of surviving cells in the sample.

The above equation may be applied to a sample of any volume provided the volumes containing a and b are equal. The sample may be considered to consist of a number of volumes or containers. In this case by increasing the number of containers the value of a may be made as large as we wish. The value of b increases also as the number of containers involved increases. Therefore, in a finite number of containers there is always one containing a survivor provided the sample size is large enough.

If the equation is solved for small samples, e.g. one container, it may be found that b may be reduced to less than one but never to zero. This is evident also by examination of the survivor curve, from which it may be seen that the number of viable cells remaining in a suspension after heating approaches but never reaches zero. A fractional value of b is interpreted as the probability of survival of a micro-organism within the specified sample. For example, if a can contains one spore of *Clostridium botulinum* with a $\mathrm{D}_{121 \cdot 1°\mathrm{C}}$ value of 0.21 minutes and it is subjected to a time-temperature combination equivalent to 2.52 minutes at 121.1°C.

Then

$$2.52 = 0.21(\log 1 - \log b)$$
$$\log b = -12$$
$$b = 10^{-12}$$

This fractional value of *b* expresses the probability that the one spore of *C. botulinum* in this particular can will survive, that is one chance in 10^{12}.

In the above example the bacterial population was reduced by a factor of 10^{12}. The level to which a bacterial population must be reduced by the heat process is one way of setting an acceptable standard. To be more specific, the reduction factor should apply to a known micro-organism, and, to be meaningful, the organism should be one of particular significance. For medium and low acid foods, that is, those with pH values greater than 4.5, *C. botulinum* is the one selected. This is because the spores of *C. botulinum* Type A and B are able to grow and produce toxin in medium and low acid foods and these spores are the most heat resistant food poisoning organisms encountered. The acceptable level to which these spores must be reduced has been established by long experience and was originally based on the maximum heat resistance obtained for these spores by Esty & Meyer (1922). In their experiments, in which more than 100 strains were tested, large numbers of spores, produced under the most favourable conditions, were heated to complete destruction in phosphate buffer at a pH of 7.0. The highest resistances obtained were:

100°C	330 minutes
105°C	100 minutes
110°C	32 minutes
115°C	10 minutes
120°C	4 minutes.

It is now generally accepted that this 'classical' standard is equivalent to a reduction in the population by a factor of about 10^{12}, and modern canning practice demands, at least, this degree of destruction in low-acid and medium-acid products (Hicks, 1961). Thus thermal death-times equivalent to those given by Esty and Meyer are used as a basis for calculating the heat process but may need to be adjusted if the D values in the product differ from those in neutral phosphate buffer.

Since the most resistant spores of *C. botulinum* have a D value at 121.1°C of about 0.21 min a 12D process is given by 2.52 minutes at 121.1°C, assuming instantaneous heating and cooling. The values of 12D are known for other temperatures, and if the logarithms of these times are plotted against the appropriate temperature the normal thermal death-time (TDT) curve results.

A unit of measurement has been devised which is used to compare the relative sterilising potential of different heat processes. This unit, equal to 1 minute at 121.1°C, is designated by the symbol F. Using ordinates as in the TDT curve, straight lines corresponding to different values of z can be drawn through this point. In order to define F completely the value of z is indicated as a superscript, for example F^8, F^9, F^{10}, etc. Since the value of z for the TDT curve of *C. botulinum* is 10°C, and, also, because this value is commonly observed for other bacterial spores, $F^{10}_{121.1\,°C}$ is more usually written as F_0. This is the unit normally used in process calculation work but in special cases where a process is being evaluated with respect to destruction of an organism with a different slope, other values, such as F^9, may be used. Since the spores of *C. botulinum* show a thermal death-time (12D) of approximately 2.5 minutes 121.1°C the heat process required to achieve a lethal value of unity with respect to this organism has an F_0 value of 2.5.

Since spoilage organisms of greater heat resistance than *C. botulinum* may be encountered, a process with a lethal value of unity with respect to this organism, while adequate from public health aspects, cannot be relied upon to ensure freedom from spoilage. From the commercial standpoint, then, the standard of processing needs to be more stringent than that required to safeguard the health of the consumer if extensive losses due to spoilage are to be avoided. An example of the type of heat-resistant organism which may be met is the National Food Processors Association Research Laboratory strain of putrefactive anaerobe. This organism, known as No. 3679, has been widely used in processing studies by numerous workers, and Townsend et al (1938), Stumbo et al (1950), Reed et al (1951), Reynolds et al (1952), Amaha (1953) and Pflug & Esselen (1954) have published results of heat resistance on this strain. Various $D_{121.1}$ values are found depending upon the substrate but maximum values up to 1.5 minutes may be anticipated. Most workers have found that in food substrates an approximate average value of z=10°C may be assumed.

With regard to thermophilic species, the phenomenal heat resistance of these organisms may, in some packs, render their destruction by traditional heat-processing techniques impracticable. Flat-sour spoilage is usually attributable to the growth of *Bacillus stearothermophilus* for which $D_{121.1}$ values of up to 5 minutes are observed. Gaseous spoilage may be caused by growth of *Clostridium thermosaccharolyticum,* some strains of which are extremely heat resistant. An organism of this type has been described by Xezones et al (1965) for which they claim a D_{124} value of 70 minutes. This is undoubtedly an exception and it is more usual to assume a heat resistance of the same order as for *B. stearothermophilus*.

The processing standard adopted will, of course, be determined by a number of factors. Where thermophilic spoilage is of no particular significance for the pack concerned (e.g. a pack not to be exported to a hot climate or a pack not normally subject to thermophilic spoilage) the standard of processing may be based on the destruction of an organism with a thermal resistance equivalent to that of anaerobe No. 3679. Since this organism will not endanger the health of the consumer, the standard is not normally based on reducing its number by a factor of 10^{12} but more often by a factor of 10^4 or 10^5. Sognefest et al (1948), discussing the thermal process requirements of non-acid canned foods, state that past experience has indicated that commercially sterile products should be obtained if processes are based upon a spore concentration (test organism No. 3679) having a resistance in neutral phosphate of 3.65 to 4.15 minutes at 121.1°C.

Where the pack is subject to thermophilic spoilage, a more rigorous heat treatment is called for. According to Lewis (1945), satisfactory processes for meat-vegetable packs are those which give lethal values of 7 to 8, that is, F_0 values of approximately 18 to 20. Bashford (1947) states that for good-quality products prepared under normal hygienic conditions, a lethal value of the order 6 to 7 is often required to ensure sterility. Quality considerations, however, frequently preclude the adoption of such heating processes, and careful selection of raw materials with a low degree of thermophilic contamination then becomes essential.

Within the medium-acid and low-acid group of foods there is evidence that the process requirements are lower for those foods in the more acid region, that is, with a pH within the range 4.5 to 5.5. Sognesfest et al (1948) carried out studies on the effect of pH on the thermal process requirements for vegetable purees using P.A. No. 3679 and *Clostridium botulinum* as the test organisms. Spoilage results indicated that the lower the pH within the range 4.5 to 9.0, the lower the sterilising value required. At a pH of about 5.5 and above, there seems to be a decided rise in the sterilising value required to prevent spoilage. It was noted that the F_0 value required, based on spoilage results, varied with different products; Alstrand & Ecklund (1952) have used these results to suggest F_0 values required for foods based on their pH and carbohydrate content. Xezones & Hutchings (1965) found also that the heat resistance of spores of *C. botulinum* in food substrates decreased with decreasing pH the effect being greater at the lower sterilisation temperatures used. In view of the diverse nature of different foods, however, it is not possible to make general recommendations. For all foods with a pH greater than 4.5 other than those containing curing salts, it is customary to process to a minimum value of F_0=3.0.

Stumbo et al (1975) have published a working guide for establishment of thermal processes for low acid canned foods. The calculations are based on certain proposals or assumptions on the number and distribution of spores initially present and on the D and *z* values to be used. A $D_{121.1}$ value of 0.2 minutes in conjunction with a *z* value of 7.8°C was used for *Clostridium botulinum* although no justification was presented for using these two values simultaneously. Indeed, the paper by Perkins et al (1975) suggests that the lower *z* values obtained in food are often associated with lower D values whilst Stumbo and his colleagues use the high classical value for D with a low value for *z*. Nevertheless the data given are in the form of practical tables of F values which may be applied to canned foods. These take into account can size, whether the product is conductive or convective heating, retort temperature and the degree of commercial sterility assurance required.

At pH values below 4.5 the risk of growth and toxin production from *Clostridium botulinum* does not exist. Furthermore, the micro-organisms capable of causing spoilage are of comparatively low heat resistance and the heat processing requirements are drastically reduced. For acid foods with a pH between 4.0 and 4.5 the heat process is usually adjusted to reduce to an acceptable level the risk of flat-sour spoilage due to the growth of *Bacillus coagulans*. This should ensure, coincidentally, freedom from spoilage by mesophiles such as *B. polymyxa*, *B. macerans* and the butyric anaerobes. A heat process of $F^{10}_{121.1} = 0.7$ may be regarded as adequate for this purpose. Products with a pH below 4.0 are normally subject to spoilage only by non-spore-forming bacteria, yeasts and moulds and consequently pressure processing is not usually applied. For such products the use of 121.1°C as a reference temperature is inconvenient and the D values of significant micro-organisms are usually given at lower temperatures such as 65° or 70°. Stumbo (1973) gives approximate $D_{65.5}$ values of 0.5 to 1.00 minutes for *Lactobacillus* spp., *Leuconostoc* spp., and yeasts and moulds. If the presence of *Byssochlamys fulva* or *B. nivea* is suspected, however, the process conditions will need to be adjusted to prevent spoilage by these moulds, the ascospores of which are exceptionally heat resistant (*see* p. 135).

Heat penetration

In evaluating the heat process applied to canned foods it is necessary to take into account the contribution made by the periods in which the can contents are heating-up and cooling down. It is necessary, therefore, to measure the rate of heat penetration into the product.

The rate of heat penetration can be determined by means of the thermo-electric couple. This instrument depends on the principle that when two dissimilar metal wires such as copper and constantan or chromel and alumel are joined at both ends to form a closed circuit and one junction is at a higher temperature than the other, a current is set up, the magnitude of which depends on the temperature difference between the two junctions. One junction (the hot junction) is inserted into the can at the point of greatest temperature lag and the second (the cold junction) is held at a steady temperature, usually 0°C (melting ice). A potentiometer is put into circuit and as the temperature of the can in the retort is raised, the difference in potential is observed.

Modern practice is to use a recording potentiometer incorporating automatic temperature compensation for the reference couple so that the cold junction in melting ice is eliminated. The scale of the instrument may be factory calibrated in temperature units for use with thermocouples of known characteristics. The instrument is calibrated by exposing both junctions, or the hot junction only in the case of instruments with a compensated reference junction, to temperatures known and verified by a standard reference mercury-in-glass thermometer. With factory calibrated equipment it is essential to use wire which corresponds to the instrument. Additionally the temperature scale should be checked regularly by the method recommended above. The system as a whole should be capable of giving reproducible readings to an accuracy of ±0.5°C and should be calibrated regularly over the range of temperatures for which it will be used. Equipment is available for converting the thermocouple signals directly into lethal rates and summing them to give a direct F_0 read-out. These readings should be supplemented with time temperature plots from the usual heat-penetration runs, and, as for all instrumental systems, checked for accuracy at regular intervals. Some of the commercial instruments that are available are reviewed by Holdsworth (1974) and Lingelbach (1974).

Thermocouples designed for measuring heat-penetration rates in canned foods are produced by several firms. One in common use is the thermocouple described by Ecklund (1949) which is a plug-in type with the socket flush to the wall of the can. The flush fitting enables the can to be closed readily on all types of seaming machines. In our laboratories we make our own thermocouples from thin gauge (0.5 mm) copper and constantan wires which are passed through diametrically opposed holes in the can wall and sealed into position with epoxy resin (Fig. 14). The advantage of this type of assembly is that the flexible wires can be reeled around the outside of the can and fixed with adhesive tape enabling it to be closed on most types of seamer. Furthermore, the thin wires held in tension across the can give little disturbance to weak convection currents in the contents. This type of thermocouple has been described by Board (1965). Metal sheathed thermocouples are viewed with disfavour and a sheath of non-conducting material is preferred as affording less risk of heat conduction along the sheath to the can centre. In conduction heating packs errors may be caused by heat being transmitted along the thermocouple wires themselves. To allow for such errors with his assemblies, Ecklund (1956) has prepared a table of correction factors to be applied to the j values (see p. 199 for an explanation of this term). Cowell et al (1959) confirm Ecklund's findings and point out that errors are largest in the early stages of heating and cooling. For accurate work, therefore, the thermocouples should be constructed from fine wires and, if necessary, correction factors should be applied.

Fig. 14 Thin wire thermocouple positioned in a can

Packer & Gamlen (1974) calculated the errors in tension wire thermocouples used for the estimation of process values in convection heating packs and provided a formula for correction of the j values. Beverloo & Weldring (1969a, 1969b) evaluated errors experimentally for various thermocouple constructions and came to the conclusion that errors introduced by the thermocouple to the process value were not serious.

Heat penetration studies on canned products to be processed on continuous sterilisers present a problem in the transfer of the thermocouple information from the probe to the recording equipment. The traditional method of carrying out such studies has been to employ a pilot plant simulator which employs slip rings for this purpose, but this has never been entirely satisfactory and it is advisable to supplement these results with trials made on production equipment. For this purpose bacteriological methods, which are described later, or in some cases radio telemetry or electronic data storage systems, may be used. The sensing element in equipment of this type often consists of a thermistor or platinum resistance thermometer rather than a thermocouple.

The basis of radio telemetry systems is the conversion of temperatures to electrical signals by a transducer, which in turn modulate a radio transmitter. This radio signal is then picked up by a receiver and converted to an electrical signal which is proportional to the measured temperature. Since the shell of the steriliser prevents transmission of the signal, looped aerials are placed inside the casing and brought through sealed glands. By this means the signal is fed to the receiver. The aerials are best fitted during construction and installation of the steriliser. It is, however, possible to install aerials on operational hydrostatic sterilisers during the period of overhaul. Several radio telemetry systems are available and one developed by the British Food Manufacturing Industries Research Association is described by Steele (1972). Multi-channel versions of this equipment are now available. In the United Kingdom, under the Wireless Telegraph Act 1949, a licence is required to operate radio telemetry equipment and care should be exercised to ensure that purchased equipment operates in the permitted frequency bands.

The electronic data storage systems consist of small self-powered measuring devices which will store the generated information in the form of a binary code. After the process the instrument is interrogated and the stored contents displayed for analysis. One version of this equipment (Felton and Guilleaume measuring instrument TME-2) consists of a platinum resistance temperature probe, a temperature evaluating and storage device and a recorder. After the process the stored data may be displayed either digitally or in the form of a time/temperature plot.

Mechanism of heat transfer

The positioning of the hot junction is dependent upon the physical properties of the product and it is necessary to consider the mechanism of heat transfer to appreciate the reason for this.

Conduction heating packs. Solid foods heat by conduction and the process is relatively slow since there is no transfer of the material from the hot to the cooler parts of the can; the rate at which the heat is conducted decreases as the temperature difference between the retort and the centre of the can becomes smaller. Jackson & Olson (1940) studied the temperature distribution in cans heating by conduction. During the initial lag period, the temperature of the material at the centre of the can remained constant while the temperature increased near the can wall. After the lag period the temperature rose on a smooth curve from the centre to the can wall, indicating that heat was conducted symmetrically from the outside of the can towards its centre. Evans & Board (1954), investigating heat transfer through the headspace in cans packed with solid foods, showed that some asymmetry was introduced due to the lower rate of heat transfer through the headspace than from the can walls. This has the effect of making the slowest heating point in the can just above the geometric centre.

Convection heating packs. Heat is transferred by convection currents caused by a change in density as the fluid heats. Any dissolved substances that increase the viscosity will reduce this effect and decrease heat transfer.

Broken heating curve packs. Some foods exhibit what is known as a broken heating curve. This refers to a distinct change or break in the rate of heat penetration, an initial period of relatively rapid heat transfer ceasing more or less abruptly and being followed by a period of much slower heat penetration. Jackson & Olson (1940) in their report on the mechanisms of heat transfer in canned foods mention thick soups, brine-packed whole-grain corn, certain tomato juice and specialities as being among those packs which give broken heating curves; they point out that the form of such heating curves may show considerable variation from can to can and from batch to batch. As a result of their investigation of the temperature distribution pattern in a bentonite suspension exhibiting the same type of broken heating curve as some foods, Jackson and Olson concluded that the 'break' is due to the product undergoing a sol-gel transition. In the first stage of heating, convection is the main factor involved in heat transfer but with the transition of the product to a gel form, heating continues mainly by conduction.

Ball (1947) has suggested that broken heating curves may result from the presence of small amounts of air in the retort during processing. The effect of an air-steam layer round the can is to prevent the condensation of steam on the can surface during the latter stages of processing and thus no latent heat is derived from the steam during this period. The temperature at which latent heat ceases to be a factor in heating the can corresponds to the 'break' in the heating curve.

Mixed conduction convection heating packs. In some products such as those with a high solids to liquids ratio or large pieces of solid food in a thin liquor the pack heats simultaneously by conduction and convection. Sometimes this happens accidently, for example if pasta clumps together to form large pieces the heat transfer may be slower than that intended due to a change in characteristics from mainly convective heating to mixed conduction convection heating. This situation can give rise to serious under-processing and spoilage. Vacuum packed products such as sweet corn and stratified products, both discussed later, transfer heat by virtue of both conductive and convective mechanisms.

Positioning of thermocouple

Thermocouples are usually mounted in the sides of the test cans, the position of the tip being determined by the location of the point of

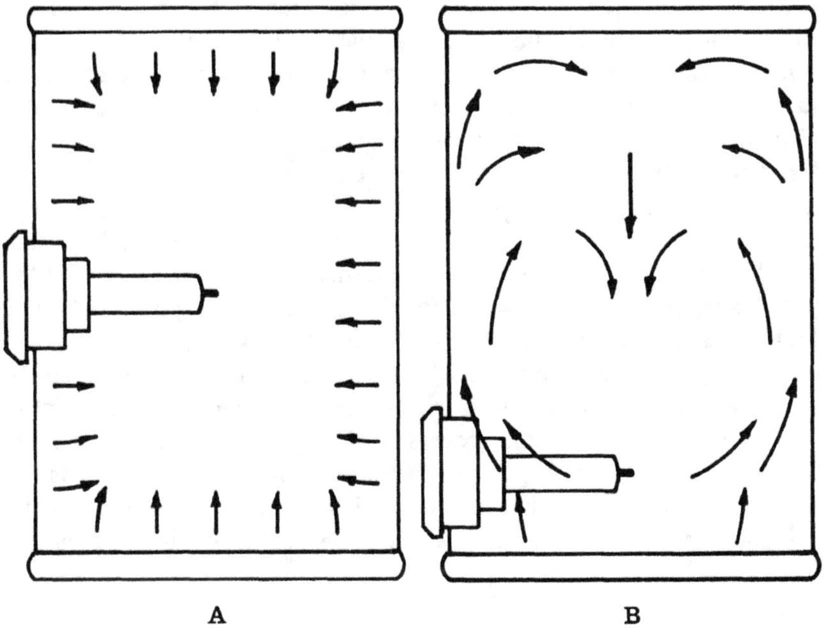

Fig. 15 Heat penetration into cans (courtesy of *Food Technology*)

greatest temperature lag. For solid or very viscous packs which heat mainly by conduction this point is usually taken as the geometric centre of the can (Fig. 15a). Evans & Board (1954) showed that the true point is just above the geometric centre but the lethal value obtained when using this point does not vary greatly from the value obtained at the geometric centre. For products such as clear soups or vegetables in brine, which heat mainly by convection, the thermocouple tip is located on the can axis at a point between the geometric centre and the bottom of the can (Fig. 15b). For No. 2 cans this point is about 1.9 cm (¾ in) from the bottom of the can and for No. 10 cans about 3.8 cm (1½ in) from the bottom when the cans are processed in an upright position. Products which exhibit broken heating curves may have the slowest heating point either at the geometric centre or near the bottom of the can axis, and an initial heat-penetration test should have thermocouples located in both positions to determine which point is the slower heating. In mixed conduction convection heating packs the slowest heating point is usually at or near the centre of the solid mass of product but in cases of doubt it should be determined experimentally.

Location of a thermocouple at the slowest heating point of a flexible pouch is more difficult since pouches do not have a fixed geometry and thickness varies with the shape and size of solid particles in the product, the differential pressure between the contents and the heat transfer medium, the disposition of pouches in the retort and the dimensions of any support racking that may be used. Variations in pouch thickness that may occur due to inadequate attention to these factors can result in significant changes in the rate of heat penetration. For this reason the thickness of pouches is generally restrained by fixed geometry racking. In practice the thermocouple tip is normally held at the anticipated pack centre or, in the case of products containing particles of appreciable size, at the centre of a selected piece of solid food. The wires may be passed through a hole in the wall of the pouch and sealed by means of a packing gland of the type described by Davis et al (1972). We have found a simple epoxy resin cement seal adequate for many applications. The tip of the thermocouple needs to be held at the correct position within the pouch and for this purpose we have used small nylon spacers of the type shown in Figure 16. These are of variable length to suit the pack size and rack dimension. They are held in position by tightening a small nylon nut on the threaded spindle of the spacer. A leak proof seal is obtained by the use of O rings and washers.

Factors affecting heat penetration

It is important to recognise the factors affecting the rate of heat transfer if the results obtained on test cans are to be reproduced in commercial production. Apart from correct retort operation and venting procedures, discussed in Chapter 5, there are a number of other factors of importance.

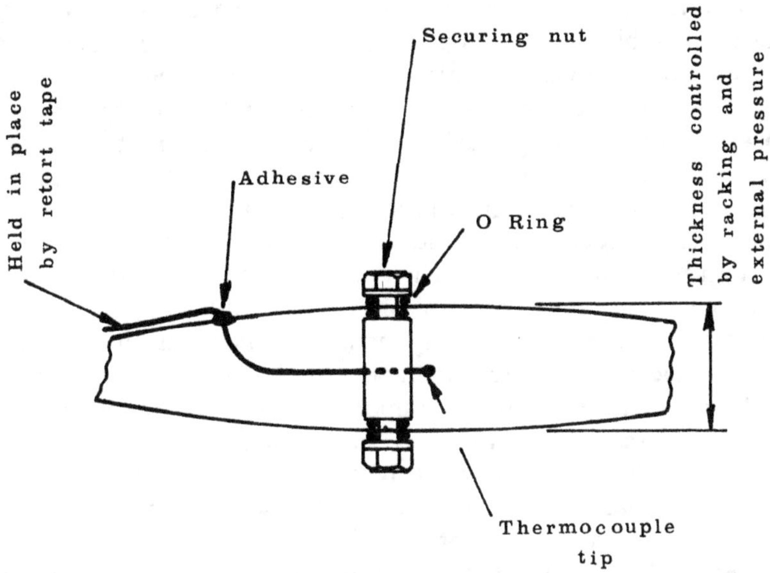

Fig. 16 Thermocouple positioned in a flexible retort pouch

Filling. For any canned food the rate of heat penetration is affected by the temperature gradient between the can and the retort, the rate becoming slower as the temperature difference decreases, so that there is an asymptotic approach to retort temperature. From this it follows that increase in the retort temperature results in more rapid heat penetration with consequent decrease in sterilising time. Due to the effect of the temperature gradient on the rate of heat penetration, a cold filled can will reach retort temperature in about the same time as a partially heated can, but this in no way detracts from the value of hot filling when processing canned foods since hot cans will reach a lethal temperature more rapidly than those cold filled.

If cans of a slow heating product are filled hot and allowed to stand before being processed the temperatures within the can will be non-uniform due to cooling near the can wall and conversely cold filled products subjected to a thermal exhaust may show higher temperatures in the areas near the can wall. Alstrand & Benjamin (1949) showed that errors are introduced into heat-penetration data due to non-uniform temperature distribution and suggested that the 'initial temperature' should be taken as the average temperature of the can contents rather than the temperature of the can centre. This has been generally accepted by the industry.

In order to allow for the worst situation likely to arise in practice the lowest anticipated filling temperature is normally used for heat penetration studies on test cans. Care must be exercised, however, to ensure that this gives a realistic representation of events. For example, pieces of farinaceous materials or unblanched meat may clump or mat together if liquored with sauce at high temperature and this gives rise to a particular problem if delays occur on lines feeding rotary cooker-coolers. In such circumstances the poor heat penetration into these larger masses would not have been anticipated from the test runs if the sauce temperature had been adjusted to a low value.

With solid materials packed in liquids, the ratio of solid to liquid and also the arrangement of the solids in the liquid will affect heat penetration; the presence of channels which permit convection currents facilitates the transfer of heat. In this connection the position of the cans in the retort may be important as in the case of packs such as celery, asparagus, etc., where vertical channels between the pieces permit the flow of convection currents when the cans are stacked in an upright position. These currents, would, however, be retarded if such packs were retorted in the horizontal position. Solids loosely packed in a liquid will heat more rapidly than if tightly packed. It follows from this that where the margin of safety on a process is small, pronounced overfilling with solid material may result in under-processing. Lewis (1945) determined the rates of heat penetration into I.T. cans containing (1) the normal fill of 16 oz (0.45 kg) and (2) 18 oz (0.5 kg) of a meat pack. The lethal value of a process of 2 hours at 115.5°C was reduced by approximately 14 per cent in the case of the over-filled can.

Ingredient related factors. A number of factors related to the product are able to affect the rate of heat penetration, for example, the turgidity of vegetables and fruits and the tightness of certain vegetables such as celery hearts. These variations may be influenced by growing conditions during the season (wet or dry), storage conditions between harvesting and preparation and the maturity of the crop. The ability of solids to absorb sauce will give rise to changes in the solids/liquid ratio in the container. The extent to which this occurs is, in the case of dry products, determined by the water uptake during soaking or blanching. The water uptake of legumes may also be affected by the hardness of the water. Since fatty tissue is a poorer conductor of heat than the lean portions the heat transfer rate is affected by the ratio of these constitutents. In soups and sauces the type of thickening agent used, small variations in the amount or changes in the thickening power from delivery to delivery, may all have an effect.

Preparation related factors. A potentially dangerous situation that may arise during production, usually not covered during test processes, is the use of partly thawed frozen raw material. This causes a considerable temperature lag in the heat penetration of the particles concerned which is particularly serious on high speed continuous lines. Mechanical handling in the factory may cause particle damage leading to different product characteristics from those tested in the laboratory or pilot plant.

Processing related factors. Since the time required to vent the retort and bring it to processing temperature ('come-up' time) may not be the same in commercial practice as with test cans, it is necessary to know the effect of this on the total process. Ball (1928) experimentally established that 42 per cent of the 'come-up' time should be added to the process time at retort temperature and Alstrand & Benjamin (1949) have confirmed that this figure should be used for conduction and convection heating packs for logarithmic or uniform rate of retort temperature rise in the 'come-up' period. Gillespy (1953) estimated the value of the 'come-up' time mathematically and gave graphs which can be used for various combinations of steam supply pressure, processing pressure, can size and initial product temperature.

Agitation of the cans during heating markedly increases the rate of heat penetration in liquid packs by transferring the heated portion of the contents to the cooler regions of the can. Clearly, agitation is only advantageous for products which are too viscous to permit convection currents but sufficiently fluid to allow mixing or for solids packed in liquids. One development, in which the mechanism of heat transfer is not immediately apparent, is the application of agitation to vacuum-packed foods (Ball, 1951; Lueck & Brighton, 1951). Vacuum packing, by which method foods are canned with little or no added liquor, is seldom used in this country. In America, however, vacuum packing is commonly used for shrimps, corn-on-the-cob, whole grain corn, etc. The mechanism of heat transfer in these packs using traditional methods of sterilisation has been studied by Jackson (1940). Based on an examination of heating curves Jackson concluded that vapour convection must be largely responsible for heat transfer. Experimental data indicated that the pressure in the cans, the amount of free liquor in the cans and the closeness of packing of the contents have marked effect on the rate of heat transfer in vacuum packed cans. Relatively long processes are necessary when using traditional methods of sterilisation but extremely rapid heat transfer is obtained if agitation is used. It has been suggested that the hot vapour cycle is supplemented by a hot liquid cycle in which small amounts of liquor trickle over the food particles. Roberts & Sognefest (1947) published data showing the advantages obtained by application of agitation to vacuum-packed products.

One of the most efficient methods of increasing the rate of heat penetration is end-over-end agitation. In this process the cans are arranged with their longitudinal axis aligned as the spokes in a wheel so that they rotate end-over-end as they move around the circumference. Clifcorn et al (1950) state that when the speed is adjusted so that the centrifugal force equals the weight of liquid contents, the headspace volume passes approximately through the centre of the can and produces maximum turbulence. Ball (1951) states that viscosity of the product has a greater effect than density alone upon the speed of rotation required for the highest rate of heat penetration. Factors affecting the process are speed and radius of rotation, viscosity of the product and the headspace in the can. An obvious disadvantage when using batch retorts is that fewer cans can be accommodated than when using the conventional method. This is partly off-set by the reduced sterilising times possible, and with mechanical aids such as automatic loading to cut down time between batches, end-over-end agitation may be a sound commercial proposition for certain products, particularly when packed in large size cans.

Another method of speeding up heat penetration, known in America as the 'Strata-Cook' process, is applicable to products which consist of discrete food particles and a finely divided component in a thin liquor, e.g. cream-style corn. The three components must be capable of being packed in a specified manner so that they are stratified within the can and the finely divided component kept separated from the discrete particles which are suspended in the liquor. After heating, the finely divided component is mixed by shaking the sealed container. Heat penetration is rapid into the bottom components due to convection currents set up in the liquor. Conduction into the top layer is speeded up because it is in contact with the bottom components as well as the can wall. US Patents 1502196 and 1502197 taken out for the 'Strata-Cook' process claim 40 to 60 per cent saving in time compared with the process for premixed products. It is obvious that the method is limited to certain types of packs.

In heat penetration, the shape and size of the container for any given product is very important since the rate of heat penetration is affected by the ratio of the surface area of the container to its volume. Small cans heat more rapidly because the surface area exposed to heat is large in relation to the volume of the can. Evans & Board's (1954) investigation of heat transfer through the headspace indicated that this can be an important factor on the rate of heat penetration into short squat cans, and care should be exercised in converting data obtained for one type of can to another of different shape because of headspace effects.

With certain products the cooling phase contributes significantly to the overall lethality of the process. Board et al (1960) investigated how the lethality of the cooling phase for conduction-heating packs was affected by a number of variables such as the type of product, can vacuum, retort pressure and delay between turning off the steam and turning on the water. There is sometimes a variation in the cooling of cans at the top and bottom of retorts in which a temperature gradient often occurs. Additionally if cans are cooled by overhead sprays and the cold water hits the tops of cans before the contents have been adequately cooled, water vapour in the headspaces condenses causing a reduction of internal pressure. Consequently the contents boil, resulting in a 'stirring' action, which greatly increases the rate of cooling. Furthermore the temperature of the incoming cooling water often varies by as much as 30°C between winter and summer or between commencement of production and normal working.

Process evaluation

In the formulation of heat processes for canned foods, the most difficult problem is to evaluate the lethal effect of the period during which the temperature of the can contents is rising to its maximum; especially where the heat-penetration rate is slow the temperature continues to rise during all or most of the process time. Since, for any organism, all temperatures above the maximum for growth may be considered as being lethal, the sterilising effect of a process commences as soon as the food reaches such a temperature. For practical purposes a significant lethal effect on spore-forming organisms in low-acid and medium-acid foods is usually assumed to commence at 93°C. In order to define the heat-penetration curve completely it is necessary, however, to start thermocouple readings at the commencement of the process. Hicks (1961) discussed the nature and extent of the uncertainties in canning process calculations and pointed out that these should be taken into account by the provision of an adequate safety factor. It is important to note that for some packs quality considerations call for a process in which the margin of safety is small, and in such cases a change in the initial temperature of the can contents, or the cooling rate, may render the process ineffective.

The main procedure used for estimating the lethal effect of the process is measurement of the temperature at the point of greatest temperature lag within the can and integration of the lethal effects at this point by graphical or mathematical procedures. Extensions of these classical methods take into account other regions in the can and methods have been proposed for calculating integrated lethality effect at all points in the container.

The increased availability of high-speed computers has led to their use for the estimation of thermal processes either directly, that is without intermediate manual calculations, or for computing parametric values used in heat process evaluation. In addition inoculation and trial process methods are used in appropriate cases, for example, where heat-penetration methods are inconvenient or for confirmation of processes evaluated by other methods.

Classical calculation methods

In the classical methods, the value of the process is computed by integrating the lethal effects of the temperatures found at the point of greatest temperature lag in the can during the heating and cooling cycle.

The general method. This method described by Bigelow et al (1920), which is a graphical method of carrying out the integration, is founded on the fact that each point on the heating and cooling curve for a can of food represents a lethal value for the organism studied and involves the construction of a lethality curve. First, the thermal death-times of the test organism are determined for several temperatures within the processing range. From the data obtained, a curve is constructed on semi-log paper using the log scale for the time and the linear scale for temperature. Figure 13 illustrates a thermal death-time curve for *Clostridium botulinum.* From such a curve can be found the thermal death-time for the organism at any temperature within the processing range. The lethal rate for any given temperature is the reciprocal of the number of minutes required to kill * the organism at that temperature. If the rate of heat penetration and cooling is known, any temperature reached during the process may be given a lethal rate value; if a curve is constructed as for a heating-cooling curve in which lethal rates are substituted for temperatures, a lethality curve is obtained. When the area beneath the lethality curve is equal to unity, the process is considered to be adequate with respect to the organism studied. A lethal rate table based on Esty and Meyer's TDT data for *C. botulinum* spores in buffer at pH 7.0 was given by Taggart & Farrow (1942) and an extended form is reproduced in the appendix.

Improved general methods. In one application of the 'general method' the lethal rates used are values equal to 1 minute at a standard temperature of 121.1°C. For example, if 4 minutes at 115.6°C is the

*As explained on p. 181 the TDT curves werebased on destruction of a large number of spores in samples of an arbitrary size. This is considered equivalent to a 12D process.

equivalent, as regards destruction of organisms, of 1 minute at 121.1°C, then the lethal rate at 115.6°C is 0.25. The area beneath a lethality curve constructed in this manner gives a value of the process in terms of time at 121.1°C and has been called the F value of the process, or the F_0 value of the process in the specific instance when the slope z of the thermal death-time curve is 10°C. The lethal rates may be obtained from tables (*see* Appendix, p. 362) or calculated by applying the formula proposed by Ball (1928):

$$L - \log^{-1} \frac{T - 121.1}{z}$$

in which T= any lethal temperature,
z= the number of degrees centigrade required for the T.D.T. curve to traverse one log cycle.

The general method of calculating process value has been criticised on the grounds that it is laborious and can be applied only when the conditions relating to can size, retort temperature and initial temperature of the can to be processed are identical with those under which the heating-cooling curve data were secured. Schultz & Olson (1940) have suggested the use of special co-ordinate paper with which less effort is entailed in making the calculations. They also give formulae for converting heat-penetration data obtained under one set of initial and retort temperatures to data corresponding to different initial and retort temperatures.

Resende et al (1969) give an improved procedure for use of this paper.

Patashnik (1953) gives a simple method of process calculation based on the 'general method' using the trapezoidal rule for estimating areas under curvilinear figures. By this rule the area is found by dividing the base into any convenient number of equal parts, erecting perpendiculars to the curve and then adding half the sum of the first and last to the sum of all the intermediate ones; the result multiplied by the common distance apart will give the area required. When applied to heat-penetration data it is necessary first to take thermocouple readings at equal time intervals and then find the lethal rates corresponding to these temperatures. It is not necessary to draw the lethal rate curve for summation of area since it can readily be obtained by adding up the lethal rates in the manner specified. Patashnik states that a 4 or 5 minute time interval between thermocouple readings is convenient and gives accurate results. In our view not more than 2 minute intervals should be used. Gillespy (1953) gives a detailed example for estimating the F value of complex processes. Hayakawa (1968) applies a Gaussian integration formula for the calculation of thermal processes. The method is claimed to give accurate results with as little as six properly spaced lethal rates during the heating and cooling cycles.

Mathematical methods. The classical methods developed by Ball (1923, 1928) may be used for calculating processes for any size of can or retort temperature provided the thermal death-times and rates of heat penetration approximate to straight lines when plotted on semi-log paper. Since this work and many other studies were carried out using the Fahrenheit scale of temperature most of the reference data to be found in the literature is on this basis. Furthermore the reference temperature which is used for the definition of the F value is 250°F which converts to the rather unwieldy value of 121.1°C with a recurring decimal place. Since from 1980 all new European temperature measuring equipment must read in degrees Celsius we have, in this text, made the metric conversions throughout. Shapton et al (1971) proposed a new unit to be known as the 'Metric F value', with the symbol MF, should be used and that this would be defined by the parameters of 120°C and a *z* value of 10°C. In spite of the obvious attractions of simplification we have not adopted this suggestion here, because, until there is international recognition of this standard American workers will continue to work with the original standard and the coexistence of two different systems must inevitably lead to confusion.

Ball defined the thermal death-time curve by reference to the coordinates of an arbitrarily chosen point (termed F) and the slope of the line (*z*). The thermal death-time at 250°F (121.1°C) was taken as the F value and *z* value as the number of degrees on the temperature scale over which the line passes in traversing one logarithmic cycle on the time scale. The hypothetical curve passing through 1 minute at 121.1°C is used for process calculations and gives the times and temperatures corresponding to F = 1. The times at any temperatures other than 121.1°C are designated F_i and are related to each other in a manner dependent upon the slope of the curve (*z*). That is:

$$F_i = \log^{-1} \frac{121.1 - T}{z}$$

The reciprocal of F_i is the lethal rate expressed as the number of F units per minute corresponding to the temperature (T) in degrees Celsius.

The heat-penetration curve is constructed by plotting values representing differences between retort temperature and product temperatures on a log scale and time on a linear scale, zero time being taken when the steam is turned on. Rather than make this subtraction each time, the paper is turned upside down, the top temperature coordinate labelled 1° below retort temperature, and the temperature plotted direct (Fig. 17) The heating curve is defined by the factors *j* and *f*. The value of *j* is obtained from $\frac{jI}{I}$ where *j*I is the difference between the retort temperature and the theoretical initial temperature and I is the difference between the retort temperature and actual initial temperature. The theoretical initial temperature is found by extending the straight portion of the heating curve until it intersects a vertical line representing the beginning of the process adjusted to include the lethal value contributed by the 'come-up' period of the retort. In practice, the vertical line is erected at a point on the linear scale representing 58 per cent of the time taken for the retort to reach operating temperature, this being in agreement with Ball's finding that 42 per cent (i.e. 100—58 per cent) of the 'come-up' time should be added to the process time at retort temperature (*see* p. 193). The factor *f* is the number of minutes on the time scale required for the straight-line portion of the curve to cross through one log cycle on the temperature scale. Similar parameters may be established for the cooling curve the appropriate subscript being used to denote whether they apply to the heating or the cooling portions of the heat-penetration curve, for example. f_h and f_c and j_h and j_c . The values of *j* and *f* may be used for the solution of various problems such as the calculation of process times or F values. Step by step details of the methods are outlined by Townsend et al (1954), Cheftel & Thomas (1963) and Stumbo (1973). A simple example of a calculation based on Ball's method is given in the Appendix, p. 359. The basic equation used in the calculation is:

$$B = f_h \log \frac{j_h I_h}{g}$$

in which B= the process time, in minutes, measured from the corrected zero of the process,

g= the difference, in degrees Fahrenheit, between retort temperature and the maximum temperature reached by the food at the geometric centre of the container.

Published tables are available (Stumbo, 1973) relating to *g* to f_h . These tables show values in °F. When working examples with metric

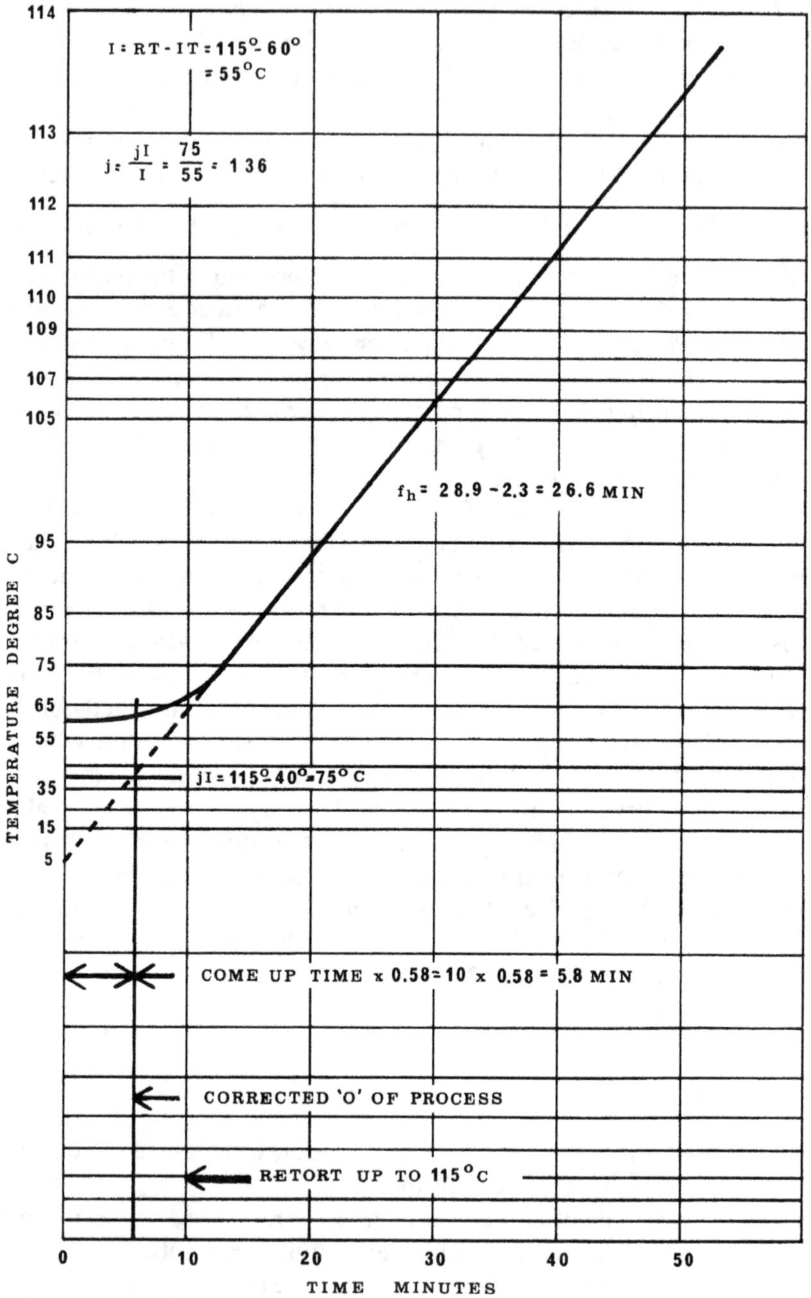

Fig. 17 Heat penetration curve

data the values obtained are in °C and must be converted by multiplying by a factor of 1.8 before using the tables.

Obviously the calculation is only valid for finite values of g and cannot be used for process calculation if the conditions are such that the centre temperature of the product reaches retort temperatures before the end of the process. The value of g is related to the ratio $\frac{f_h}{U}$ where $U = FF_1$ (the time in minutes at the retort temperature equivalent to the F value of the process). Ball (1928) used a value of 1.41 for the j value of the cooling curve but Stumbo & Longley (1966) showed that the relationship between g and $\frac{f_h}{U}$ is dependent upon j (as well as z) and they produced limited tables (z values ranging from 12° to 22°F) giving $\frac{f_h}{U}{:}g$ relationships which allow for this. Jen et al (1971) updated these tables with the aid of a computer. Recalculated and refined tables are given by Stumbo (1973) in which the values of U account for all the lethal heat during heating and cooling. Skinner & Urwin (1971) developed an empirical formula for the data contained in the table for $z = 10$°C of Stumbo and Longley enabling the procedure to be carried out on a programmable desk calculator. Hayakawa (1977) reviews various computer programs for estimation of the parametric values used in thermal process evaluation.

Integrated lethality methods

The classical methods of process evaluation are based on the destruction of spores at the slowest heating point in the can. Stumbo (1948) pointed out that these methods do not properly take into account the influence that the number of bacteria in the food has on the process required to sterilise the entire contents of the container.

Convection heating packs. In a convection heating pack, or a fluid pack subjected to mechanical agitation, product movement in the container occurs and all points receive approximately equivalent heat treatments. Thus the heat treatment at the geometric centre (F_c) is close to the effective mean throughout the container (F_s) and may be obtained from the following relationship:

$$F_c = F_s = D_r(\log a - \log b)$$

in which D_r = The DRT at the reference temperature
a = The initial no. of spores and vegetative cells in the entire container
b = The number of survivors in the entire container after receiving a heat treatment of the designated F value.

A worked example showing the evaluation of process time and F values is given by Stumbo (1973).

Conduction heating packs. Stumbo (1948) visualises the contents of the can divided into a series of concentric cylinders ranging from one the size of the container itself to one the size of a short piece of thin pencil lead at the can axis. With uniform distribution of bacteria throughout the can, the area of each of these cylinders is taken to represent the number of bacteria at its surface and this progressively increases from the central cylinder to that coincident with the food surface adjacent to the can wall. Because of the logarithmic thermal death-rate of bacteria, the initial number of bacteria influences the heat treatment required to reduce the survival to a given level; consequently, according to Stumbo, the F value required at the surface of each of these cylinders increases with its area, which is directly dependent on its distance from the can axis. Although all the other cylinder surfaces receive greater heat treatments than the one at the can axis, Stumbo states that with convection heating packs the heat treatments may, at some of these surfaces, be insufficient to compensate for their increased F requirements and the location of the 'greatest probability of spore survival' is not at the centre of the can but may be quite close to the can wall. Stumbo's concept has been criticised by Ball (1949) and Jakobsen (1954) on the grounds that the 'probability of survival' is loosely defined and should strictly be taken as the ratio of the number of survivors to the initial number of organisms in the same volume, location, etc. Thus, in the ideal case of instantaneous heating throughout the can, the probability of survival will be the same at all points in the can. Since in practice the point of greatest temperature lag receives less heat than any other, the probability of survival at this point must be greater than elsewhere. Ball and Jakobsen indicate that if Stumbo had used the term 'distribution of survivors' instead of 'probability of survival' their objections would not apply.

Nevertheless, the fact that regions in the can other than the point of greatest temperature lag should be considered in process evaluation had been established. Stumbo (1949), Gillespy (1951), Hicks (1951) and Ball & Olson (1957) went a stage further by suggesting methods of process evaluation in which they added up, or integrated, the chances of survival over the whole volume of the can. Each of these workers carried out the calculations in two stages, first determining the survival ratios for particular points or small regions in the can, and then integrating over the whole volume. These methods acknowledge that fractional values of spores obtained by the application of logarithmic formulae relating to thermal destruction do not represent sterility but must be interpreted as the probability of spore survival; this should be taken into account for all regions of the can. These methods often indicated that processes for canned foods should be slightly more severe than those calculated by the classical methods. It is interesting to note that they may be judged to give a theoretical reason for the commercial practice, where quality considerations permit, of using higher F values for large cans than for small cans of the same convection heating product. Heat-penetration tests and inoculation studies by Sognefest & Benjamin (1943) and Roberts (1949) confirm the need for a higher sterilising value in the case of large cans of convection heating products. Schack et al (1959), as a result of a bacteriological survey of canned ham pieces before and after processing, concluded that the traditional F value underestimated the true spore killing potential of the heating process. With non-comminuted, conduction heating products such as these, where the infection is probably on the outer surface, Stumbo's layer concept more adequately explains the true sterilising effect.

The variable factors influencing the thermal stability of micro-organisms in foods has necessitated, in practice, the adoption of processes in which a margin of safety has been incorporated. These processes, which have been based on the results of heat penetration to the slowest heating point within the can, have been validated by the results of many years' experience in practical canned food processing. For this reason the classical procedures continue to be the most widely used. Methods of process calculation based on integration of lethal effects throughout the can may be used in special situations. The reader is referred to Stumbo (1973) for details on methods of applying these procedures. Hayakawa (1977) should be consulted for the application of computers to thermal process evaluation.

Microbiological methods

Normally these methods are supplementary to the usual process evaluation procedures outlined previously but, in certain circumstances, where heat penetration is not readily obtainable they may form the most reliable method of estimating process lethality. For example, when it is difficult to simulate the agitation imposed by certain continuous retorts and commercial equipment is not suitable for the passage of radio telemetry equipment. There are two methods that may be used:

1. Inoculated pack methods in which the spores are added directly to the food

2. Methods in which spores, suspended in buffer and contained in a small bulb or capillary tube, are placed in the can of food.

Inoculated pack method. The canned food is inoculated with the spores of a heat-resistant organism and subjected to the normal heat process. The extent of the survival of the organism is estimated by culture or incubation tests and from these results the heat process may be estimated. The subject has been discussed by Tjaberg (1974).

The spores used as the inoculum must have an appropriate heat resistance for the study in hand. Normally interest lies in exploring low levels of spore survival which means that in-can incubation of the processed cans is used for estimating the level of survival. Thus the heat resistance and amount of inoculum should be such that a low incidence of spoilage may be anticipated when the cans are subjected to the usual heat process. Needless to say the spores should be capable of germination and growth in the food concerned and ideally spoilage should be accompanied by gas production so that the result can be ascertained by sorting of the sample for blown cans. The most used organism for this work is P.A. 3679. The level of inoculum is usually between 10 000 and 100 000 spores per can. The thermal death-time of the spore suspension should be checked and is normally between 12 and 16 minutes at 115.5°C in phosphate buffer (Townsend et al, 1954).

As regards the method of introduction and location of the inoculum within the can, this is governed by the product involved. Fluid products, such as soups, are usually batch inoculated by dispersing the spore suspension with the aid of a mechanical stirrer. Solid products, packed in brine or syrup, are usually inoculated in the can by means of pipette. For those products where uniform mixing is not possible the inoculum should be injected at a point near the geometric centre of the can by means of a hypodermic syringe.

The inoculated cans should be processed at various levels so that they show, after incubation, almost 100 per cent spoilage at the lowest sterilisation level and none at the highest. This normally necessitates four or five different processes. Sufficient samples should be prepared to enable a prediction of the true level of spoilage to be made with some degree of certainty (*see* p. 313). Control cans containing uninoculated material must be included. With rotary cookers, other factors, such as the speed of rotation of the can and the headspace, have profound effects on the heat penetration (*see* Fig. 18) and therefore care should be taken to see that these are adjusted to the practical working values when carrying out tests.

The spore bulb method. The bulbed capillaries used by us for carrying

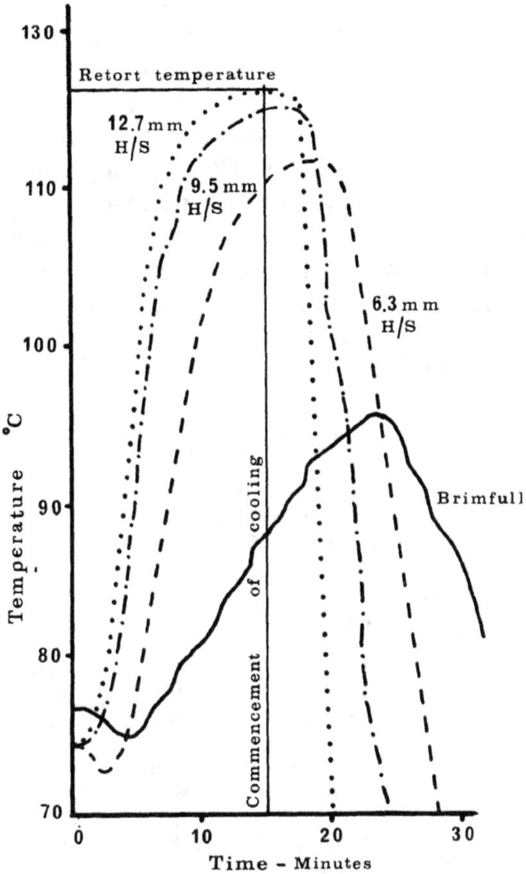

Fig. 18 Effect of headspace on heat penetration rate in rotary cooker-coolers

out studies on the heat resistance of micro-organisms were described on page 169. The same bulbs may be implanted in solid pieces of food or positioned at the slowest heating point of solid packs without difficulty. With convection heating packs, or those processed with agitation in continuous retorts, the bulbs must be fixed in the appropriate position by means of a support and this introduces a problem since the presence of the support may interfere with the normal product movement. We have used a rod of Tufnol 6 mm wide and 3 mm deep, cut so that it fits diagonally across the can from top to bottom. The ends of the Tufnol are tapered to provide a good fit and the length is adjusted so that when the cover is seamed in place the rod is held firmly in position. The stem of the capillary is inserted through a small hole drilled in the rod and the bulb is held in place by short collars of silicon rubber tubing. Some adjustment of position is possible by regulating the position of the hole and the length of the collars. Approximately 10^6 spores of *Bacillus stearothermophilus* suspended in citrate buffer are used in each bulb and they are normally activated before use by subjecting them to a heat treatment of between 5 and 10 minutes in an oil bath at 120°C. By comparing the number of survivors that can be recovered after normal processing with the number of spores initially present, plus a knowledge of the heat resistance of the suspension used, which is conveniently carried out with replicate bulbs, the lethality of the process may be evaluated. A method similar to the above but using capillary tubes has been described by Michiels (1972).

REFERENCES

Alstrand D V, Benjamin H A 1949 Food Res 14: 253
Alstrand D V, Ecklund O F 1952 Food Tech 6: 185
Amaha M 1953 Food Res 18: 411
Ball C O 1923, Bull Nat Res Cncl Washington 7: Part 1, No 37
Ball C O 1928 Univ Calif publ Publ Hlth 1: 15
Ball C O 1947 Food Ind 19: 174
Ball C O 1949 Food Tech 3: 116
Ball C O 1951 2nd Int Congress on Canned Foods (CIPC) Paris
Ball C O, Olson F C W 1957 Sterilization in food technology. McGraw-Hill, New York
Bashford T E 1947 J Roy San Inst 67: 519
Beverloo W A, Weldring J A G 1969a Conserva 18 (4): 91
Beverloo W A, Weldring J A G 1969b Lebensmittel Wissenschaft und Technologie 2 (1): 9
Bigelow W D, Bohart G S, Richardson A C, Ball C O 1920 Nat Canners Assoc Bull 16L
Board P W 1965 Circular No. 7-P, Commonwealth Scientific and Ind Res Org, Australia
Board P W, Cowell N D, Hicks E W 1960 Food Res 25: 449
Burton H, Perkin A G, Davies F L, Underwood H M 1977 J Food Technol 12: 149
Busta F F 1967 Appl Microbiol 15: 640
Cheftel H, Thomas G 1963 Res Lab J J Carnaud & Forges de Basse Indre, Bull No 14. (Published for US Dept Agric & Nat Sci Found Washington by Israel Programme for Scientific Translations Jerusalem 1965)
Clifcorn L E, Peterson G T, Boyd J M, O'Neil J H 1950 Food Tech 4: 450
Cowell N D, Evans H L, Hicks E W, Mellor J D 1959 Food Tech 13: 425
Davies F L, Underwood H M, Perkin A G, Burton H 1977 J Food Technol 12: 115
Davis R B, Long F E, Robertson W F 1972 Food Technol 26 (8): 65
Ecklund O F 1949 Food Tech 3: 231
Ecklund O F 1956 Food Tech 10: 43
Edwards J L, Busta F F, Speck M L 1965 Appl Microbiol 13: 851
Esty J R, Meyer K F 1922 J Infect Dis 31: 650
Evans H L, Board P W 1954 Food Tech 8: 258
Gillespy T G 1951 J Sci Food Agric 2: 107
Gillespy T G 1953 J Sci Food Agric 4: 553
Hayakawa K 1968 Food Technol 22: 905
Hayakawa K 1977 Adv in Food Res 23: 75
Hicks E W 1951 Food Tech 5: 134

Hicks E W 1961 J Food Sci 26: 218
Holdsworth S D 1974 Food Manuf 49 (11): 35
Jackson J M 1940 Proc Inst Food Tech p 39
Jackson J M, Olson F C W 1940 Food Res 5: 409
Jakobsen F 1954 Food Res 19: 66
Jen Y, Manson J E, Stumbo C R, Zahradnik J W 1971 J Food Sci 36: 692
Lewis F C 1945 Food 14: 32
Lingelbach H 1974 Fleischwirtschaft 54 (7): 1160
Lueck R H, Brighton K W 1951 2nd Int Congress on Canned Foods (CIPC) Paris
Michiels L 1972 Industries Alimentaires et Agricoles 89 (9/10): 1349
Packer G J K, Gamlen J L B 1974 J Food Sci 39: 739
Patashnik M 1953 Food Tech 7: 1
Perkin A G, Burton H, Underwood H M, Davies F L 1977 J Food Technol 12: 131
Perkins W E, Ashton D H, Evancho G M 1975 J Food Sci 40: 1189
Pflug I J, Esselen W B 1954 Food Res 19: 92
Reed J M, Bohrer C W, Cameron E J 1951 Food Res 16: 383
Resende R, Stumbo C R, Francis F J 1969 Food Tech 23: 325
Reynolds H, Kaplan A M, Spencer F B, Lichtenstein H 1952 Food Res 17: 153
Roberts H L 1949 Processing Meeting of the Nat Canners' Assoc Jan Quoted by Alstrand and Ecklund Food Tech 6: 185
Roberts H L, Sognefest P 1947 Supplement to NCA Inf Letter Jan 31st 1947
Schack W R, Greenberg R A, Blodgett G A, Silliker J H 1959 Food Res 24: 112
Schultz O T, Olson F C W 1940 Food Res 5: 399
Shapton D A, Lovelock D W, Laurita-Longo R 1971 J appl Bact 34 (2): 491
Skinner R H, Urwin S 1971 Process Biochem 6 (11): 35
Sognefest P, Benjamin H A 1943 Processing Conf of the Nat Canners' Assoc Quoted by Alstrand and Ecklund Food Tech 6: 185
Sognefest P, Hays G L, Wheaton E, Benjamin H A 1948 Food Res 13: 400
Steele D J 1972 Food Manuf 47 (11): 35
Stumbo C R 1948 Food Tech 2: 115
Stumbo C R 1949 Food Tech 3: 126
Stumbo C R 1973 Thermobacteriology in food processing, 2nd edn. Academic Press, London
Stumbo C R, Longley R E 1966 Food Tech 20: 341
Stumbo C R, Murphy J R, Cochran J 1950 Food Tech 4: 321
Stumbo C R, Purohit K S, Ramakrishnan T V 1975 J Food Sci 40: 1316
Taggart R, Farrow F D 1942 Food 11: 13
Tjaberg T B 1974 SIK Rapport (359) 167
Townsend C T, Esty J R, Baselt F C 1938 Food Res 3: 323
Townsend C T, Somers I I, Lamb F C, Olson N A 1954 A Laboratory manual for the canning industry. Nat Canners' Ass Washington
Xezones H, Hutchings I J 1965 Food Tech 19: 1003
Xezones H, Segmiller J L, Hutchings I J 1965 Food Tech 19: 1001

10

Continuous flow sterilisation and aseptic processing

With the conventional method of processing canned foods at temperatures up to about 120°C the relatively lengthy heat treatment necessary to destroy spoilage organisms often results in deterioration of the quality of the food; appearance, flavour and vitamin content may each be impaired to a greater or lesser degree. It is therefore, not surprising that over many years a vast amount of research has been expended on the problem of achieving preservation by means of heat without loss of quality.

The fundamental aim of this work has been to bring about more rapid heating of foods in order that higher sterilisation temperatures and correspondingly reduced times of heating may be used. Ball (1938) observed that foods into which heat penetrates rapidly usually show less deterioration when sterilised at temperatures above 121°C than when sterilised at lower temperatures. To explain this it is necessary to consider the effect of various time and temperature treatments on the organoleptic and nutritive properties of the food and to compare these with the thermal death-time data for bacterial spores.

The data for bacterial spores is well documented in the form of the various thermal death-time curves. The slope of these is defined by z, the value of which approximates to 10°C. The extent to which the z value may be regarded as constant in the higher temperature regions is discussed in Chapter 8. Similar curves may be constructed for nutritional or quality indicators such as thiamin destruction (Jackson et al 1945; Feliciotti & Esselen, 1957; Mulley et al, 1975), degradation of chlorophyll (Schanderl et al, 1962; Gupte et al, 1964) and the oxidation of ascorbic acid (Evenden & Marsh, 1948). It is found that the slope of these curves, measured in the same way as for the thermal death-time curves for spores, is approximately 33°C. Because of the widely differing slopes between the bacterial destruction curves and curves showing the loss of desirable constituents in the food it is obviously better that sterilisation is achieved by means of a high temperature short time (HTST) process than by traditional means. The symbol C_0 has been used to express equivalent heat (cooking)

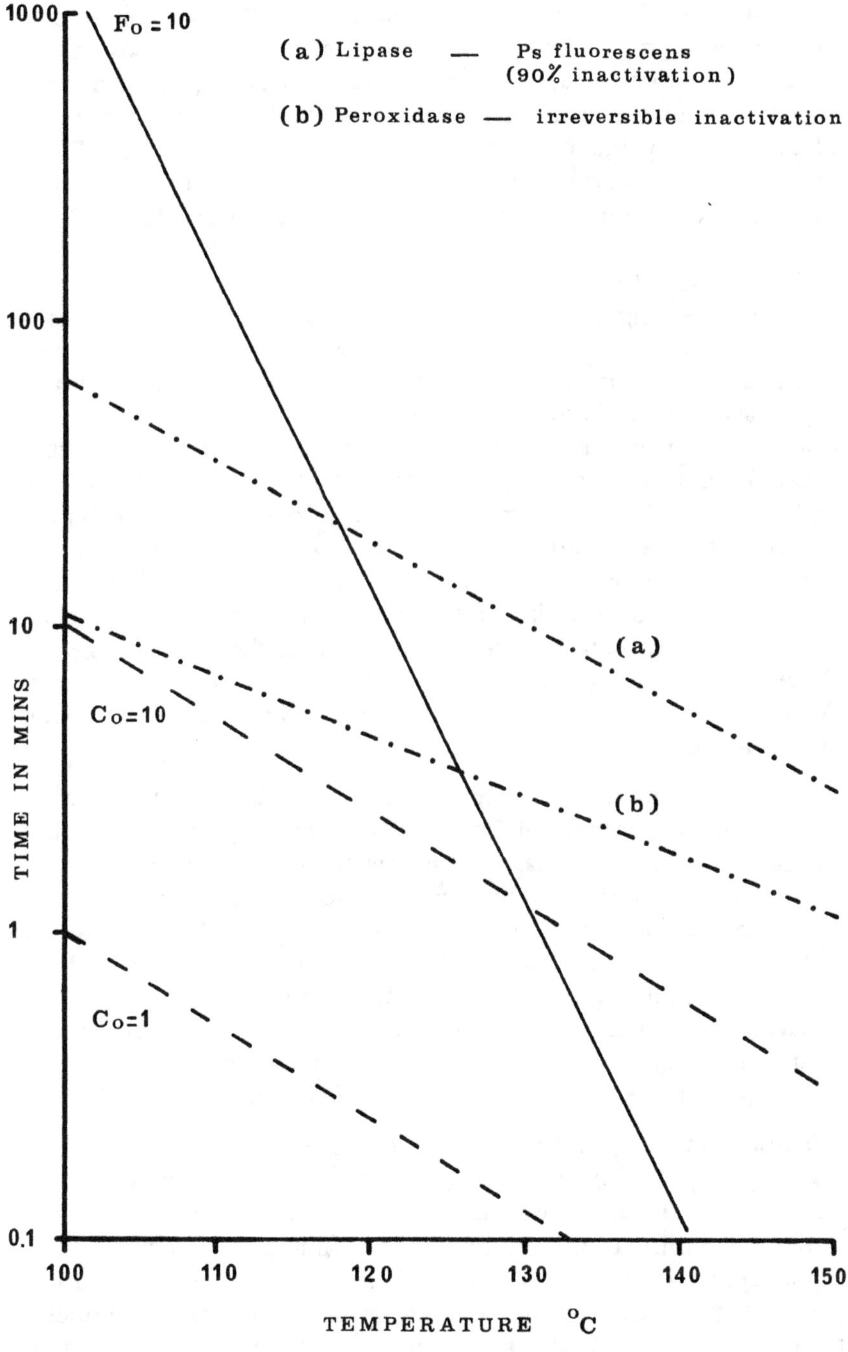

Fig. 19 Comparative cooking and sterilisation curves

processes of 1 minute at 100°C. The C_0 line may be given the slope appropriate to the cooking effects, taken as 33.2°C by Leonard et al (1964), and a graph drawn to illustrate the comparative cooking and sterilisation effects (Fig. 19). Whilst a certain degree of licence has been taken in the construction of these cooking curves, they do illustrate the principles involved in the application of HTST processes. For more detailed information on the effect of heat processing on the organoleptic quality of the food see Ball & Olson (1957).

Ball & Olson (1957) have defined HTST processes as sterilisation by heat for times ranging from a few seconds to 6 minutes. This definition covers certain in-can methods of heating achieved in flame sterilisers and some forms of agitating cooker-coolers but is more generally appropriate to continuous flow sterilisation in which more efficient heat transfer methods may be used. In these systems process times generally range from a few seconds to about 1 minute. In the UK the milk industry uses the term UHT for continuous milk sterilisation carried out at temperatures above 135°C. In this text we are using HTST in accordance with the definition of Ball and Olson and regarding UHT as referring only to continuous sterilisation processes carried out at temperatures in the upper part of the HTST range, i.e. above 135°C.

Excluding the in-can processes there are two methods for applying HTST techniques:

1. Heating the food to high temperatures in bulk, or on-stream, after which it may, or may not, be partially cooled before filling. The hot product is filled and sealed in the can at a temperature sufficiently high to sterilise the container and lid and then cooled in the can in the normal manner. For products with a pH above 4.5 this involves filling at temperatures above 100°C and the operation is carried out under pressure to prevent the product boiling
2. Heating the food to high temperatures and holding in bulk or on-stream, to effect sterilisation, after which it is cooled, filled and sealed in sterile containers under aseptic conditions.

These methods involve carrying out the whole or part of the sterilisation procedure before canning. In the first method residual heat in the product is used to sterilise the container and lid thus avoiding the difficult procedure of aseptic filling. By this method the filling temperature is regulated according to the nature of the product, particularly hydrogen ion concentration. For some products, temperatures below 100°C are adequate and for others temperatures above 100°C are necessary so that a pressure filling and closing technique is required to prevent boiling of the product. For example, the method for tomato juice (Sognefest & Jackson, 1947) involves heating the juice rapidly to 120° to 122°C, cooling to 90°C at which temperature it is filled, closed and the cans inverted and held for 3 minutes to sterilise the headspace and covers of the can. In the Smith–Ball (1955) process, the product is pre-sterilised by flash heating, partially cooled to 121.1° to 126.7°C at which temperature it is introduced into a pressurised room (about 1.3 bar) where it is filled into cans which are allowed to stand for a short period in order to destroy spoilage organisms which may have gained access during the filling and closing operations. The principle has been refined in a full-scale commercially operating unit. This process, known as the 'Flash 18' process, has been described by Ziemba (1964) and is covered by US patent No. 3232770 and British Patent No. 1069853.

The second method involves filling the food under aseptic conditions. The equipment used for this purpose depends on the nature and type of package involved. The aseptic filler must sterilise the empty container or the material from which it is formed, fill the product aseptically, sterilise the lidding material and apply it hermetically to the container. The production of the sterile product by continuous flow sterilisation involves (1) heating the product by pumping it through a heat exchanger so that its temperature is raised to a controlled and predetermined value; (2) passing the product directly through a holding section to give a pre-determined minimum residence time in order to effect sterilisation and (3) cooling it at least to a temperature of 35°C if aseptic filling is intended. Because of the nature of the equipment used the heat exchange process is suitable only for liquids or liquids containing small particles. Liquids containing particles with cross-section in excess of a few millimetres give special problems which have not been fully resolved.

The objective must be to attain a sufficiently high temperature in the heat exchanger so reducing residence time requirements in order that a special holding tube is unnecessary. This idealised situation is more or less achieved in the UHT treatment of milk in which temperatures in the order of 140°C and transit times of a few seconds are common. Burton (1977) is of the opinion that processes at temperatures above 150°C should not be attempted even for easily processed products such as milk because the short holding times involved are very difficult to control and reproduce. Unfortunately it is not possible to use high temperatures and short residence times for many products because: (1) viscous products are difficult to heat uniformly and rapidly to the operating temperature; (2) the presence of small particles make it desirable to impose an unheated holding section for temperature equilibration purposes and (3) the products may contain heat resistant enzymes which are more likely to survive processes of the UHT type than those carried out at lower temperatures.

For the transfer of heat into liquid foods the alternative types of heat exchange generally employed are:

1. *Indirect heat exchange;* the product and the heat exchange fluid are separated from each other by the heating surface. There are three types:
   a. Tubular heat exchanger
   b. Plate heat exchanger
   c. Scraped surface heat exchanger
2. *Direct heat exchange*; steam is condensed into the product for heating or vapour is removed from it by evaporation for cooling. There are two types:
   a. Steam injection (steam into product)
   b. Steam infusion (product into steam).

The choice of method generally depends on the nature of the product. The direct heat exchange methods are more efficient and give fewer problems associated with scale on the heating surface. The indirect systems lend themselves to better heat recovery, tend to give a more stable output temperature and are not prone to contamination from condensibles in the steam. For the more viscous products and those containing small particles the scraped surface heat exchanger is generally used.

Evaluation of the effective thermal process in continuous flow sterilisation is complicated by the fact that all components of the medium may not spend the same time in the steriliser and holding tube. The distribution of the fluid velocities in round pipes approximates to one of the three types of flow shown in Figure 20. In piston flow, the mean time in the holding tube is equal to the residence time of the components. Unfortunately this idealised flow pattern is rarely achieved. For turbulent flow an accurate calculation of the velocity is not possible but there are empirical equations which relate average and maximum velocity. The relationship varies with the Reynolds number but the maximum velocity is generally assumed to be 1.25 times the average velocity. For laminar flow the velocity profile is parabolic so that the average velocity is one half the maximum. According to Holdsworth (1969) many of the products suitable for continuous flow sterilisation, such as soups, purees, sauces and concentrates, are non-Newtonian fluids of high apparent viscosity exhibiting laminar flow.

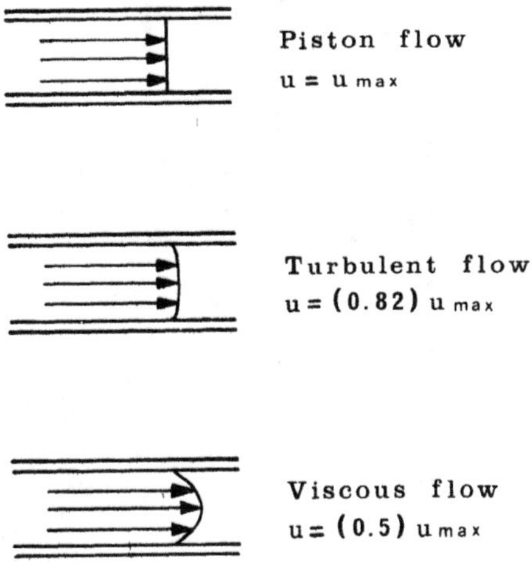

Fig. 20 Distribution of velocities in fluids exhibiting different types of flow inside round pipes. u — mean velocity of the fluid

Several mathematical analyses have been made of continuous sterilisation processes. Those by Deindoerfer & Humphrey (1959) and Richards (1965) apply to turbulent flow conditions; those by Aiba et al (1965), Charm (1966) and Leniger & Beverloo (1969) apply to laminar flow situations under isothermal conditions such as those existing in lagged holding tubes. Simpson & Williams (1974) have developed a design for non-isothermal behaviour such as that existing in a tube steriliser in which the heating and cooling are carried out indirectly without interpositioning of a holding tube. De Ruyter & Brunet (1973) and Manson & Cullen (1974) have developed computer models for products containing particulates processed in a scraped surface heat exchanger used in conjunction with a holding tube.

For design purposes the holding tube length may be calculated by reference to a nomogram reproduced by permission of the James Dole Corporation (Fig. 21) from which it is possible to compute the holding time required. With a knowledge of the holding tube capacity and average flow rate the tube length for piston flow conditions may be calculated. However if the pattern deviates from piston flow a correction must be applied to account for the fastest moving stream. For example, if the flow pattern is laminar, as it is likely to be except for thin liquids, then the length should be doubled. Residence time in the heat exchanger is usually ignored but the extra lethal effect attributable to this contributes about a 10 per cent safety factor.

An indication of transit times may be obtained by allowing a marker dye, or other indicator such as salt solution or radioactive tracer, to

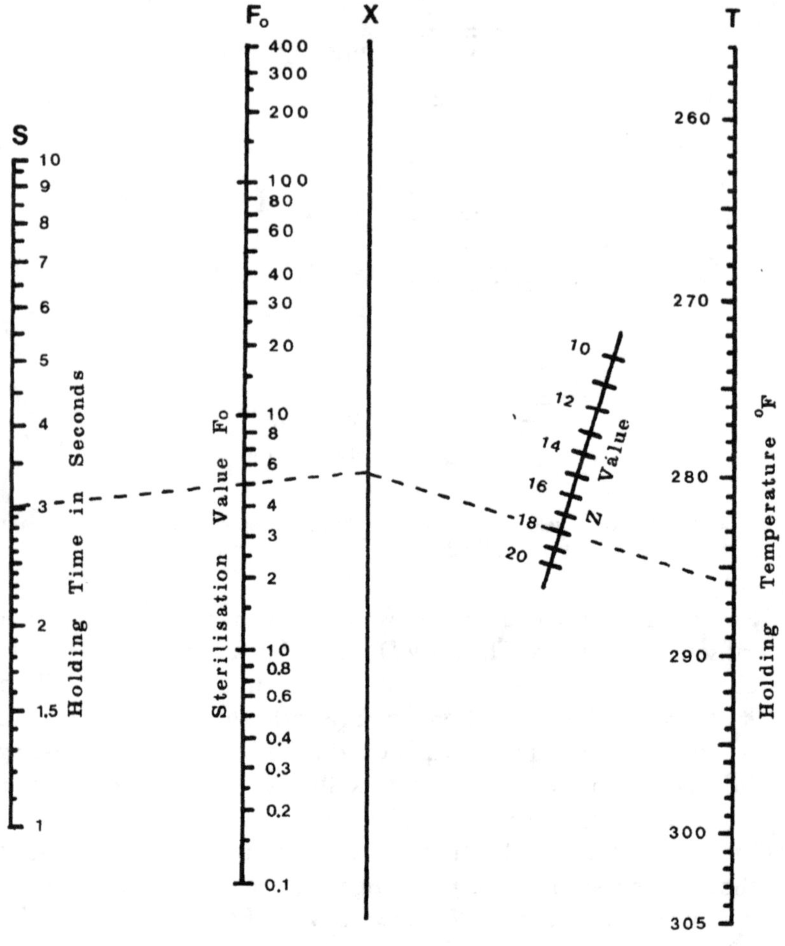

Fig. 21 Nomogram for obtaining holding tube length (courtesy of James Dole Corporation)

pass through the system noting its distribution at the end of the holding tube. A check on sterilising efficiency may also be made by running the plant on a product containing a known number of heat resistant spores such as *Bacillus stearothermophilus* and checking the reduction obtained. A knowledge of the heat resistance of the spores in the product is required for interpretation of the results, but since the temperature is practically constant in the holding tube the decimal reduction time at the operating temperature only is required since the *z* value is not involved in the calculation. Dallyn et al (1977) have suggested immobilising the spores in alginate gel beads of 1 or 2 mm diameter. These beads may be passed through the system and the surviving beads containing spores recovered. Using this technique it is suggested that information on extent and distribution of lethal effect may be obtained.

When continuous sterilisation is used as a prelude to aseptic filling the product is cooled by one of the methods mentioned earlier, that is, by passing it through a heat exchanger of the indirect type or allowing it to flash into a vacuum vessel so that the temperature is reduced to a level suitable for filling. The temperature selected depends on the nature of the product and/or the type of packaging, but normally it should not exceed 35°C. The product is often filled directly into the container the filling speed being adjusted to correspond to the production rate. Alternatively the cooled sterile product may be collected and held in a sterile reservoir thus providing a greater flexibility in filling rate.

Whilst products containing small particles may be handled by one of the heat exchanger systems mentioned earlier, generally of the scraped surface indirect type, products containing larger pieces do not respond well to this treatment. Particles having cross-sections of, say, 1.5 cm require process times of 3 or 4 minutes whilst the sauce component in which the particles are suspended may be treated in 30 seconds or less. Thus the preferred system would handle the solid and liquid components separately recombining the components after processing.

Sterilisation of the container and the lidding material may be carried out thermally, a method acceptable to all authorities, or chemically, a method becoming more widely accepted and extensively used in Europe. Other methods of sterilisation such as UV irradiation (Maunder, 1977) IR irradiation (Molin, 1976) or gamma irradiation (Sturm & Gilliand, 1974) have been proposed but not yet sufficiently proven to meet widespread demand.

Thermal sterilisation

This is used mainly for cans and can ends and is applied in the Dole aseptic system which was pioneered by W.M. Martin in the late 1940s and early 1950s. The system is described by Swords (1973). It consists of five components as follows:

1. A tunnel or tunnels, heated by superheated steam, through which pass the empty cans. The metal reaches a temperature of up to 220°C which is maintained for 40 seconds. Temperature control is critical since loss of solder binding will take place if the temperature is in excess of 235°C
2. A cooling section in which the hot metal is cooled by sprays of sterile water directed on to the outside of the can

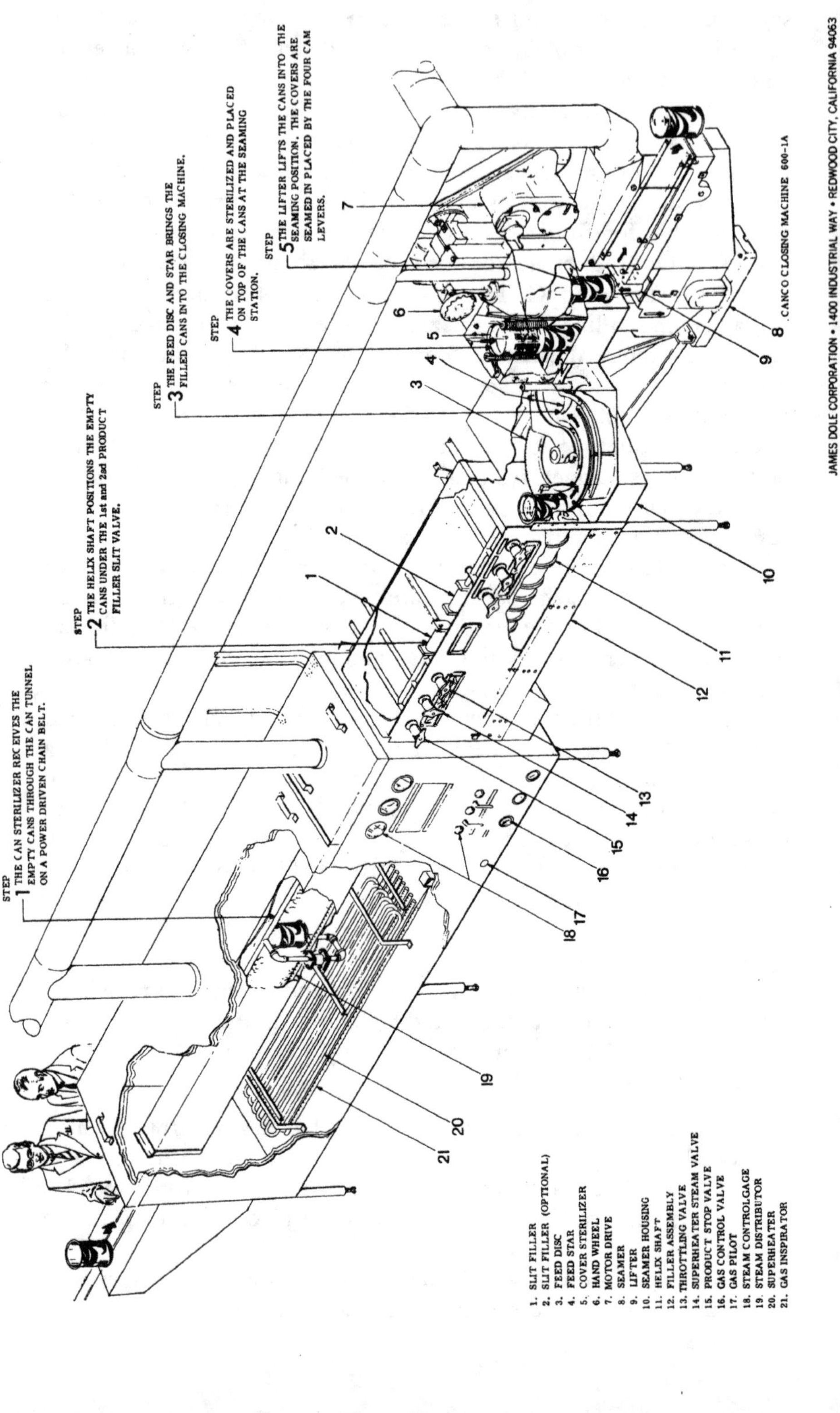
STEP 1 THE CAN STERILIZER RECEIVES THE EMPTY CANS THROUGH THE CAN TUNNEL ON A POWER DRIVEN CHAIN BELT.
STEP 2 THE HELIX SHAFT POSITIONS THE EMPTY CANS UNDER THE 1st and 2nd PRODUCT FILLER SLIT VALVE.
STEP 3 THE FEED DISC AND STAR BRINGS THE FILLED CANS INTO THE CLOSING MACHINE.
STEP 4 THE COVERS ARE STERILIZED AND PLACED ON TOP OF THE CANS AT THE SEAMING STATION.
STEP 5 THE LIFTER LIFTS THE CANS INTO THE SEAMING POSITION. THE COVERS ARE SEAMED IN PLACED BY THE FOUR CAM LEVERS.
1. SLIT FILLER
2. SLIT FILLER (OPTIONAL)
3. FEED DISC
4. FEED STAR
5. COVER STERILIZER
6. HAND WHEEL
7. MOTOR DRIVE
8. SEAMER
9. LIFTER
10. SEAMER HOUSING
11. HELIX SHAFT
12. FILLER ASSEMBLY
13. THROTTLING VALVE
14. SUPERHEATER STEAM VALVE
15. PRODUCT STOP VALVE
16. GAS CONTROL VALVE
17. GAS PILOT
18. STEAM CONTROLGAGE
19. STEAM DISTRIBUTOR
20. SUPERHEATER
21. GAS INSPIRATOR
CANCO CLOSING MACHINE 600-1A
JAMES DOLE CORPORATION • 1400 INDUSTRIAL WAY • REDWOOD CITY, CALIFORNIA 94063

3. A cover steriliser in which the can ends are heated by superheated steam, the temperature once again reaching 220°C
4. A filling section in which the cool sterile product is filled into the cans. These then pass continuously under the filler which is often of a simple type known as a slit filler. The product flows continuously through the slit at a fixed delivery rate so as to maintain the correct filling weight
5. A seamer which is a standard can seaming machine modified for aseptic operation. Superheated steam is used both for pre-sterilisation and to maintain asepsis during operation.

The system (Fig. 22) is fully instrumented with controls, recorders and alarm systems. The secret of successful operation is to maintain continuity of supply of undamaged containers and ends and to ensure that no can-jams or stoppages occur up-stream of the seamer. Line efficiencies of up to 95 per cent are claimed if careful attention is paid to design and maintenance.

Methods for determining the sterilisation requirements for the equipment, the product, the containers and covers and the operating temperatures necessary to maintain aseptic conditions are given by Townsend & Collier (1955).

Until comparatively recently the largest container size that could be handled by the Dole system was the 603 × 700 soldered tinplate can but there is now available a model handling containers from 2 US gallons (7.571 litres) to 6 US gallons (22.712 litres) capacity. The 2 and 3 gallon pails are drawn two-piece containers without side seams. The normal double seam lid closure is used.

A relatively large volume of seasonal food is packed aseptically into 55 gallon (208.192 litres) drums mainly for use by processors as an intermediate either for repacking or for use in other products. There are three widely used systems:

1. The No-Bac Fifty Five Filler from the Cherry-Burrell Corporation
2. The Fran Rica Filler
3. The Hambart Sterile Filler.

In the Cherry-Burrell system both sterilisation and filling of the drum are carried out in a retort. Sterilisation is by wet steam at a pressure of 0.66 bar. The idea originated from the Rheem Manufacturing Company and Cherry-Burrell developed the filling equipment. The drum is of special construction, electrolytically coated inside with tin and with top and bottom ends double-seamed on to the drum body. After the filling operation, whilst the drum is still in the retort, it is closed by applying a sterilised cap.

In the Fran Rica system a sterilising nozzle introduces wet steam at 1 bar into the drum for 1.5 minutes with a venting nozzle. This is subsequently retracted and a filling nozzle lowered. After filling a cap is swaged into place. Throughout these operations the top double seam is in contact with the head of the equipment in order to maintain pressure.

The Hambart system is suitable only for acid products since the drums are subjected only to a low temperature flowing steam treatment at atmospheric pressure. The system uses two parallel tunnels each containing three chambers separated by air-actuated sliding doors. In the first section the drum is heated in flowing steam; in the second section the steamed drum is filled aseptically and in the third section a steamed cap is applied.

Operational rates on the above systems are reported to be between 14 and 20 drums per hour.

Chemical sterilisation

The efficiency of chemical sterilising agents has been discussed by Lubieniecki-von-Schellhorn (1970) and Toledo (1975). The ideal chemical sterilant should have a good and rapid sporicidal activity, be easily applied, be compatible with the packaging machinery and the packaging material and be easily removed leaving no harmful residues to contaminate the food. The 'sterilising' agent may be gaseous, in which form it is suitable for use on preformed carton blanks where long contact times are permissible, or liquid for on-line application to the contact surface of the packaging material or container.

Gaseous sterilisation was reviewed by Bruch (1961). Of the substances considered, the only one which has been seriously studied is ethylene oxide. Toledo (1975) has summarised in his paper the work carried out on its microbiocidal properties. This information is reproduced here in Table 7. However, Doyle & Ernst (1967) and Mullican & Hoffman (1968) found that spores occluded in crystals or in insoluble solids were protected from the action of the gas. Thus in spite of the slow but effective sterilising action of the ethylene oxide, there must remain some doubt as to whether it is possible to obtain a truly sterile condition as required in a fully aseptic system. In the Pure-Pak system the gaseous sterilisation of the blanks is supplemented by means of hydrogen peroxide fogging of the erected carton. The products packed, mainly milk, are distributed through the chill chain, making it difficult to assess the extent of sterility achieved.

The most widely used liquid sterilant is hydrogen peroxide which is applied either by immersion of the web from which the container is

Table 7 Microbiocidal properties of ethylene oxide (courtesy of R.T. Toledo)

Organism	Temp °C	RH %	Conc Mg/l	D (min) on type of surface: Paper	Glass	Reference
B. coagulans [a]	30	33	700	10.4	13.5	Blake & Stumbo (1970)
B. coagulans	40	33	700	7.0	9.02	Blake & Stumbo (1970)
B. coagulans	50	33	700	3.53	4.05	Blake & Stumbo (1970)
B. coagulans	60	33	700	3.07	3.33	Blake & Stumbo (1970)
L. mesenteroides	30	33	700	3.45	1.69	Blake & Stumbo (1970)
L. brevis	30	33	700	5.88	2.32	Blake & Stumbo (1970)
B. subtilis var. niger ATCC 9374	54.4	40	500	4.3	6.66	Kereluk et al (1970)
Cl. sporogens ATCC 9584	54.4	40	500	3.67	3.30	Kereluk et al (1970)
Cl. sporogens ATCC 7955	54.4	40	500	3.25	2.80	Kereluk et al (1970)
B. stearothermophilus ATCC 7953	54.4	40	500	2.63	2.63	Kereluk et al (1970)
B. pumilus ATCC 7061	54.4	40	500	2.81	2.21	Kereluk et al (1970)
M. radiodurams	54.4	40	500	3.00	2.25	Kereluk et al (1970)
M. phlei ATCC 11728	54.4	40	500	2.40	2.25	Kereluk et al (1970)
Cl. botulinum 62A	40	47	700	11.5	—	Savage & Stumbo (1971)
Cl. botulinum 62A	40	33	700	11.8	—	Winarno & Stumbo (1971)
Cl. botulinum 62A	40	23	700	7.4	—	Winarno & Stumbo (1971)
Cl. botulinum 62A	40	11	700	6.0	—	Winarno & Stumbo (1971)

(a) z values for B. coagulans 83.3°F on glass and 94.2°F on paper

ultimately formed, as in the Tetra-Pak or Tetra-Brik systems, or by spraying on to preformed containers.

The microbiocidal properties of hydrogen peroxide have been widely investigated in recent years. At ambient temperatures the sporicidal activity is too slow to enable it to be used effectively in high speed on-line systems, but the 'D' value rapidly drops as the temperature is raised. According to the work of the Food Processors Association (Denny, 1973, quoted by Toledo, 1975) the most resistant botulinum strain, *Clostridium botulinum* 169B, has a 'D' value of only 0.03 minutes at 88°C in 35 per cent hydrogen peroxide. This gives an effective immersion time of 0.36 minutes to achieve a 12D process which is probably an overkill since the packaging material is generally not heavily infected (less than 0.05 bacteria per cm^2). The resistance of various strains of *Bacillus subtilis* notably *globigii*, has been widely investigated largely because the spores of this organism appear to be exceptionally resistant to the action of hydrogen peroxide. Toledo et al (1973) found a 'D' value of 7.3 minutes at 25°C for one strain when tested in 26 per cent hydrogen peroxide. Various *z* values ranging from 24°C to 52°C have been reported. Cerny (1975) has recently reported that spores of *B. subtilis* which had apparently been killed by hydrogen peroxide treatment could be reactivated by subsequent heat treatment for 2 to 10 minutes at 80°C. Since no reactivation was observed below 50°C this observation would not appear to present a hazard in aseptic filling provided the product is filled at temperatures below this value. A general review of the sporicidal properties of hydrogen peroxide was made by Bockelmann & Bockelmann (1972). Toledo (1975) has discussed the use of hydrogen peroxide for package sterilisation.

In our experience the effectiveness in commercial systems varies according to the method of application. A properly designed system of spraying pre-formed containers, followed by hot sterile air drying, can be extremely effective but in view of the criticality of temperature and the need for complete removal of the hydrogen peroxide before filling, a very fine balance has to be maintained between the application rates and heating/drying times and temperatures. With such systems occasional failures have been reported and these have been attributed to the presence of an occasional dirty package with adherent oxidisable organic matter present which protects the micro-organisms from the hydrogen peroxide. A high degree of sterility is achievable from a properly designed form/seal operation in which the unformed web is totally immersed in hot peroxide for an adequate time. The main criticism which can be levelled at this system is the possibility that undesirable substances may be leached from the packaging material and these may accumulate in the bath of hydrogen peroxide leaving residues behind which may contaminate the food. Johnson & Toledo (1975) have also remarked on the deleterious effect of hydrogen peroxide vapour in the headspace where it has been shown to oxidise ascorbic acid in orange juice at a faster rate than attributable to air alone. We have also noted some storage defects which could be attributed to headspace effects and for this reason the use of a gas flushing system before sealing the container is advisable.

Various systems use a combination of hydrogen peroxide and heat for the sterilisation of packaging materials, mainly plastic-coated board materials intended for UHT milk. These systems form the carton from a continuous web of material which is treated in various ways.

Tetra-Pak uses immersion in cold peroxide followed by radiation heating of the surface; Tetra-Brik, immersion in hot peroxide (60° to 80°C followed by drying in hot sterile air); Zupack, immersion in a bath of hydrogen peroxide followed by drying in hot sterile air. The Pure-Pak system used largely in the USA is different in that the cartons are formed from preformed blanks which have been pretreated by ethylene oxide gas and then hydrogen peroxide is used as a secondary treatment on the erected carton.

Due to the increasing availability of thermoplastic laminates with improved barrier properties against oxygen and water vapour transmission, many aseptic, or so-called aseptic systems, based on the use of rigid cups or pots, are being offered. Some rely on the use of preformed cups which are treated by hydrogen peroxide sprays before drying and filling. The lidding material, usually in reel form, may be treated by immersion in a bath of hydrogen peroxide, dried and heat sealed to the filled cup. Others use a continuous form—fill—seal technique, the hydrogen peroxide being applied to the web of material before the forming operation. The general principles employed in machines of this type are shown in Figure 23.

The Seres-Thimonnier system for the aseptic packaging of sachets uses a combination of hot alcohol (70°C) followed by a bactericidal UV treatment. Alcohol alone is not a highly efficient sporicidal agent and the efficiency of UV light has been questioned in some circumstances (but see Ch. 11). Although, the combined washing and disinfectant action of the alcohol in conjunction with the UV treatment is probably quite effective in practice, there is insufficient evidence to enable us to reach a definitive opinion on the safety of such a method for use in a fully aseptic system.

A number of difficulties arise in connection with HTST processes, one of which is connected with the problem of enzyme inactivation and regeneration. In conventional processing at temperatures up to

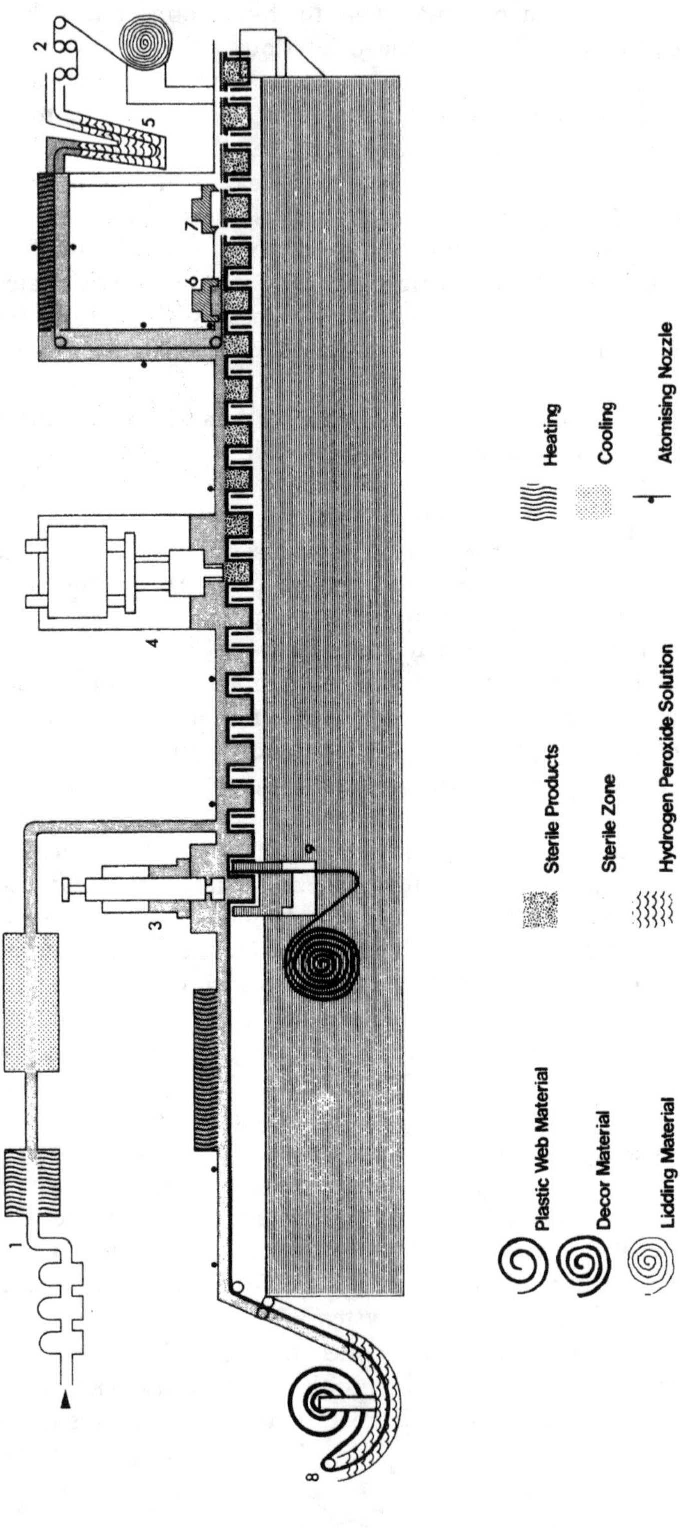

Fig. 23 Schematic representation of an aseptic form-fill-seal machine

about 120°C, the time – temperature relationships are determined mainly by the heat resistance of spoilage bacteria and are such that enzymes in the food are destroyed. With foods processed by the HTST method, however, the heating may be insufficient to inactivate enzymes although quite adequate to destroy the spoilage organisms. This is due to the differing slopes of the enzyme inactivation curve and the thermal death-time curve so that HTST processes favour enzyme survival (Fig. 19). Esselen & Anderson (1956) studied the thermal destruction of peroxidase in vegetables and found *z* values in the order of 28° to 44°C whilst that for bacterial spores is in the region of 10°C. Immediately after heat processing some enzymes show little or no activity, but after some hours activity increases or returns. This is known as enzyme regeneration and it may be controlled by increasing the severity of the heat treatment. Adams (1978) investigated the inactivation and regeneration of horseradish peroxidase using two purified grades of the enzyme. Although the different purities of peroxidase did not give significantly different $D_{121.1\,°C}$ and *z* values, the presence of either pea or green bean puree significantly reduced the D value and, in the case of pea puree, increased the *z* value. Adams pointed out that data obtained for one vegetable cannot be applied to another.

Lu & Whitaker (1974) investigating the factors affecting the rates of heating activation and reactivation or horseradish peroxidase, found the rate of inactivation to be pH dependent with maximum stability at pH 6.0 and an increasingly rapid rate of deactivation below pH 4.5.Reactivation did not occur at temperatures below 0°C or pH values below 5.0. Guyer & Holmquist (1954) and Leeson (1957) review the literature on this subject and Board (1961) reviews in detail the problems of stability in canned foods due to enzyme survival.

Comparatively recently it has been realised that certain enzymes, including proteases and lipases from Gram-negative psychrotrophs such as *Pseudomonas* spp., are capable of surviving the UHT treatment. The formation of heat stable proteases in raw milk has been reported by Mayerhofer et al (1973), Bengtsson et al (1973), Malik & Swanson (1974) and Adams et al (1975). Heat stable lipases from *Pseudomonas fluorescens* have been noted by Driessen & Stadhouders (1974). The extreme heat resistance of lipase from this source can be seen by referring to Figure 19. The subject has been reviewed by Speck & Adams (1976) and by Cogan (1977).

In connection with colour it is found that the green pigments in vegetables withstand the HTST process well but unfortunately in products such as peas and spinach puree the colour is not retained during storage. In a few weeks at ambient temperature these products are only marginally better in colour than conventionally processed packs. It has been shown that the fall of pH on storage of HTST processed spinach (Clydesdale, 1966) and green peas (Buckle & Edwards, 1970) was much greater than in conventionally processed versions of the same vegetables. This probably accounts for the loss of colour due to the acid causing hydrolysis of chlorophyll and chlorophyllide pigments. The reasons for the pH drop was investigated by Jara (1974) who found that acetic acid and pyrollidone carboxylic acid, in particular, increased on storage in HTST puree.

REFERENCES

Adams J B 1978 J Food Technol 13: 281
Adams D M, Barach J T, Speck M L 1975 J Dairy Sci 58: 828
Aiba S, Humphrey A E, Millis N F 1965 Biochemical engineering. Academic Press, New York, ch 8
Ball C O 1938 Food Res 3: 13
Ball C O, Olson F C W 1957 Sterilisation in food technology. McGraw-Hill, New York
Bengtsson K, Gardhage L, Isaksson B 1973 Milchwissenschaft 28: 495
Blake D F, Stumbo C R 1970 J Food Sci 35: 26
Board 1961 In: 4th Int Congr on canned foods Berlin, p 39
Bockelmann I von, Bockelmann B von 1972. Lebensmittel Wissu Tech 5 (6): 221
Bruch C W 1961 Ann Rev Microbiol 15: 245
Buckle K A, Edwards R A 1970 J Food Technol 5: 173
Burton H 1977 J of the Soc of Dairy Tech 30: 135
Cerny G 1975 Naturwissenschaften 62 (6): 299
Charm S E 1966 Food Technol 20: 665
Clydesdale F M 1966 Chlorophyllase activity in green vegetables with reference to pigment stability in thermal processing. Thesis. Abstract only in Dissertation Abstracts B 27 1180
Cogan T M 1977 Irish J Fd Sci Technol 1: 95
Dallyn H, Falloon W C, Bean P G 1977 Lab Practice 26: 773
Deindoerfer F H, Humphrey A E 1959 Appl Microbiol 1: 264
Doyle J E, Ernst R R 1967 Appl Microbiol 15: 726
Driessen F M, Stadhouders J 1974 Neth Milk Dairy J 28: 10
Esselen W B, Anderson E E 1956 Food Res 21: 322
Evenden W, Marsh G L 1948 Food Res 13: 244
Feliciotti E, Esselen W B 1957 Food Technol 11: 77
Gupte S M, El-Bisi H M, Francis F J 1964 J Food Sci 29: 379
Guyer R B, Holmquist J W 1954 Food Technol 8: 547
Holdsworth S D 1969 Proc Biochem 4. (10): 15
Jackson J M, Feaster J F, Pilcher R S 1945 Proc Inst Food Technol 81
Jara D G 1974 Changes in the colour pH and organic acids of green vegetables on heat processing and storage. Thesis. Abstract only in Food Sci Tech Abs 6: 10J 1385
Johnson R L, Toledo R T 1975 J Food Sci 40: 433
Kereluk K J, Gammon R A, Lloyd R S 1970 Appl Microbiol 19: 146
Leeson J A 1957 Inactivation of enzymes by heat. Sci Bull No 2 Fruit and Vegetable Canning and Quick Freezing Res Assn
Leniger H A, Beverloo W A 1969 Dechema-Monographien 63 (1125-1143): 219
Leonard S, Luh B S, Simone M, Everson G 1964 Food Technol 18 (1): 81
Lu A T, Whitaker J R 1974 J Food Sci 39: 1173
Lubieniecki-von-Schellhorn M 1970 Die Ernahrungswirtschaft 10: 810
Malik A C, Swanson A M 1974 J of Dairy Sci 57: 591 (Abs)

Manson J E, Cullen J F 1974 J Food Sci 39: 1084
Maunder D T 1977 Food Technol 31 (4): 36
Mayerhofer H J, Marshall R T, White C H, Lu M 1973 Appl Microbiol 25: 44
Molin G 1976 Proc European Meeting of Meat Res Workers Part 22 P J8: 3
Mulley E A, Stumbo C R, Hunting W M 1975 J Food Sci 40: 985
Mullican C L, Hoffman R K 1968 Appl Microbiol 16: 1110
Richards J W 1965 Brit Chem Eng 10: 166
deRuyter P W, Brunet R 1973 Food Technol 27 (7): 44
Savage R A, Stumbo C R 1971 J Food Sci 35: 182
Schanderl S H, Chichester C O, Marsh G L 1962 J Org Chem 27: 3865
Simpson S G, Williams M C 1974 J Food Sci 39: 1047
Smith H L, Ball C O 1955 Food Manuf 30: 139
Sognefest P, Jackson J M 1947 Food Technol 1: 78
Speck M L, Adams D M 1976 J Dairy Sci 59 (4): 786
Sturm W, Gilliand A 1974 Verpackungs-Rundschau 25 (4): 298
Swords B J 1973 A I Ch E Symp Series 69 (132): 100
Toledo R T 1975 Sterilants for aseptic packaging. In: Trends in packaging 9th Annual Symp issued as Special Report No 18, New York State Institute of Food Technologists p 6
Toledo R T, Escher F E, Ayres J C 1973 Appl Microbiol 26: 592
Townsend C T, Collier C P 1955 Establishing sterilisation procedures for aseptic canning units. In: Information Letter No 1526 (Convention Issue). National Canners Assoc, Washington
Winarno F G, Stumbo C R 1971 J Food Sci 36: 892
Ziemba J V 1964 Food Engineering 36 (3): 122

11

The use of radiations in food preservation

Two forms of radiation energy, namely ionising energy and ultraviolet light (UV) energy are known to show microbiocidal properties to varying degrees. The former comprises energy from gamma rays, X-rays and accelerated electron beams. The UV energy is attributable to that portion of the spectrum between visible light and soft X-rays, in particular, the frequencies near 2537 Å (253 nm).

Ionising radiations

The bactericidal properties of ionising radiations have been known for many years (Goldblith, 1966) but only since the 1940s has serious thought been given to the use of ionising radiations for the purpose of food preservation.

The radiations of interest for sterilising food are generally either gamma rays, originating from radioactive isotopes (Cobalt-60 or Caesium-137), or electron beams, generated by particle accelerators. Gamma rays are very penetrating and can therefore be used for sterilisation in reasonable depth. Foods may be packed and sealed in normal-size containers, which are usually treated with half the required dose from either side. Electron beams have a low penetrating power and this limits their use for treatment of many packaged foods. However, when surface treatment only is required, then low-energy electrons would be the natural choice.

The application of ionising radiations is controlled by regulating the rate of energy output from the source, adjusting the physical relationship, mainly distance, between the source and the target and altering the time of exposure. The amount of energy absorbed is known as the dose and has usually been measured in rads where 1 rad is defined as the absorption of 100 ergs per gram of absorber. The International System of Units (SI) replaces the rad with the gray (Gy) which is equivalent to one joule/kg, i.e. 100 rad.

For the treatment of foods, the energy level must be limited in order to avoid nuclear change and induced radioactivity. Thus the limiting energy level for electromagnetic radiations (gamma or X-rays) is 5

MeV and for accelerated electrons is 10 MeV. In the latter case the effective depth of the radiation will be limited to about 5 cm assuming a two-sided irradiation process is used.

Factors affecting the radiation resistance of micro-organisms

The main effect of ionising radiations on living organisms is damage to DNA or a closely related functional component of the reproductive process. An inadequate dose may not prevent reproduction but may give rise to mutants in the progeny. Whilst damage to the cell is the main cause of 'death', reproduction and growth may be affected also by secondary chemical changes in the growth medium and by the suitability of the environment in general.

Oxygen effect. It seems to be well established that the presence or absence of oxygen during irradiation affects the radiation resistance of micro-organisms. This is most marked with vegetative cells. For example, it has been demonstrated that complete removal of oxygen from the cell suspension may increase the resistance of *Escherichia coli* cells three-fold (Hollaender et al, 1951). Also it has been demonstrated by Wheaton & Pratt (1962) that spore suspensions of *Clostridium botulinum* in neutral phosphate buffer were more resistant when irradiated under nitrogen than under air.

The possibility of peroxides being involved in the inactivation of the cell is indicated by the discovery of Williams & Kempe (1957) that catalase, which decomposes hydrogen peroxide, reduces the lethal action of gamma rays on the spores of *Clostridium botulinum* and P.A. 3679. The presence of certain chemicals, many of them reducing agents, has been shown to afford protection against irradiation (Burnett et al, 1952; Williams-Walls, 1960; Kohn & Gunter, 1960).

It was reported by Hollaender et al (1951) and Biagini (1955) that the concentration of cells of *Escherichia coli* in the suspending medium influenced their radiation resistance; the cells became more sensitive with decreasing concentration. Gunter & Kohn (1956) confirmed this effect and showed that the resistance conferred upon a cell suspension of concentration above 2×10^9 per ml could be completely eliminated by shaking or bubbling with oxygen before irradiation. The explanation advanced for this phenomenon is that the modification in radiosensitivity is brought about by development of a state of endogenous anoxia in proportion to the cell concentration in the suspension.

Epp et al (1968) irradiated *Escherichia coli* B/r in mixtures of oxygen and nitrogen using high intensity electron beams applied as short pulses of 30 nsec duration, the effect of which was to cause oxygen depletion within the irradiated cells. A plot of the log survivors against the irradiation dose showed a tailing effect probably because of the shift to anaerobic conditions. The significance of this when comparing the radiosensitivities of micro-organisms is obvious and Underdal & Rossebo (1972) included these tailing effects in making regression analyses of D_{10} values for *Salmonella seftenberg*.

Water effect. There is evidence that vegetative bacteria are more resistant when irradiated in the dry state (Bellamy & Lawton, 1954; Davies, 1976). The removal of water by freezing can perhaps be considered analogous and, in fact, Bellamy and Lawton demonstrated reduced sensitivity of *Staphylococcus aureus* in the frozen state. Similarly *Escherichia coli* is protected by freezing (Stapleton & Eddington, 1956; Fuld et al 1957). Freezing and drying, through the physical removal of water, probably diminish the contribution of the indirect effects of irradiation (Goldblith, 1971).

Bacterial spores sometimes behave differently and Pepper et al (1956) have reported increased sensitivity in the dry or frozen state. Fuld et al (1957) confirmed that *Bacillus subtilis* is more resistant in the non-frozen state, but found that *B. coagulans* and *Clostridium sporogenes* showed no difference in the frozen and unfrozen conditions. *C. botulinum* has been reported more resistant in phosphate buffer when frozen (Denny et al, 1959) but in meat Ingram & Thornley (1961) found similar resistance at O°C and -75°C.

Harnulv & Snygg (1973) investigated the radiation resistance of the spores of *Bacillus subtilis* and *B. stearothermophilus* at various water activities. The experiments were made under anoxic conditions in water vapour and in aqueous solutions of glucose and glycerol at various water activities. In water vapour, resistance of both types of spores increased slightly with decreasing water activity. In glycerol solutions an initial rapid increase in resistance with decreasing water activity was followed by a slower increase on further reduction of water activity. A similar initial phase of a rapid increase in radiation resistance with reducing water activity was observed with glucose solutions.

According to Wheaton & Pratt (1962), freeze-dried spore suspensions of *Clostridium botulinum* were less resistant in the dry state than when resuspended in the neutral phosphate buffer.

It would seem therefore that the main effect of limitation of water by freezing or drying is to increase the radiation resistance of vegetative cells. The evidence regarding the effect on spores is conflicting but it appears that their radiation resistance is not affected to anything like the same extent.

Hydrogen ion concentration. The effect of hydrogen ion concentration does not appear to have been studied extensively. According to Williams-Walls (1960), the sensitivity of the spores of *Clostridium botulinum* and P.A. 3679 to gamma radiation was not altered by changing the pH within the range of 3.2 to 8.4. On the other hand, Proctor et al (1958) noted that the spores of *Bacillus coagulans* tended to be more sensitive at lower pH values. Farkas & Andrassy (1968) irradiated aerobic spores at varying pH values. *Bacillus cereus* showed no change in radiation resistance in the pH range 5 to 8 but at values of less than 5 the radiation resistance decreased. The indications are that effect of pH is not marked.

Temperature. The temperature applied during irradiation influences the radiation sensitivity of bacterial cells (Bridges & Horne, 1959; Goldblith, 1971). Licciardello (1964) noted that the death rate of salmonellas was increased if the irradiation temperature was raised to between 45° and 50°C. Tulis et al (1973) showed that the spores of the radiation sterilisation indicator, *Bacillus pumilis,* could be inactivated using a 60 per cent reduced dose of irradiation if the temperature was raised to 60°C. The effect of freezing has already been noted but this may be attributed to a reduction in the availability of water in addition to the temperature effect.

The composition of the medium. Hollaender et al (1951) found that the cultural conditions used affected the resistance of *Escherichia coli.* Growth in media containing glucose conferred some resistance to the cells. Stapleton & Engel (1960) investigated this effect and found that the cells became radio-resistant if the pH was allowed to decrease by formation of acid during growth. The level of resistance was related to the final pH of the culture in the stationary phase of growth.

The composition of the medium in which the cells are irradiated has a marked influence on their destruction. Hollaender and Stapleton, cited by Desrosier & Rosenstock (1960), have demonstrated the protective effect of protein (Table 8).

Table 8 Survival of *Staphylococcus aureus* in water, dilute gelatins and nutrient broth (Desrosier & Rosenstock, 1960)

Substrate	Percentage cells surviving irradiation dose of 20 000 rad
Water	12
Dilute gelatin	22
Nutrient broth	42

Generally micro-organisms are more resistant in foods than in buffered suspensions. Proctor et al (1958) have shown this to be the case with *Bacillus coagulans* in tomato juice and Wheaton & Pratt (1962) found similar results using pork-pea infusion medium.

Anderson et al (1960) have reported that sodium chloride, at the concentrations used in foods and, certain spices, such as mustard, may enhance the lethal effect of irradiations. Other spices, however, had no effect, or actually increased the dose required to bring about sterilisation. Anellis et al, (1965) found that when salt was present smaller doses of radiation were needed to sterilise canned meat which had been inoculated with spores of *Clostridium botulinum*. Whether the effect was caused by sensitisation of the spores to irradiation or by the irradiated spores becoming less tolerant to sodium chloride was not indicated. This point was clarified by Roberts et al (1965), who showed that radiation renders the spores more sensitive to sodium chloride.

It seems clear that when micro-organisms are 'inactivated' during irradiation most of the cells retain certain metabolic activities whilst being incapable of producing macro-colonies on plating media. It is not surprising, therefore, to find that the method of recovery of irradiated cells has an influence on the apparent lethal effects of ionising radiations. Stapleton et al (1955) have demonstrated that *Escherichia coli* shows maximum survival when incubated at a temperature well below that which is normally considered optimum for this organism. Furthermore, *E. coli* appears to be more fastidious in its nutritional requirements after the cells have been irradiated. A medium of inorganic salts and glucose, which normally suffices for non-irradiated cells, was not fully satisfactory for irradiated ones, the latter requiring the addition of yeast extract.

A more detailed discussion concerning environment and the radiation resistance of micro-organisms is given by Kelner et al (1955), Bridges & Horne (1959), de Serres (1961), Goldblith (1971) and Davies (1976). Nevertheless it can be seen from the above that environment during and after irradiation may have a profound influence upon the resistance of micro-organisms to ionising radiations. Therefore, caution should be exercised in application of results to food processing and, as with heat sterilisation, whenever possible the survival of spoilage micro-organisms should be investigated in the food itself.

Species of micro-organism. It is generally considered that micro-organisms are inactivated in an exponential manner with increasing increments of radiation. In other words, the log of survivors plotted against dose normally yields a straight line curve although there are a number of exceptions in which a shoulder is exhibited on the survival curve. Thus to compare the resistance of various species of potential food spoilage organisms, some common proportional reduction in the population should be used as a measure. Either the D value, which is the radiation dose required to reduce the population by 90 per cent, or the value, designated LD99, which is the radiation to reduce it by 99 per cent, form a convenient standard for comparison. Silverman & Sinskey (1977) have produced a table showing the radiation resistance of several micro-organisms part of which is reproduced in Table 9.

Generally, the smaller the organisms the greater is their resistance to the effects of irradiation. Thus the dose needed to kill bacteria and moulds may be insufficient to inactivate viruses, enzymes and toxins. Wagenaar & Dack (1960) found the toxin of *Clostridium botulinum* to be very resistant to gamma irradiation. In broth, with initial toxin levels of 1000 M.L.D. a reduction to below 20 M.L.D. required 4×10^6 rad. Kempe & Graikoski (1962), in experiments on the gamma ray sterilisation of ground beef, found residual botulinus toxin in incubated cans which had been inoculated with more than 2 670 000 spores of *C. botulinum* per gram. Samples with lower densities of inoculation were free from residual toxin and the authors suggest that toxin was introduced with the inoculum. Thus in any process of sterilisation by irradiation there must be some complementary procedure to inactivate enzymes and destroy toxins. This is probably most conveniently achieved by a comparatively mild heat treatment.

There has been a significant discovery by Anderson et al (1956) in the isolation of a *Micrococcus*, now known as *Micrococcus radiodurans*, which is very resistant to radiation. The organism is similar to *M. roseus* or *M. rubens tetragenus* and has been isolated on more than one occasion from irradiated samples of ground beef and pork. Duggan et al (1963) investigated the radiation resistance of four strains of *M. radiodurans*. In raw beef the most resistant strain was reduced by a factor of about 10^5 by a dose of 3.0 megarad, that is, it showed a higher radiation resistance than spore-forming spoilage bacteria. The organism is highly aerobic, sensitive to heat and inhibited by sodium chloride above 2 per cent, and therefore is easily controlled by means other than irradiation.

Sterilisation doses for foods

Any process to be considered for food sterilisation must be able at least to destroy spores of *Clostridium botulinum* or render them permanently inactive in the food in question. In practice *C. botulinum* is more radiation resistant than common spoilage organisms, and, unlike heat-sterilised canned foods, an irradiation process designed to

Table 9 Radiation resistance of micro-organisms (Silverman & Sinskey, 1977). Reproduced by permission of Lea and Febiger

Species	D_{10} (Mrad)	Presence of a 'Shoulder' (Mrad)	Irradiation Menstruum
Anoerobic spore formers			
Clostridium botulinum			
Type A NCTC 7272	0.12	0.9-1.0	Water
Type A 36	0.33	0.4	Buffer
Type B 213	0.11	0.9-1.0	Water
Type B 53	0.33	0.4	Buffer
Type D	0.22	0.25-0.35	Water
Type E Beluga	0.08	0.25-0.35	Water
Type F	0.25	0.25-0.35	Water
Clostridium sporogenes			
PA 3679/S_2	0.22	0.25-0.35	Water
NCTC 532	0.16	0.25-0.35	Water
Clostridium welchii (perfringens)			
Type A	0.12	0.25-0.35	Water
Type B	0.17	0.25-0.35	Water
Type C	0.18	0.25-0.35	Water
Type E	0.12	0.25-0.35	Water
Type F	0.20	0.25-0.35	Water
Clostridium tetani	0.24	0.25-0.35	Water
Aerobic spores formers			
Bacillus subtilis	0.06		Saline + 5% gelatin
Bacillus pumilus E601	0.17	1.1	Water

Vegetative bacteria			
Salmonella typhimurium	0.02		PO_4 buffer
Salmonella typhimurium R6008	0.13		PO_4 buffer
Escherichia coli	0.009		PO_4 buffer
Pseudomonas species	0.003-0.006		PO_4 buffer
Flavobacterium species	0.003-0.006		PO_4 buffer
Lactobacillus brevis NCDO 110	0.12	0.02-0.05	PO_4 buffer
Staphylococcus aureus	0.02		PO_4 buffer
Streptococcus faecium	0.28		Dry state
Micrococcus radiodurans R.	0.22	1.2	PO_4 buffer
Micrococci (smooth strain 248)	0.52	4.0	Water
Moraxella osloensis	0.58		Ice
Yeast			
Saccharomyces cerevisiae	0.05		Saline + 0.5% gelatin
Torulopsis candida	0.04		Saline + 0.5% gelatin
Moulds			
Aspergillus niger	0.047		Saline + 0.5% gelatin
Penicillium notatum	0.02		Saline + 0.5% gelatin
Viruses			
Foot-mouth	1.3		Frozen at —60°C
Vaccinia	0.17		In vacuo
Process requirements			
Trichina inactivation	0.02-0.05		
Enzyme inactivation	2.0-10.0		
Insect deinfestation	0.1-0.5		

safeguard against *C. botulinum* should also be adequate against other exceptionally heat-resistant anaerobes such as *C. sporogenes*. An international committee proposed the term *radappertisation* to describe such processes (Goresline et al, 1964).

The food canner has used the classical studies of Esty and Meyer on the heat resistance of *Clostridium botulinum* spores as a standard for formulating heat processes. This standard, which has gained wide acceptance, requires that an initial population of 1×10^{12} spores is reduced to less than 1 survivor. According to Desrosier & Rosenstock (1960) it has been established that, in foods with pH value greater than 4.5, a radiation dose of 400 000 rad is required to give a 90 per cent reduction in population. Thus, using a similar standard of safety as with heat-processed canned foods, the sterilisation dose can be taken as $12 \times 400\,000$ or 4.8 million rad. According to the work of Wheaton et al (1961) there is a striking difference between the resistance of different strains and also survival in different food products. Since the freezing of meat diminishes off-flavours during irradiation and because Denny et al (1959) had demonstrated reduced sensivity of *C. botulinum* spores in frozen phosphate buffer, Ingram & Thornley (1961) investigated the effect of low temperature on the irradiation resistance of *C. botulinum* in meat. They concluded that there is no evidence of a significant difference in the sterilising doses required at 0°C and at -75°C. The difference in sensitivity of different strains, the influence of the medium on resistance and the fact that processes which are marginal from a spoilage point of view must carry a greater hazard from *C. botulinum* than with heat processing, impose a particular responsibility in the selection of a safe dose. On the basis of current knowledge, provided that botulinus toxin has been inactivated by a complementary process, the value of 4.8×10^6 rad for food with pH above 4.5 appears to be reasonable.

Verification of the safe process is usually based on inoculated pack studies. According to Urbain (1978) a spore level of 10^7 per unit (can) should be used. Determination of the minimum radiation dose (MRD) is based upon (1) the presence or absence of viable botulinal cells in the cans, regardless of their ability to outgrow and produce toxin and/or can swelling; (2) a single most resistant strain of *Clostridium botulinum* and (3) a shifted exponential (an initial shoulder followed by a semilog decline) rate of spore death. An experimental design for an inoculated pack study on a meat pack is given in Urbain's paper. Whilst it is convenient to assume that the rate of spore inactivation is exponential, as mentioned earlier, an initial shoulder is often observed on the survivor curve. Anellis et al (1975) and Ross (1974) used improved methods employing two interrelated functions operating simultaneously, (1) a spore inactivation rate which is not necessarily exponential and (2) a sterilisation rate for the container which is dependent upon inactivation of the most resistant spore present. This treatment provides a greater margin of safety than the conventional method and gives the maximum and most conservative dose required for obtaining sterility.

The large radiation dose required to inactivate the spores of *Clostridium botulinum* usually produces off-flavours in foods. Irradiation at temperatures of -20°C or below largely prevents or limits such off-flavours without unduly increasing the resistance of bacterial spores. However, in frozen food at these temperatures there is a corresponding increase in the resistance of vegetative cells, some of which may become as resistant as spores. Thus the practicability of sterilisation of foods solely by the use of ionising radiations is problematical. Furthermore, if food is to be sterilised in this way the post-process control procedures will undoubtedly be more difficult than with heat-sterilised foods. With the latter the presence of 'blown' cans in the stack is often an indication of underprocessing due to the fact that gas producing micro-organisms of greater heat resistance than *Clostridium botulinum* exist and are likely to be present in the food. Sterilisation by means of ionising radiations would eliminate these indicator organisms as readily as the spores of *Clostridium botulinum* and the presence of 'blown' cans would no longer act as an indication of a border line process.

Complementary effects of radiation and heat

It was noted by Morgan & Reed (1954) that the spores of P.A. 3679 were more easily destroyed by heat if they were first subjected to a sub-lethal dose of gamma radiation. This effect has been noted and confirmed also by Kan et al (1957). Conversely, heating of spores before irradiation does not affect their radiation sensitivity. The mechanism by which the spores are sensitised to the effects of heat is uncertain but may be due to rupture of DNA by irradiation and an unwinding action by the heat thus preventing repair.

Kempe (1955) noted that the heat treatment required for sterilisation of a spore suspension of *Clostridium botulinum* could be reduced four-fold by pre-irradiation in phosphate buffer. The effect was less in gelatine and nutrient broth. This work was extended by Kempe et al (1957, 1958, 1959) using ground beef. In these experiments a pre-irradiation dose of about 1×10^6 rad was required to reduce the F value for *C. botulinum* three-fold. The results were similar with P.A. 3679 although the combined process would require larger F values than for *C. botulinum*. Kempe (1961) points out that different foods will show different degrees of sensitisation and may, therefore, need different combined irradiation-heat processes. With a combined process of this type the irradiation and the heat treatment would probably have to be adjusted to deal with heat tolerant types such as the spores of P.A. 3679, and this would carry with it a greater safety factor against the survival of *C. botulinum* than a straight irradiation process.

Except for certain specialised products the double process, both parts of which are reasonably severe, does not appear to have outstanding advantages over the conventional process. However Grecz et al (1971) simultaneously heated and irradiated the spores of *Clostridium botulinum* 33A suspended in ground beef. They found that in the temperature range of 65° to 100°C there was a dramatic reduction in D values. Navanugraha (1973) carried out a similar experiment using the spores of *C. sporogenes* P.A. 3679. The D value decreased from 0.210 Mrad at 20°C to 0.087 at 95°C. With dry spores of *Bacillus subtilis* heated to 95°C the radiation dose could be reduced to one-third of that required without a simultaneous heating process (Reynolds et al, 1970; Reynolds & Brannen, 1973). Irradiation at high temperatures would be expected to increase the development of off-flavours but Urbain (1978) suggests that it might be possible to optimise the thermal and irradiation processes to an extent that would yield products of better sensory quality than that obtained by either used individually.

Low dose procedures

It is not the intention here to deal in any detail with the use of ionising radiations for applying pasteurisation doses since the canning microbiologist is likely to be interested mainly in the possibility of the use of radiations for producing 'sterile' foods as is customary with heat-processed canned goods. However, since it is the field of low dose application, perhaps in combination with refrigeration or the use of antibiotics, that probably offers the best chance of success, this aspect cannot be ignored completely.

At least two approaches have been made to this problem — namely, irradiating at levels designed to eliminate particular non-sporing pathogens such as bacteria of the *Salmonella* group (Nickerson et al, 1957; Hannan & Thornley, 1957; Thornley, 1962) and irradiating at levels of about 200 000 rad with a view to lowering the microbial count and eliminating sensitive spoilage types. These processes have been termed *radicidation* and *radurisation* respectively (Goresline et al, 1964). The latter procedure tends to alter the normal spoilage pattern. For example, pre-packaged fresh meats held at refrigerator temperatures usually become spoiled in a few days due to the growth of psychrophilic pseudomonads. A gamma ray dosage of 100 000 rad will inactivate all pseudomonads and extend the storage life at 2°C considerably but spoilage eventually takes place due to the growth of bacteria similar to *Microbacterium thermosphactum* (Wolin et al 1957). Thornley (1957) demonstrated an increase in the refrigerated life of minced chicken meat held anaerobically and subjected to 250 000 rad; the eventual spoilage organisms were *M. thermosphactum* and *Streptococcus faecium* the latter appearing to be unusually radiation resistant. According to Laycock & Regier (1970) irradiation of temperate climate fish causes the pseudomonads to be replaced by the more resistant *Achromobacter*. In tropical regions Gram-positive species form part of the microflora and these are not eliminated by *radurisation*.

Ingram & Farkas (1977) reviewed the microbiology of foods pasteurised by ionising radiations. They confirmed that irradiated foods, except for those treated with relatively low doses, tend to spoil in an unusual manner but saw no cause for alarm since this is a familiar pattern, for example, in cured foods. They found no evidence of mutation giving rise to new pathogens or increasing the pathogenicity of known organisms.

Because spoilage of refrigerated foods is normally caused by bacteria which are sensitive to irradiation, and because most pathogenic bacteria normally do not develop below 5°C, the combined use of refrigeration and irradiation would seem to offer some promise of success. Certain products however, such as fish, may be contaminated with *Clostridium botulinum* type E, the spores of which will survive the irradiation treatment and may give rise to toxic outgrowth at refrigeration temperatures. Hobbs (1977) reveiwed the literature and concluded that at a storage temperature of 10°C toxin commonly accumulates before the end of the shelf life. This rarely occurs at 5°C. Thus fish, particularly fatty fish such as herring, which has been subjected to low dose irradiation must be kept refrigerated at temperatures below 3.8°C to avoid the risk of botulism. More detail on the radurisation of fresh meats, poultry, fish, fruit and vegetables and cereal grains and baked products is given by Urbain (1978).

The effect of ionising radiations on food

It is possible to inactivate the potentially harmful and spoilage microorganisms by subjecting them to an appropriate dose of ionising radiation. However, this normally results in changes with respect to flavour, colour, odour and texture. With certain products there is a further loss of quality on storage due to the premature onset of oxidative rancidity of the fat component.

The sensitivity of animal protein foods to off-flavour development during irradiation varies with the species of the animal from which the protein originated (Sudarmadji & Urbain, 1972). Coleby (1959) concluded that, although meats varied very considerably in their response to irradiation, the changes in colour, odour and flavour which resulted from sterilising doses (4.8 Mrad) were usually sufficient to render the process of doubtful commercial value. Roberts (1971) took the same view. Shewan (1959) reached a similar conclusion with regard to the irradiation sterilisation of most sea-foods. In spite of these difficulties attempts to overcome the problems continued to be investigated, mainly under the aegis of the US Army Natick laboratory. Various methods of reducing the side effects have been investigated the most successful of which appears to be irradiation at low temperatures (Coleby et al 1961). Since the lethal effect of irradiation on micro-organisms is reduced by lowering the temperature the dose required to obtain sterility is increased. In spite of this it appears that the gain in sensory quality is not completely off-set by the higher dose requirement since the Natick laboratory has selected an irradiation temperature of $-30° \pm 10°C$. This temperature gives an optimum balance of quality cost and irradiation dose (Wierbicki et al, 1975).

It is extremely difficult to reach any definite conclusion on the acceptability of radiation sterilised foods. Unfortunately many workers, whilst investigating microbiological aspects, fail to emphasise or note flavour and physical changes in the food. Others give optimistic forecasts of the effect of radiations on certain foods without specifying accurately the doses or conditions of irradiation. The acceptability of the food may be influenced by the method of preparation, for example, certain defects such as irradiation odours in meat may be dispersed in the cooking process. Finally, what is an acceptable product immediately after irradiation may rapidly become unacceptable on storage. Our interpretation of the present position is as follows:

(1) The destruction of enzymes requires an irradiation dose that renders the food unacceptable. Therefore some pre- or post-irradiation treatment for the destruction of enzymes is indicated.

(2) True sterilisation procedures (4.8 Mrad), even on foods in which enzymes have been inactivated, usually render the food unacceptable.

(3) The combined use of refrigeration and low-dose irradiation is capable of extending the storage life of certain products whilst retaining a reasonable level of acceptability.

A primary consideration is the clearance of irradiated food as safe, wholesome products which are not injurious to the health of the consumer. The areas of concern are general toxicology, carcinogenicity, teratology, mutagenicity, hepatic microsomal enzyme function, induced radioactivity, microbiology, packaging and nutrition (Urbain, 1978). No loss of macro-nutrients are observed even at *radappertisation* doses (Anon, 1964; Raica et al, 1972) although, as in heat processing, certain vitamins may be partially destroyed. Induced radioactivity may be controlled by limiting the energy of the radiation source as explained earlier. In order to be effective, the radiations must penetrate the product to be sterilised. Although the penetration of gamma rays is greater than electrons of practical velocities, the exposure times with gamma rays are much greater. With electrons, the greater the velocity the greater the penetration; the upper limit to velocity being set by the economics of the plant and the possibility of the food becoming radioactive. As far as microbiological safety is concerned the danger would appear to come from *Clostridium botulinum*, either in *radappertised* foods because of the absence of the normal spoilage indicators such as *C. sporogenes* or *Bacillus stearothermophilus* which are present with heated sterilised foods, or in *radurised* foods from *C. botulinum* type E because of the alteration in the out-growth pattern of the various contaminating organisms. For more detailed information on the wholesomeness of irradiated foods the reader is referred to Urbain (1978) and Elias (1976).

Several governments have regularised the legal position on irradiation of foods. In the UK the irradiation of food for human consumption has been prohibited except where approval has been obtained for a specific food and its treatment (The Food (Control of Irradiation) Regulations, 1967). An exception is the production of sterile diets for patients who need it as an essential factor in their treatment. In the USA and Canada irradiation is dealt with under the regulations applying to food additives and permission is granted or refused on petitions concerning acceptability. Currently only disinfestation of flour and sprout inhibition of potatoes is acceptable in the USA. In Europe usage is, on the whole, restricted to sprout inhibition of potatoes or onions. It is possible that wider applications for the *radurisation* of strawberries and chicken may be permitted since these already have clearance from the World Health Organisation.

Discussions relating to the economics of any projected process are outside the scope of this book. There is no lack of literature on the subject and the reader is referred to Desrosier & Rosenstock (1960),

Anon (1968), Brynjolfsson (1973), Deitch (1973), Gard & Warland (1973) and MacQueen (1973).

Summarising, it can be said that theoretically irradiation processes are capable of producing sterile foods, probably in conjunction with a mild heat treatment since enzymes, viruses and toxins may not be eliminated by irradiation alone. The hazard from the production of toxic substances or induced radioactivity, whilst still under investigation, at this stage appears to be small and controllable. Nevertheless, irradiation preservation of food must still be regarded as being in the development stage and if it is to be applied more widely some way will have to be found of eliminating or considerably reducing the chemical and physical changes it induces. In the meantime there is hopeful promise for the application of pasteurisation doses as a means of increasing the shelf life of refrigerated foods or for destroying specific pathogens in certain foodstuffs.

Ultraviolet irradiation

The lethal action of ultraviolet light on micro-organisms has been investigated for many years and extensive literature has been built up on the subject. In some cases, the results of laboratory experiments and industrial application are in poor agreement, and this appears to be due to the use of different sources of radiation, methods of estimating lethality, etc., by various investigators. The bactericidal effect is probably due to a photo-chemical reaction involving the nucleic acid, although other mechanisms have been suggested such as indirect lethal action resulting from the formation of peroxides and the activation of bacteriophages.

The penetrating power is very low, so that, lethal action is confined to organisms on or near the surface unless the medium is transparent to UV radiation as is the case with air, water and some thermoplastics. The sterilising effect occurs only in direct UV light and not in the shade of obstructions such as opaque dust particles or contours of the surface, e.g. in holes or pores.

It is generally agreed that the wavelength for maximum germicidal effect is 2600 Å (260 nm). Low pressure mercury vapour lamps fortunately have a high output at 2537 Å (253 nm) — a value close to the peak bactericidal wavelength. Until recently the industrial use of UV lamps was limited by the available radiation intensity but the newly developed high intensity Brown-Boveri lamp described by Brandli (1975) and Bachmann (1975) is claimed to offer new and interesting applications for UV irradiation. Whereas the radiation intensity of conventional tubes falls into the range of 1μ to 10μ W/cm^2 (at a few centimetres exposure distance), the high intensity Brown-Boveri UV-C lamps give radiation intensities of 0.1 to 1 W/cm^2 (Maunder, 1977).

Several factors influence the sensitivity of micro-organisms to UV light. Nagy (1947) pointed out that both laboratory and field tests indicated that short exposures, given over a long period of time to cover one or more life cycles of the organism are more efficaceous than higher intensities of radiation for short periods. This is due to the fact that during certain stages of the life cycle susceptibility to UV light is increased. Other influential factors include pH, temperature and the number of organisms per unit area exposed. The relative humidity affects the death rate of bacteria suspended in air, this being more noticeable at relative humidities greater than 50 per cent, when an increase in relative humidity results in a decreased death rate (Koller, 1952). Riley & Kaufman (1972) noted the sharpest decline in death rates as the relative humidity levels were increased from 50 per cent to 70 per cent followed by a reactivation at levels above 80 per cent.

Spores of bacteria are usually more resistant to ultraviolet light than the corresponding vegetative form, but reports vary as to the extent. Some workers have claimed that bacterial spores are only slightly more resistant whilst others report factors between 5 and 10. According to Koller (1952), moulds and yeasts are much more resistant than vegetative bacteria although this is not in complete agreement with Ewell (1948), who states that *Mucor* species are about 6 times and *Penicillium* 5 to 15 times as resistant as bacteria, but that yeast have the same or slightly higher resistance than bacteria. Janicki et al (1960) and Sarkin (1977) have compiled tables showing the energy doses required for 'killing' a fixed percentage of a population of various micro-organisms. These tables clearly show the relatively high resistance of some mould spores. This may be accounted for by the protection afforded them by fatty or waxy secretions (Ellis & Wells, 1941). Pigment also apparently affords some protection, dark pigmented spores being more resistant to irradiation than non-pigmented types. There seems to be only limited information on the effect of UV light on clostridial spores and we could find no published information of the effect on the spores of *Clostridium botulinum*. Using the spores of *Bacillus subtilis* and *B. stearothermophilus*, Maunder (1977), compared the efficiency of conventional lamps with the high intensity Brown-Boveri lamp for the sterilisation of plastic packaging film. Using a distance of 0.75 inches, several minutes exposure were required from the conventional lamp to reduce the count to a low level whereas according to the information given by Bachmann (1975) a 5 second exposure from the high intensity lamp would have given sterility in respect of these organisms.

Notable among the processes in the food industry for which UV irradiation has been applied are the tenderising or ageing of meat, the curing and wrapping of cheeses, prevention of surface mould growth on bakery products and air purification in bottling and food processing establishments. After slaughter, meat becomes tender on storage as a result of enzyme activity. The process is speeded up at relatively high temperatures and UV light has been used to control the growth of surface micro-organisms (Moulton, 1939). Of more immediate interest these days is the possible use of UV irradiation for water sterilisation or for the sterilisation of packaging materials. For the former application the water may be passed through a tube of quartz which is subjected to bactericidal radiations generated by UV lamps which are arranged around its circumference. The subject is discussed in detail by Summer (1975).

The on-line sterilisation of packaging material for aseptic filling has been proposed. The development of the high intensity lamps make this a more practicable operation than formerly since exposure times in the order of 5 secs are reported to be effective on a wide range of micro-organisms. There would still appear to be some doubt as to whether this procedure would be acceptable for a truly aseptic operation since there is a danger of shadowing from dust particles which readily deposit on many thermoplastic materials due to an acquired electrostatic charge. Furthermore in our view, there is a need for more investigations into the effects of UV radiations on clostridual spores, particularly those of *Clostridium botulinum*, and on the relatively UV resistant micrococci. The sterilisation of material intended for aseptic packing has been studied by Sturm & Gilliand (1974) and Cerny (1977).

REFERENCES

Anderson A W, Cain R F and Elliker P R 1960 20th Ann meeting inst food tech (171). Food Tech 14 Suppl p. 43

Anderson A W, Nordan H C, Cain R F, Parrish G, Duggan D 1956 Food Tech 10: 575

Anellis A, Grecz N, Huber D A, Berkowitz D, Schneider M D, Simon M 1965 Appl Microbiol 13:37

Anellis A, Shattuck E, Rowley D B, Ross E W Jr Whaley, D N, Dowell V R Jr 1975 Appl Microbiol 30: 811

Anonymous 1964 Report of the working party on irradiation of food. Min Health HMSO London

Anonymous 1968 The commercial prospects for selected irradiated foods. Rep Tid-24058. US Dept Commerce, Washington DC

Bachmann R 1975 Brown Boveri Rev 62 (5): 206
Bellamy W D, Lawton E J 1954 Nucleonics 12: No 4 54
Biagini C 1955 Arch Biochem Biophys 56: 38
Brandli G 1975 Brown Boveri Rev 62 (5): 202
Bridges B A, Horne T 1959 J appl Bact 22: 96
Brynjolfsson A 1973 In: Factors influencing the economic evaluation of food irradiation. IAEA, Vienna p 13
Burnett Jr W T, Morse M L, Burke Jr A W, Hollaender A 1952 J Bact 63: 591
Cerny G 1977 Verpackungs-Rundschau 28 (10): 77
Coleby B 1959 J Appl Radiat Isot 6: 115
Coleby B, Ingram M, Shepherd H J, Thornley M J, Wilson G M 1961 J Sci Food Agric 12: 483.
Davies R 1976 In: Skinner F A, Hugo W B (eds) Inhibition and inactivation of vegetative microbes. Soc for Appl Bact Symp Series no 5. Academic Press, London, p 239
Deitch J 1973 In: Radiation preservation of food. IAEA, Vienna (Proceedings of a Symposium held in Bombay), p 593
Denny C B, Bohrer C W, Perkins N E, Townsend C T 1959 Food Res 24: 44
Desrosier N W, Rosenstock H M 1960 Radiation Technology in food, agriculture and biology. An. Publishing Co Inc
Duggan D E, Anderson A W, Elliker P R 1963. Appl Microbiol 11: 398
Elias P S 1976. IAEA Bulletin 18 (Suppl): 19
Ellis C, Wells A A 1941. The chemical action of ultra-violet rays. Reinhold Publ Corp New York
Epp E R, Weiss H, Santomasso A 1968. Radiation Res 34: 320
Ewell A W 1948 Refrig Eng 55: 466
Farkas J, Andrassy E 1968 Elelmiszertudomany 2 (1/2): 59
Fuld G J, Proctor B E, Goldblith S A 1957 Int J Appl Rad and Isotopes 2: 35
Gard C W S, Warland H M F 1973 In: Radiation preservation of food. Proceedings of a Symposium held in Bombay, IAEA, Vienna, p 629
Goldblith S A 1966 In: Food irradiation. Proceedings of a Symposium held in Karlsruhe, IAEA, Vienna, p 3
Goldblith S A 1971. In: Hugo W B (ed) Inhibition and destruction of the microbial cell. Academic Press, London, p 285
Goresline H E, Ingram M, Macuch P, Mocquot G, Mossel D A A, Niven C F Jr, Thatcher F S 1964 Nature 204: 23
Grecz N, Walker A A, Anellis A, Berkowitz D 1971 Can J Microbiol 17: 135
Gunter S E, Kohn H I 1956 J Bact 72: 422
Hannan R S, Thornley M J 1957 Food Manuf 32: 457 559
Härnulv B G, Snygg N G 1973 J Appl Bact 36: 677
Hobbs G 1977 Food Irradiat Inf (7): 39
Hollaender A, Stapleton G E, Martin F L 1951 Nature, Lond, 167: 103
Ingram M, Farkas J 1977 Acta Alimentaria 6 (2): 123
Ingram M, Thornley M J 1961 J appl Bact 24: 94
Janicki J, Sobkowska E, Wierzbowski J 1960 Acta Chim Hung 23: 483
Kan B, Goldblith S A, Proctor B E 1957 Food Res 22: 509
Kelner A, Bellamy W D, Stapleton G E, Zelle M R 1955 Bact Rev 19: 22
Kempe L L 1955 Appl Microbiol 3: 346
Kempe L L 1961 Canad Food Industr 32: 45
Kempe L L, Graikoski J T, Bonventre P F 1957 Appl Microbiol 5: 292
Kempe L L, Graikosi J T, Bonventre P F 1958 Appl Microbiol 6: 261
Kempe L L, Graikoski J T, Bonventre P F 1959 Appl Microbiol 7: 133
Kempe L L, Graikoski J T 1962 Appl Microbiol 10: 31
Kohn H I, Gunter S E 1960 Radiation Res 13: 250
Koller L R 1952 Ultra-violet radiation. Chapman & Hall Ltd, London
Laycock A, Regier L W 1970 In: Preservation of fish by irradiation. IAEA, Vienna, p 13
Licciardello J J 1964 J Food Sci 29: 469

MacQueen K F 1973 In: Factors influencing the economic evaluation of food irradiation. IAEA Vienna p 1
Maunder D T 1977 Food Technol 31 (4): 36
Morgan B H, Reed J M 1954 Food Res 19: 357
Moulton C R 1939 Food Manuf 14: 232
Nagy R 1947 Food Ind 19: 102
Navanugraha U 1973 Simultaneous heating and irradiation of Clostridium sporogenes spores in raw ground beef. M S Thesis, Michigan State Univ
Nickerson J T R, Charm S E, Brogle R C, Lockhart E E, Proctor, B E, Lineweaver H 1957 Food ech 11: 159
Pepper R E, Buffa N T, Chandler V E 1956 Appl Microbiol 4: 149
Proctor B E, Goldblith S A, Fuld G J, Oberle E M 1958 Radiation Res 8: 51
Raica N, Scott J, Nielsen W 1972 Radiat Res Rev 3: 447
Reynolds M C and Brannen J P 1973 In: Radiation preservation of food. IAEA, Vienna (Proceedings of a Symposium held in Bombay), p 165
Reynolds M C, Lindell K F, Laible N 1970 A study of the effectiveness of thermoradiation sterilization. Res Rep SC-RR-70-423 Sandia Lab Albuquerque, New Mexico
Riley R L, Kaufman J E 1972 Appl Microbiol 23: 1113
Roberts T A 1971 Proc Inst Food Sci Technol 4: (2)
Roberts T A, Ditchett P J, Ingram M 1965 J appl Bact 28: 336
Ross E W Jr 1974 J Food Sci 39: 800
Sarkin R 1977 South African Food Rev 4 (Feb): 100
deSerres F J 1961 Microbial reaction to environment. Cambridge Univ Press, p 196
Shewan J M 1959 Int J Appl Rad and Isotopes 6: 143
Silverman G J and Sinskey T 1977 In: Block SS (ed) Disinfection sterilization and preservation. Lea and Febiger, Philadelphia, p 542
Stapleton G E, Eddington C W 1956 Rad Res 5: 39
Stapleton G E, Engel M S 1960 J Bact 80: 544
Stapleton G E, Sbarra A J, Hollaender A 1955 J Bact 70: 7
Sturm W and Gilliand A 1974 Verpackung-Rundschau 25 (4): 298
Sudarmadji S, Urbain W M 1972 J Food Sci 37: 671
Summer W 1975 Process Biochem 10 (Nov): 43
Thornley M J 1957 J appl Bact 20: 286
Thornley M J 1962 Radiation control of salmonellae in food and feed products. Tech Report No 22 Int Atomic Energy Agency 81
Tulis J J, Fogarty M G, Sliger J L 1973 Developments in Ind Microbiol 14: 49
Underdal B, Rossebo L 1972 J Appl Bact 35: 371
Urbain W M 1978 Adv Food Res 24: 155
Wagenaar R O, Dack G M 1960 Food Res 25: 279
Wheaton E, Pratt G B, Jackson J M 1961 J Food Sci 26: 345
Wheaton E, Pratt G B 1962 J of Food Sci 27: 327
Wierbicki E, Brynjolfsson A, Johnson H J, Rowley D B 1975 Proc Eur Meeting Meat Res Workers No 21
Williams N J, Kempe L L 1957 Appl Microbiol 5: 366
Williams-Walls N J 1960 Dissert Abstr 20: 1975
Wolin E F, Evans J B, Niven Jr C F 1957 Food Res 22: 682

12

Types of spoilage

Canned foods are classified as spoiled when for any reason the food has undergone a deleterious change or the condition of the container renders such change possible. Spoilage may arise from a variety of causes which include the activity of micro-organisms, chemical reaction between the contents and the container, faulty technique in canning practice, rough handling and poor storage conditions. Spoiled cans may present a perfectly normal external appearance or they may show obvious abnormalities such as varying degrees of distortion. In certain instances, for example, where distortion of the can has arisen as the result of faulty canning practice or retort manipulation, the contents may be quite wholesome. In spite of this such cans are considered to be spoiled since they are visually indistinguishable from cans which have spoiled as the result of microbial activity and are, therefore, unsaleable.

The ends of a normal can with a good vacuum are slightly concave or flat. A can with its ends bulged by positive internal pressure due to gases formed as the result of microbial or chemical activity is termed a 'swell' or 'blower'. A 'hard swell' is one in which the ends of the can are permanently and firmly distended; if the ends can be moved by thumb pressure but cannot be forced back to the normal position the can is referred to as a 'soft swell'. The term 'springer' is used to describe a can in which one end is bulged but can be forced back into the normal position whereupon the opposite end bulges. A 'flipper' is a can of normal appearance in which an end 'flips' out when the can is struck against a solid object; the end snaps back to the normal position when very light pressure is applied. It should be noted that all swelled cans pass successively through the 'flipper' and 'springer' stages. Internal pressure may be generated by the growth of micro-organisms or from the formation of hydrogen gas resulting from internal corrosion. The latter results in hydrogen swells. These are not normally apparent for some months after packing whereas gas production due to microbial growth is usually evident within 15 days. Defects may be due to other causes such as faulty retort operation, incorrect filling or exhausting or the use of cans of inadequate substance.

The chief defects or causes of spoilage may be listed as follows:

Microbial	Under-processing
	Inadequate cooling
	Leaker infection
	Pre-process spoilage
Chemical	Internal corrosion giving rise to hydrogen swells or pinholing
Physical	Faulty retort operation
	Under-exhausting
	Over-filling
	Internal vacuum too high
	Use of cans of inadequate substance
Miscellaneous	Rust
	Damage.

The microbial causes of spoilage are reviewed below and summarised on pages 364-5. Other forms of spoilage are also briefly considered in this chapter since an appreciation of the various defects which may arise is essential if the microbiological examination of canned foods is to be properly carried out.

Under-processing

In under-processed packs the activity of the surviving organisms may result in gas production which causes the can to become a 'swell' or the contents may undergo acidification or some other undesirable change affecting quality but gas is not produced. When growth of micro-organisms occurs without gas production, the affected cans have a normal appearance externally and spoilage is only detectable after the can has been opened.

Any pack which suffers spoilage as a result of the activities of organisms surviving the heating process can, in a sense, be termed under-processed, but it should be noted that the primary cause of spoilage may not always lie in the formulation of the processing conditions. For a given product packed in accordance with correct canning procedure, i.e. clean plant, sound raw materials, etc., a process may be adequate to ensure keeping quality. Under other conditions where the packing plant may be inexpertly run, the same process may not be sufficient to prevent spoilage. It is important therefore, to make some distinction between spoilage resulting from a mass infection with a heat-resistant organism which has arisen through some fault in canning practice and that occurring in a pack for which the heating conditions are below generally accepted standards. While both packs may in the strict sense of the term be regarded as under-processed, an increase in the heating conditions is not the proper remedy in both cases.

Unless the pack has been grossly under-processed, it will generally be found that the spoilage is due to a single type of organism. In low-acid or medium-acid products this is usually a spore-forming species. In acid products which have not been subjected to a pressure process, yeasts or moulds or aciduric sporing or non-sporing bacteria may be found; non-aciduric spore-formers may also be present, but these organisms have no significance since they are normally incapable of growth in an acid environment. There are, of course, exceptions, but as a general rule the presence of a single species in a can free from obvious mechanical defects is indicative of under-processing.

The first step in the bacteriological investigation of spoilage of this kind is the identification of the responsible organism. It should be appreciated that the biochemical activities of an organism may not be the same in canned food and laboratory media, and an organism which is not normally gas producing may cause gaseous spoilage in canned food. This fact is well founded on the reports of many workers, examples being the spoilage of chopped ham by species of the genus *Bacillus* (Jensen et al, 1934), the blowing of cans of rice pudding by an organism resembling *Streptococcus lactis* (Morris, 1940) and the production of gas in canned peas and corn by a strain of *Staphylococcus aureus* (Segalove et al, 1943). In the initial stages, complete identification is not required, but it is at least essential to determine to what group the organism belongs in order that its origin may be traced without delay. Thermal death-time estimations carried out in the food involved may enable adjustments to be made to the processing conditions if quality considerations permit.

In attempts to trace the source of the organism and, if possible, to effect its elimination, all raw materials should be investigated. In addition to the main ingredients, materials such as colouring solutions and spices should not be overlooked. Canned products such as peas, asparagus, evaporated milk, etc., used as raw materials may be heavily infected with 'flat-sour' or other types of spores and their bacteriological quality should be checked before they are used.

In acid products, such as canned fruit, spoilage may arise due to the growth of *Byssochlamys fulva* or related organisms and this gives rise to textural breakdown of fruit usually without gas production (*see* p. 135).

Cameron (1938) has emphasised the importance of contamination from plant and equipment; where thermophilic 'flat-sour' type spores are concerned he indicates that this is the most important single factor.

He remarks that the chances of spoilage are greater in complicated plants in which there are heated blending tanks, mixers, reservoirs, etc., wherein thermophiles may proliferate and continually inoculate the materials passing through them. When spoilage due to such contamination build-up occurs, cans packed during the first hour or so of the day's run are unaffected since the contamination does not reach a significant level during this period.

Improperly cleaned plant in which overnight development of heat-resistant organisms has occurred has also been noted as a factor in spoilage outbreaks. Here it is sometimes found that spoilage is confined to those cans packed at the beginning of the shift; with the continual flow of material through the plant the contamination may be reduced or eliminated with the result that cans produced after the plant has been in operation for some time are not affected. In the investigation of spoilage outbreaks it may therefore be helpful to determine, if possible, at what stage during the run the affected cans were packed.

Where anaerobic organisms are concerned in spoilage outbreaks, it will usually be found that the infection is a direct one derived from a raw material; favourable conditions for the development of these organisms do not often present themselves in plant, equipment, etc.

Under-processing caused by faulty operation or poor design of retorts may be responsible for spasmodic outbreaks of spoilage. Outbreaks of this type are usually confined to part of a retort load of cans which become spoiled as the result of localised under-heating due to the presence of water or pockets of air in the retort during the process. In severe cases, this under-heating is reflected in the semi-cooked condition of the food but when the degree of under-processing in the affected area is slight, the lack of uniform heat distribution in the retort may not be suspected since cooking will appear normal although the lethal effect on the cans has been reduced. Also if a process relies upon agitation to promote heat transfer into the product an adequate headspace is essential to maintain the normal stirring action in the can or under-processing may result. Clumping or matting of solid particles may lead to inadequate heat transfer. Under-processing has been known to occur due to the addition of dehydrated vegetables or unblanched pasta direct to a product. In this case a combination of clumping and inadequate rehydration leads to spoilage through survival of microbial spores.

It has been noted elsewhere that agreement between pressure gauge and temperature recorder or thermometer does not of necessity indicate complete elimination of air from the retort. This point has been stressed by Somers (1946) in reporting tests in which temperatures among cans in inadequately vented retorts were as much as 55.5°C below the indicated retort temperature although there was perfect temperature-pressure agreement. Alstrand & Blair (1955) drew attention to the fact that dividers in retort baskets, used with automatic loading systems, may lead to the presence of air pockets. Felmingham & Leigh (1966) have attributed outbreaks of spoilage due to bad venting practice and the use of divider plates. They recommend a modification of the venting procedure when such plates are used.

The manner in which cans are stacked in the retort may also result in under-processing. Ecklund, et al (1942) found temperatures between tightly packed 6-lb rectangular cans during processing were considerably less than the indicated processing temperature, bulging of the walls of the can preventing free access of steam to all parts of the cans. As mentioned earlier in connection with heat penetration, the alignment of cans in the retort may, with certain types of pack, retard convection currents and so result in under-heating. Heat penetration may also be retarded in solid-liquid packs when the weight of solid filled per can is excessive.

In products processed by HTST methods spoilage may arise due to the survival and regeneration of enzymes. For example canned peas which were processed to F values in the range of 5 to 12 at 127°C or 3 to 7 at 121°C developed viny off-flavours within 8 months' storage at ambient temperatures but not when processed to similar F values at 115°C (Guyer& Holmquist, 1954). This topic is discussed further in Chapter 10.

Inadequate cooling

Since the heat resistance of the thermophiles is such that thermal processing sufficient to ensure their destruction often results in some deterioration in the quality of the food, their control chiefly consists of the elimination or reduction of infection and the rapid cooling and storage at relatively low temperatures of those products susceptible to thermophilic spoilage.

Flat-sour thermophiles multiply rapidly in the range 48° to 70°C and failure to cool cans immediately after processing to a temperature below this may lead to serious spoilage. It is customary to cool cans to an average temperature of about 35°C; the cans will then retain sufficient heat to dry quickly. Efficient drying of the cans is important in reducing post-process infection and corrosion of the tinplate.

In connection with thermophilic spoilage resulting from inadequate cooling of cans, an investigation by Pearce & Wheaton (1952) is of interest. They found that thermophilic flat-sour spores died off when stored in non-acid food substrates at 21°C; thermophilic anaerobe spores were unaffected. With the flat-sour spores, increasing the sterilising value of the process increased the rate and completeness of the auto-sterilisation. Schmidt & Nank (1957), who used the term spore deactivation to describe this phenomenon, found that the rate of deactivation was temperature dependent. In cream-style corn containing surviving spores of the flat-sour strain 1518, storage for 14 days at 36.7°C was more effective than storage for 64 days at 21.1°C. We have found that the rate of deactivation depends also on the nature of the suspending medium, particularly in respect of hydrogen ion concentration, deactivation proceeding more rapidly as the pH is decreased. Practical application of these findings can be made to reduce flat-sour spoilage by (1) reducing the number of flat-sour spores in the product by packing control and the use of maximum processing conditions preferably with the highest temperature permitted by quality consideration, and (2) storage at a suitable temperature for several weeks.

Leakage through seams

Among the causes of spoilage in canned foods, that arising from leakage of can seams after processing is undoubtedly of greater economic importance than any other. The industry's annual wastage through leaker spoilage must be very considerable, but improvements in canning technique and the growth of technical supervision should enable the losses to be reduced.

The micro-organisms infecting canned foods as the result of post-process leakage of the containers may be of widely varying types including cocci, non-sporing rods and sporing rods; yeasts are only rarely found and moulds are present only when the container is grossly defective. Cameron & Esty (1940) point out that this is not the case with acid products where acidity exerts a selective action on the organisms gaining access and permits only the growth of certain types. The main source of the organisms is the cooling water. Airborne organisms are probably of little significance as a cause of leaker spoilage, air-cooled cans invariably showing a lower proportion of spoiled cans than those cooled in water unless the latter has been disinfected by chlorination or some other means. Bryan & Morris (1932) found the ratio of spoilage in air-cooled cans to that in cans cooled in water to be 1:10. Bean & Everton (1969) demonstrated that certain 'flavobacteria' isolated from chlorinated can cooling water were capable of causing spoilage in canned milk products.

General experience seems to be that post-process infection of cans results in the formation of 'swells', indicating that the original leak becomes sealed by the can contents or that the defective seam has some valvular action. The type of spoilage, however, will obviously depend on the nature of the organisms present and if these are non-gas-producing types the cans will remain normal in appearance; similarly, cans will not swell if the defect in the seam is so severe that it permits free passage of the gas as it is formed by the contaminating organisms. The nature of the food is also able to influence gas production, for example, *Staphylococcus aureus* usually produces gas in canned garden peas but in processed peas (canned dried peas) it often develops without gas production.

In cans which have leaked there is often present more than one type of organism, but they may, at times, be contaminated by a solitary species such as a sporulating organism. This may apply particularly where cans have been cooled in chlorinated water in which there will be a tendency for the bacterial content to be reduced to resistant spore-formers. In pressure-processed cans the presence of viable cocci or non-sporing rods is generally regarded as evidence of post-process leakage since these organisms are unable to survive the processing (Cameron & Esty, 1940 — but see Ch. 13).

In connection with leaker investigations, especially of low-acid packs, microscopic examination of the contents may reveal cocci which, however, may have been present in the raw material at the time of canning and subsequently destroyed by the heating process. Microscopic examination is invaluable in the examination of spoiled cans but should always be supplemented by cultures to determine whether the organisms observed are viable. Since the majority of the organisms present in cooling water and air have an optimum growth temperature of 20° to 30°C, cultures at this temperature should always be included in the examination of suspected leakers. Some organisms found in water and air grow poorly or not at all at 37°C, and it is quite possible that cultures prepared from infected cans and incubated at this temperature may fail to reveal the presence of viable organisms in some cases.

In solid products, such as meat rolls, the infecting organisms are more likely to be on the surface of the contents and cultures prepared with inoculum carefully taken from the centre, may prove to be sterile. It is important, therefore, when examining this type of product for leaker spoilage, to ensure that the inoculum includes material from the surfaces.

In all leaker examinations, the use of solid media in addition to the liquid type is advisable since it facilitates better assessment of the degree to which the various organisms are present, and also obviates the possibility in liquid media of overgrowth of a slow-growing species by rapidly growing types. Another advantage which accrues from the use of solid media is the elimination of doubt concerning the possibility of contamination of the cultures during their preparation. The finding of numerous colonies of cocci or non-sporing rods on solid media makes it almost certain that they originated from the sample.

Leakage in cans is sometimes encountered where there is no bacterial contamination of the contents. Defects in the seams may be so small that bacteria are unable to pass through, but the vacuum is slowly depreciated until equilibrium with the atmosphere is reached. Adam & Dickinson (1944) have remarked that 1 to 3 per cent of cans may suffer heavy fall in vacuum due to minute leaks. Partial loss of vacuum is probably more common than is usually appreciated and its extent may be influenced by a number of factors, notably, the magnitude of the initial vacuum, the weight and type of seam lining compound, the standard of the can seam and the type of heat process applied. We have noted an incidence of vacuum loss in a product treated by a tomato juice process in which the product was filled very hot (95° to 100°C) into the can, closed and cooled. The can seams conformed to the normal standards and the failure was attributed to the high initial vacuum plus an inadequacy of the seam lining compound under the processing conditions applied. In such cases the cans remain normal in appearance unless subjected to temperature fluctuations such as removal from a cool to a warm storage place or to change in altitude whereupon the ends of the cans may 'flip'. Analysis of the headspace gases is very useful in the laboratory examination of cans of this type; it is usually possible to find oxygen present in such cases to the extent of several per cent of the total gases, whereas in normal airtight cans it is entirely absent or present in very small amounts. In lacquered cans, absorption of the residual oxygen by the metal is delayed by the lacquer film, but in plain cans it is usually removed within a few days.

Pre-process spoilage

Spoilage of this type is an example of faulty canning practice whereby bacterial development in the food is permitted during the preparation period. The extent to which such growth may occur is illustrated in a report by Gunderson, et al (1947). In the preparation of chicken meat, they found that the total bacterial count increased 520-fold and the coliform count 1600-fold during the 145-minute interval between cooking and chopping of the boned meat.

Particularly during warm weather, development of fast-growing organisms may occur if there is any undue delay between filling the container and heat-processing. This especially applies to cold-filled packs such as meat pastes and similar products, which are finely chopped or milled by machine; this process tends to raise the temperature of the mixture with the result that growth of the organisms is stimulated. Processing may subsequently sterilise the pack, but the liberation of gas produced by the organisms during the lag period before processing may cause swelling or flipping of the cans. A typical case of pre-process spoilage in canned meat has been described by McClung & Wheaton (1936). The spoilage organism was *Clostridium welchii*, which survived the preliminary cooking of the meat. Rapid growth of the organism in the warm meat resulted in the production of sufficient gas to swell the cans. No viable organisms could be isolated from the cans after processing. In spoilage of this type, cultures prepared from processed cans are invariably sterile, but microscopically, numerous organisms can be demonstrated.

Similar bacteriological findings may be obtained with cans which have swelled following under-processing or leakage and in which the bacteria have died out. Cultures prepared from such cans are sterile but numerous organisms may be seen microscopically, although the cells occasionally disintegrate and disappear entirely. Thermophiles are particularly liable to undergo rapid auto-sterilisation. It is sometimes difficult to distinguish these two conditions, but it is usually possible to arrive at a correct diagnosis by noting the condition of the organisms in stained smears. Where organisms have undergone auto-sterilisation the cells show uneven staining and other degenerative changes such as gross alterations in morphology. Where death has resulted from heat-processing, the cells usually stain crisply and do not appear degenerate.

Hydrogen swells

This description is applied to cans which bulge as the result of the formation of hydrogen following internal corrosion of the can. Varying degrees of bulging may be observed ranging from the 'flipping' to the 'hard-swell' stage and under certain conditions corrosion may result in actual perforation of the can.

Hydrogen swells, the contents of which may be quite harmless, are chiefly associated with foods such as fruits containing organic acids, and their formation depends upon a number of factors including the efficiency of the exhaust, the size of the headspace in the can and storage temperature. Considerable losses due to hydrogen swelling have been experienced, especially with canned fruits, and the corrosion of tinplate has been extensively investigated.

In the examination of hydrogen swells, inspection of the interior of the cans may reveal little since the degree of corrosion often appears as great in normal cans. Analysis of the headspace gases is the only reliable method, the proportion of hydrogen in hydrogen swells usually being at least half of the total volume of gas present. Very useful diagnostic methods for determining the causes of hydrogen swell outbreaks have been reported by Adam & Dickinson (1944). For detailed information concerning tinplate corrosion publications by Hoare & Hedges (1945), Hartwell (1951), Britton (1975) and Mahadeviah (1976) should be consulted.

Some foods corrode tinplate by direct chemical dissolution of the tin giving the product an abnormally high tin content without evolution of hydrogen. Once the tin coating is removed the unprotected steel may corrode rapidly giving a hydrogen swell. The presence of inorganic nitrate in medium-acid canned foods has been found to give rise to this defect. We have noted rapid internal de-tinning of cans containing beans in tomato sauce with bacon, the nitrate in this case came from the curing ingredients used on the bacon. Other instances are noted by JFK (1966), Cultrera (1967), Farrow et al (1971) and Gersons (1971).

Incorrect retort operation

Too rapid reduction of steam pressure at the conclusion of processing results in the setting up of high pressure inside the cans, and this may cause such severe straining and distortion that when cooled the cans have the general appearance of 'swells'. It will be found that strained cans produced in this manner have no positive internal pressure and the ends can be forced back more or less to their normal position. A diagnostic feature of cans which have been subjected to abnormal strain in processing is the 'peaking' or permanent deformation of the panelling or expansion rings on the ends of the cans. Whilst moderately strained cans may not warrant classification as spoiled, every attempt should be made to eliminate them, since such strain is highly conducive to leaker spoilage. The necessity for less rapid reduction in steam pressure, or better still, pressure cooling, is indicated where straining occurs. Straining may also result when ends made from unduly thin tinplate are used; they are unable to withstand the pressures set up in the cans during processing.

Under-exhausting

During heat processing, improperly exhausted cans may suffer severe strain due to the excessive internal pressures set up by the expansion of the entrapped gases. The appearance after processing of seriously under-exhausted cans may vary from slight flipping to gross distortion, depending on the amount of residual air and gases evolved from the contents and the size of the headspace, but even in grossly distorted cans the internal pressure after processing and cooling seldom exceeds a few pounds per square inch. Such cans, of course, are unmarketable. Moderate under-exhausting, resulting in a low vacuum in the cans, may not make itself apparent by causing serious container distortion during heat-processing. Difficulty may arise, however, when such cans are shipped to hot climates or high altitude areas where the increased temperature or reduced atmospheric pressure may result in bulging of the can ends. According to Jarvis (1943), loss of vacuum amounts to approximately 25 mm per 300 m rise in altitude.

Since oxygen acts as a depolarising agent and hastens the process of internal corrosion, due care must be paid to maintain proper exhausting techniques and fill weights in order to reduce the amount of oxygen in the headspace. The influence of storage temperature on corrosion rates should be borne in mind since many studies have indicated the deleterious effects of high temperatures.

Usually, most of the gas entrapped in the headspace of under-exhausted cans is air, exceptions being found with vegetable and to a lesser extent with meat packs from which carbon dioxide is liberated during processing and may be found in the headspace in large amounts. Normally, the bulk of the headspace gas present after processing is nitrogen. The proportion of oxygen found in canned foods is always very much less than that present in air and is normally less than 2 per cent. The determination of headspace gas composition and typical analyses are given by Lopez & Krebs (1967). The volume of air originally present can be computed on the basis of the amount of nitrogen found and, if the headspace volume is known, the original vacuum can be calculated.

Over-filling

Due to expansion of the contents, over-filled cans become strained during retorting. The absence of vacuum in such cans results in flipping or, in severe cases, springing of the ends. Over-filling of cans is unlikely to occur in efficiently heat-exhausted packs, since any excess contents will overflow from the can as the result of expansion.

With vacuum-closed or cold-filled products such as meat pastes, rolls, etc., this safeguard is absent, and careful adjustment of the filling weight is necessary for such materials.

Panelling

This condition (Fig. 24) usually arises in the larger sized cans in which there is a very high vacuum, the bodies of the cans being forced inwards by atmospheric pressure. A similar effect is produced in pressure-cooled cans where the air pressure has been excessively high or when cans are composed of thin tinplate. The diminution of volume caused by the partial collapse of the cans results in a decrease in the vacuum so that panelled cans which have received only a moderate exhaust may become flippers. In very severe cases panelling may result in seam leakage, especially if the seam is of sub-standard quality.

Rust

The classification of cans showing external rust formation requires very careful consideration. If, after removal of the rust, inspection

Fig. 24 Panelling

with a hand lens reveals that the iron plate is definitely pitted, it is advisable to classify such cans as spoiled since the danger of perforation is great. Cans which are slightly rusty without noticeable pitting of the iron may be considered fit for immediate sale and consumption.

The application of an external lacquer film will minimise rust formation, and cans intended for long storage or exposure to severe climatic conditions are lacquered externally as a routine measure. Rust formation is particularly liable to occur under labels when the label or the label adhesive contains hygroscopic substances. Rust formation here becomes apparent through staining of the label. Packaging of wet cans into cases is a major cause of external rusting. This arises if the cans are cooled excessively. Provided excess water is removed mechanically by air blasts, tipping of crates or other means and the cans retain sufficient heat, drying occurs and the difficulty should not arise. An equilibrium temperature of about 35°C after cooling should be the aim.

Difficulties may arise due to 'sweating' of the cans which occurs if the temperature of the can is less than the dew point of the surrounding air. This condition may arise if undue fluctuations in the temperature of the warehouse occur. Proper heating and ventilation of the store, along with regular checks on temperature and relative humidity, are desirable.

Trouble has arisen due to the packaging of cans in cases of unseasoned wood or damp fibreboard cases particularly if held in large stacks so that drying is inhibited. Information on canned food storage is given by Davis (1959).

Damage

Apart from other considerations, considerable bacteriological significance should be attached to cans damaged by rough handling. This has been clearly shown by Blackwood (1943). The important points where such cans are concerned are the extent and location of the damage. Marked deformation of the seam will be attended by considerable risk of leakage and cans with this type of damage should be classified as spoiled. Similarly, where dents on the can body are so severe that seam distortion has occurred, cans should be rejected; slight indentations on the can body are, however, permissible. A seam of normal mechanical quality can sustain slight indentation without leaking and the effects of slight damage to the seam should therefore be considered in conjunction with the quality of the seam.

REFERENCES

Adam W B, Dickinson D 1944 Chem Ind p 426
Alstrand D V, Blair B W 1955 Nat Canners' Ass Inf Letter No 1526 (Convention Issue), p 67
Bean P G, Everton J R 1969 J appl Bact 32:51
Blackwood G W 1943 Food Manuf 18: 325
Britton S C 1975 In: Tin versus corrosion. Publication No 510. International Tin Res Inst Greenford, Middx, p 30
Bryan J M, Morris T N 1932 Food Invn Bd Ann Rept, p 178 H.M.S.O., London
Cameron E J 1938 Food Res 3: 91
Cameron E J, Esty J R 1940 Food Res 5: 549
Cultrera A 1967 Industria Conserve 42 (3): 170
Davis R B 1959 Res and Dev Dept Continental Can Co Reprinted from Penn Packer Oct 1959 7549 47 P
Ecklund O F, Roberts H L, Benjamin H A 1942 Canner 94: No 1393
Farrow R P, Johnson J H, Gould W A, Charbonneau J E 1971 J Food Sci 36: 341
Felmingham J D, Leigh R J 1966 J Fd Technol 1: 345
Gersons L Jr 1971 Bull Inacol 22 (7-8): 391
Gunderson M F, Rose K D, Henn M J 1947 Food Ind 19: 1516
Guyer R B, Holmquist J W 1954 Food Technol 8: 547
Hartwell R R 1951 Adv Food Res 3: 327
Hoare W E, Hedges E S 1945 Tinplate. Arnold London
Jarvis N D 1943 Principles and methods in the canning of fishery products. U.S. Fish and Wildlife Service Res Rept No 7 U.S. Govt Prntg Off Washington
Jensen L B, Wood I H, Jansen C E 1934 Ind Eng Chem 26: 1118
J F K 1966 Fd Preserv Quart 26 (2-4): 45
Lopez A, Krebs B S 1967 Food Tech 21: 365
Mahadeviah M 1976 Indian Food Packer March-April p 9
McClung L S, Wheaton E 1936 Food Res 1307
Morris C S 1940 Proc Soc agric Bact (Abs) 52
Pearce W E, Wheaton E 1952 Food Res 17: 487
Schmidt C F, Nank W K 1957 Food Res 22: 562
Segalove M, Davison E, Dack G M 1943 Food Res 8 54
Somers I I 1946 Canner 102: No 9 13

13

Microbiology of sound canned foods

'Commercial sterility'

From what has been said in preceding chapters, especially in relation to spore-bearing bacteria, it will be appreciated that commercially sound canned foods are not invariably sterile. For acid products such as fruits, heating processes are not designed to deal with micro-organisms other than yeasts, moulds and non-sporing bacteria; the presence in such packs of viable spore-bearing bacteria is not, with the exception of certain aciduric species, considered to have any significance from a spoilage or public health point of view since their development is restrained by the high acidity. With some exceptions, processes in general use for low or medium-acid packs are theoretically adequate for the destruction of all micro-organisms but the most heat-resistant thermopiles and possibly some species of aerobic mesophilic spore-formers. The exceptions include cured or pickled meats containing curing salts which, because of their preservative action, permit the use of less stringent processes than are necessary for fresh meat products. Other exceptions are those products, which, because of their low water activity are unable to support the growth of spore-forming organisms for which a mild heat treatment, or indeed no heat treatment at all, may suffice to provide a shelf stable product. Canned sweetened condensed milk is such a product the keeping properties of which depend on its high sugar concentration. Such products are usually unsterile and the residual organisms found in them tend to be more varied than those found in other packs.

Under normal conditions of storage, the spores which survive commercially adequate processing are incapable of development, and this fact has led to the adoption of the term 'commercial sterility' to indicate the bacteriological condition of unsterile but marketable canned foods. In some quarters, objection has been raised to the use of the word 'sterility' in this connection. Clearly the description 'commercial sterility' is an unsatisfactory one, since 'sterility' implies the absence of all living micro-organisms. The term has, however, come to have a specific meaning for canning technologists and to avoid confusion it is used here. 'Commercially sterile' cans then, may be defined as cans which have been so processed that the food, under ordinary storage conditions, will neither spoil nor endanger the health of the consumer. This definition represents the minimum bacteriological standard acceptable from the public health point of view and requires that for all low- and medium-acid packs except those mentioned previously, the processing of the food should have been adequate for the destruction of *Clostridium botulinum* spores; organisms of lower heat resistance such as cocci and non-sporing rods should be eliminated in all heat-processed canned products.

While it is generally acknowledged that the spores of thermophilic organisms and other exceptionally heat-resistant types are often present in canned foods, there appears to be a divergence of view concerning the incidence of the relatively non-heat resistant organisms such as non-sporing rods and cocci in sound cans. Apart from other aspects, the question of whether or not such organisms can survive pressure processing has great practical significance inasmuch as the presence of these types in canned foods is regarded by most workers as an indication of post-process contamination or gross under-processing (Cameron & Esty, 1940; Bashford, 1942; Tanner, 1944; and others).

Results of examinations of canned foods

Since the early days of the application of microbiology to food canning, the bacteriological condition of marketable canned foods has attracted the attention of many investigators. An extensive paper by Weinzirl (1919) contains an historical survey of the results of earlier workers, including those of Russell (1895), Prescott & Underwood (1897-98), Vaillard (1900) and others. Vaillard reported that 70 to 80 per cent of the cans (presumably sound) he examined were unsterile, the organisms present including cocci, sporing and non-sporing rods and moulds. These results were criticised on the score of faulty technique by Deichstetter (1901) who, found canned foods to be sterile except for two cases in which the cans were under suspicion.

Weinzirl's examination covered 782 commercial or marketable samples consisting of 104 cans of fruit, 370 cans of vegetables, 273 cans of meat and 35 cans of milk. Of these samples, which gave no evidence of spoilage, a total of 179 (23 per cent) contained viable organisms. Spore formers were practically the only type of organism present. Most of the cans examined were tested for leaks by air pressure and seam examination, with the result that the seams of 35.9 per cent were suspect. Spores were, however, isolated from 19.2 per cent of the cans with satisfactory seams. An investigation on a similar scale was carried out by Cheyney (1919), who examined 725 samples of apparently sound canned foods (meat, fish, vegetables and fruit) for bacteria, moulds and yeasts. Before analysis, all cans were incubated at 37°C for 10 days. Of the 725 cans, 58 (8 per cent) were unsterile, all types of food being represented in this number. Cheyney states that the bacteria found constituted a sharply limited group of resistant spore-bearing rods. Staphylococci were isolated from one can of sausages and one of sardines, but one of these cans leaked when tested. Mould spores were isolated from cans of fruit. These results could not be correlated with the quality of the can seams.

The results of extensive studies on the sterility of canned foods were reported by Savage (1923). The proportion of unsterile cans found among apparently sound shop samples varied between 18.2 per cent for unsweetened condensed milk to 100 per cent for crab. A summary of the organisms found is given in Table 10 taken from Savage's book.

Table 10 Micro-organisms found in sound cans (Savage, 1923)

Sample	Yeasts	Anaerobes Obligate	Sporing	Thermo-philes	Micro-cocci	Non-sporing rods
Meat	0	0	7	6	5	1
Salmon	0	0	3	3	4	1
Sardines	0	0	4	1	1	0
Herrings	0	0	3	3	0	0
Crab	0	2	2	9	3	0
Lobster	0	0	3	2	0	0
Crayfish	0	0	1	0	0	0
Fruit	1	7	10	0	0	0
Unsweetened condensed milk	1	0	5	1	4	0

Many of these organisms were capable of causing spoilage in the products and their failure to grow in unspoiled cans was attributed by Savage to lack of available oxygen in the can and the fact that when few organisms only survive processing, they are unable to initiate growth.

Among other contributions to the subject made by Savage and his co-workers, that dealing with evaporated milk (Savage & Hunwicke, 1923) is of special interest. Of 44 apparently sound shop and factory samples, 8 were found to be unsterile. The bacterial types present in these cans were as follows:

1. Sporing aerobe, a yeast and a diplococcus
2. Sporing aerobe and diplococcus
3. Sporing aerobe
4. Sporing aerobe
5. Diplococcus
6. Sporing aerobe

7. Thermophilic bacterium
8. Diplococcus.

Savage and Hunwicke remarked that the acidity and other characteristics of the contents indicated that the samples were normal and not undergoing spoilage. They also pointed out that the sporing bacilli were all present in very small numbers and usually diplococci were scanty. On re-inoculation into cans of milk some of the cocci were found to be capable of growth and spoilage production.

Nichols (1939) examined 315 samples of evaporated milk and 253 cans of cream, both of which receive processes which can be expected to leave a residual bacterial flora of heat-resistant spore-formers only. The samples were taken direct from three factories and from general sources. One hundred per cent of the cans of milk examined from general sources (shops, etc.) were sterile, as were the samples from one factory; 94 and 93 per cent of samples from the other two factories were sterile. Of the cream samples, those from general sources were 100 per cent sterile as were those from one factory; 85 and 76 per cent of the samples from the other two factories were sterile. The organisms isolated were apparently chiefly spore-forming aerobes and anaerobes, although this is not clearly stated in the report. Cocci were isolated from some cans which appeared normal; these organisms were believed to be post-process contaminants. The lower percentage of sterile cans from factory sources appeared to be associated with some special circumstance of operation, and Nichols states that a sterile pack should ordinarily be quite attainable.

Several workers have recorded seasonal variations in the numbers of mesophilic spores found in raw milk; both Candy & Nichols (1956) and Ridgway (1958) reported that in sterilised milk the dominant surviving spores are those of *Bacillus subtilis*. Candy & Nichols (1956) showed that thermophilic spore-formers may frequently be present in sterilised milk and isolated the gas producing *Clostridium thermosaccharolyticum*. Several workers have suggested that UHT treatment of milk may not destroy but merely stimulate germination of some *Bacillus* spp. Martin (1974) came to the conclusion that a UHT treatment of about 1 second at 137.8°C did not completely destroy spore-formers such as *B. licheniformis*, *B. laterosporus* and *B. megaterium*. Grosskopf & Harper (1974) found spores of psychrotrophic organisms, under some circumstances capable of outgrowth at refrigeration temperatures, in pasteurised and sterilised aseptically packaged milk. These organisms were tentatively identified as *B. circulans*, *B. coagulans*, *B. lentus*, *B. licheniformis*, *B. macerans*, *B. megaterium*, *B. pantothenticus*, *B. subtilis*, *B. pumilus* and *B. cereus* the latter organism being particularly significant in view of its involvement as a food-poisoning organism. Thus the safety of refrigerated long-life milk is questioned. However, the temperature and time of the heat treatments that had been imposed were not stated in Grosskopf and Harper's paper and it could be that these were inadequate. Martin carried out inoculation trials on soft-serve frozen dessert mix with *B. cereus* using one million spores per ml. The mix was heated at 137.7°C for 3 seconds. He found that 99.5 per cent of the spores were destroyed and the rest remained dormant in the product for 6 weeks and then declined in numbers provided that it was stored at 4.4°C. At storage temperatures above 10°C, however, growth occurred, indicating a risk of *B. cereus* in unrefrigerated commercial samples of made-up milk products, such as desserts, in which a large number of *B. cereus* spores could be contributed by the ingredients. We point out however, that a 3 second heat treatment at 137.7°C would be regarded as an unacceptably low process for such products since it corresponds only to an F_0 of 2.3. In addition, according to Martin, although *B. cereus* is found in raw milk, the organisms are in the vegetative state and do not constitute a major risk.

Sweetened condensed milk is not heat treated after filling and closing the can and its keeping properties are attributable to its content of lactose and other soluble sugars which serve to reduce the a_w to 0.85 or lower, a value which prevents the growth of most bacteria. Spoilage by yeasts and moulds is avoided by pre-pasteurisation of the milk, attention to good hygiene and minimisation of headspace in the can. Micrococci and aerobic spore-formers are the commonest flora but the presence of coagulase-positive staphylococci should be regarded as undesirable. A total aerobic count of less than 200 organisms per g should be attainable.

An extensive survey was that made by Fellers (1926) on canned salmon. Of 5276 cans of normal quality, 3.4 per cent were considered to be unsterile. The organisms encountered included sporing aerobes, obligate sporing anaerobes, non-sporing rods, cocci and moulds. Fellers states that of the cans containing aerobes, 70 per cent showed less than 10 organisms per gram by plate count. Over 50 per cent of the spore-forming aerobes grew in canned salmon under nearly perfect anaerobic conditions. Golicz (1969) sampled 6360 cans of fish products comprising, fish in tomato sauce, fish in oil, fish in natural juice and fish in other liquids. The period of testing extended over 4 years. In all cases some cans, ranging between 2.3 per cent and 5.0 per cent, of all groups showed positive cultures the organisms isolated including *Micrococcus albus*, *Streptococcus faecalis* and other Gram-positive cocci.

Ruyle & Tanner (1935) reviewed the literature dealing with the bacteriology of canned meats and meat products, and also carried out an examination of these products. From their perusal of earlier work they drew the following conclusions:

1. Canned meats show an appreciable incidence of bacteria
2. Viable bacteria frequently fail to cause spoilage even though the cans are stored under conditions favourable to spoilage
3. Non-heat resistant forms such as cocci have been found in cans which have been processed under conditions which would normally be expected to destroy such types
4. Few investigators who have examined meat products have given adequate attention to the condition of the seams of containers.

The 559 normal cans examined by Ruyle and Tanner did not include spiced ham, a product processed at below 100°C and which was found to give 100 per cent unsterile cans. They found 31 cans (5.5 per cent) to be unsterile, the bacterial types consisting of obligate anaerobes (2), sporing aerobes (17), facultative thermophiles (1), non-sporing rods (9) and micrococci (15). The micrococci were most prevalent in sausage and potted meat. These authors discounted the theory that these non-heat resistant types were air-borne contaminants and remarked that such organisms have been recovered in many carefully carried out investigations. To account for the survival of cocci in meat products, Ruyle and Tanner advanced several possibilities. It was suggested that the cocci may have been a stage in the life cycle of an organism more resistant than spores; that the organisms were protected by fat, or that the cocci may have represented a stage in the life cycle of an organism which at other times appeared as a rod. They regarded the presence of viable organisms in sound canned foods as of minor significance.

One of the most extensive investigations was that carried out by Crossley (1938) on potted meats. Of 14 365 apparently sound jars from various sources, 12.1 per cent were found to be unsterile, the organisms isolated being spore-forming aerobes or anaerobes. Non-spore-forming types were found only in association with defective samples.

In another report (Anon., 1944) it is stated that 90 per cent of over 5000 sound cans of meats and meat products (types not specified) contained organisms of the *Bacillus subtilis* group. No other organisms are mentioned although the method of examination could be expected to reveal other types. These findings are so unusual that it seems likely that some peculiar circumstance was in operation during the tests or the packing of the products.

Theory of fat protection

Although the foregoing results are somewhat lacking in uniformity, it is clear that many able and experienced workers believe that normally non-heat resistant bacteria are frequently present in sound canned foods which have been pressure-processed. Assuming this view to be correct, the question arises as to the mechanism whereby these organisms survive heat treatment far in excess of that which might reasonably be expected to prove lethal. Jensen (1954), discussing fat protection, refers to the work of Mudd & Mudd (1924), who originated a method of determining the behaviour of various cells, including bacteria, at an oil-water interface. These workers found that cells with hydrophobic surfaces readily pass from the aqueous into the oil phase whereas cells with hydrophilic surfaces usually remain at the interface. The behaviour of bacteria at an oil-water interface can be influenced by various means. For example, organisms normally unable to enter the oil phase can be made to do so with the aid of mechanical work of extreme degree. The presence of 0.06 per cent sodium oleate allows such organisms to pass into the oil with the aid of less extreme work. According to Jensen (1954), fatty acids and soaps enhance the passage of bacteria into oil. He determined the interfacial behaviour of eight cultures of bacteria (cocci and bacilli) isolated from canned foods and found all were wetted by oil. After 15 transfers on nutrient agar, the strains were wetted only by water.

The effects of heat on bacteria suspended in fat or oil have been discussed earlier, and while published results on the subject are not in entire agreement it seems well established that at least some types of bacteria are protected to a marked degree, especially when the oil or fat is water-free. In this connection, Ingram (1955) remarks that experiments in which dry organisms have been suspended in dry fat are of doubtful significance since complete dryness of bacteria or substrate are conditions which rarely are likely to occur in practice. The explanation of the mechanism of fat protection offered by Senhaji & Loncin (1977) and Senhaji (1977) which was reported in Chapter 8 seems a plausible one. It supports the view that in the presence of water the a_w within the fat phase will equilibrate with the aqueous phase thus reducing the protective effect of the oil. The indeterminate rate of this process would undoubtedly give rise to variable results which is in line with the observed facts.

If the fat protection theory is accepted it must be ceded that where organisms of low heat resistance have been reported in canned fruit and vegetables, some other factor must have been responsible for their survival or the cans must have leaked or the organisms have been airborne contaminants. Hirst (1944) believed that faulty technique was often responsible for non-heat resistant organisms being reported in canned fruits and vegetables. In his view non-heat resistant types do not survive sterilisation of these packs.

With regard to fatty or oily foodstuffs, a question of the greatest practical importance concerns the fate of any fat-protected organisms which may survive the process. With an organism preferentially wetted by oil, it seems reasonable to suppose that once having entered the oil phase it will remain there and will be unable to develop for lack of moisture, nutrients, etc., Lang's work (*see* p. 150) indicates that oil-trapped spores can be made to pass from the oil to the aqueous phase by vigorous agitation but under ordinary conditions of handling canned foods it appears unlikely that oil-wetted organisms are capable of entering the aqueous phase. It may not be without significance that in the few reports which deal with the quantitative aspects of the subject, the numbers of non-heat resistant organisms found in the food tend to be quite low, in many cases a few per gram. Significant also may be the fact that several workers have reported that organisms found in canned products in very low numbers are capable of growth and spoilage production on re-inoculation into the same type of food from which they were isolated. In our view, the practical importance of oil- or fat-protected organisms in pressure-processed canned foods is small since it is highly probable that such organisms remain dormant in the oil or fat. Yesair et al (1946) concur in this. They state that there is no evidence that fat-protected cocci grow in canned products and evidence thus far acquired supports the belief that the factor which protects the organisms also prevents their subsequent growth.

Crossley (1944) believes that the presence of cocci in pressure-processed foods indicates post-process infection, except in the case of meat packs containing much fat or oil where the question of their survival is open to doubt. Crossley's opinion is if survival of cocci does occur it does not lead to deterioration of the food.

Air-borne contamination

Undoubtedly many of the earlier reports dealing with the sterility of canned foods are misleading due to the possibility of contamination having occurred during the culturing operation. Attention was drawn to the importance of air-borne contamination by Mudge & Foord (1940) and further stressed in a report by Williams & Clark (1942) which referred to the bacteriological condition of canned salmon. In order to soften the bones of the fish, canned salmon receives a heavy process which, on the basis of heat-penetration data, is in excess of that required for the destruction of spoilage organisms. Based on the results of cultural, pH and organoleptic tests, Williams and Clark concluded that all the cans they examined were sterile. Liquid, tissue and oil were examined (the latter after emulsification) by plating and by tube culture, using the usual precautions against contamination. As a check on air-borne contamination, plates were exposed in the vicinity of the cans during sampling and as a control on the pouring of the inoculated plates, sterile media was poured into sterile dishes at the same time. A high proportion of the inoculated plates and the exposure and pouring controls showed growth. Of the inoculated and the pouring control plates, those which were positive showed chiefly one to three colonies per plate, usually situated on the surface of the medium. When more than one colony was present they were frequently of dissimilar types and included aerobic spore-formers of the subtilis group, cocci and moulds. The exposure controls showed heaviest growth, colony counts as high as 24 per plate being found. With regard to the liquid cultures, aerobic and anaerobic growth was obtained in 5 and 0.5 per cent of the tubes respectively. These positive tubes showed no correlation with the plate cultures; the organisms were similar to those found on the positive plates. Williams and Clark state: 'In any sampling and cultural operation according to the procedures commonly used in bacteriological laboratories there is a greater chance of air-borne contamination on test plates than on control plates. The various manipulations all afford an opportunity for contamination so that both the percentage of non-sterile plates made from a truly bacteriologically sterile material, and the average number of colonies per plate, should be higher among inoculated plates than among the controls. We do not have sufficient information to justify a statement as to the range of difference, but we feel that no exception can be taken to a demand that there be a substantially higher number of positive plates among the test series than among the control series before a statement imputing lack of sterility to a material is justified.' Referring to their culture tests on canned salmon, they state: 'The fact that the bacterial flora of all plates, both inoculated and control, showing growth was similar in all respects except amount, and also that this flora was the typical air-borne type, indicates that the origin of the organisms was in fact air'. Williams and Clark suggest that the difficulty of eliminating air-borne contamination offers a rational explanation for certain of the results of bacteriological examinations of canned foods which are explainable only by theoretical reasoning or assumptions. Foord et al (1943) have also demonstrated the importance of air-borne contaminants in absolute sterility determinations.

Johansson & Ferris (1946) have published data which emphasise the difficulty of preventing contamination. By means of high speed photographs and air-sampling methods, they demonstrated that the common laboratory operations, such as pipetting, pouring and agitation of dilution bottles, frequently produce bacterial contamination of the surrounding air and environment. Photographs revealed droplet aerosols formed by blowing the last drop out of pipettes or removing the cap of a dilution blank which had just been vigorously agitated. In the experiments using a trace organism, 20 out of 37 samples of air were found to be positive. It is clear from these results that sterility tests should not be carried out in a place habitually used for manipulating bacterial cultures.

In the light of the evidence available at the present time, it is difficult to reach a definite conclusion concerning the bacteriological condition of sound canned foods in general. While the results of Mudge & Foord (1940), Williams & Clark (1942), and others are very striking, it would be rash to disregard the results of other workers on the bacteriology of sound canned foods without more extensive tests. There seems little doubt, however, that some of the earlier results, at least for some types of canned products, do not apply at the present time. Tanner (1944) points out that as the result of improvements in plant sanitation, processing, etc., the findings of earlier workers no longer hold for the modern product.

With modern methods of controlling the environmental air there should be no difficulty in carrying out reliable sterility studies. Criteria for clean work areas have been laid down and classifications are based on the number and size of particles in a set volume of air. This is possible since dust-free air does not possess a natural flora, the micro-organisms being resident in the dust particles. The American classification given in Federal Standard 209a (Anon, 1966) is also used

Table 11 Clean room classification

Maximum number of particles per cu ft 0.5 μm and larger (per litre)	Class according to English system (metric system)	Maximum number of particles per cu ft 5.0 μm and larger (per litre)
100 (3.5)	100 (3.5)	<1
10 000 (350)	10 000 (350)	65 (2.3)
100 000 (3 500)	100 000 (3 500)	700 (25)

in the UK. It designates air as belonging to one of the three categories as shown in Table 11.

Air designated as Class 100 can be maintained by use of high efficiency particle filters and a laminar air flow system. A monitoring system for checking the level of particulates in the air should be applied and can be achieved by using a method based on the microscopic examination of the particles removed by membrane filter in a vacuum line (Anon, 1968a, 1968b). A qualitative assessment of the cleanliness of the air may be obtained by exposing plates of nutrient agar in the work area for up to four hours. The subject of the environmental air in the aseptic processing of foods is reviewed by La Rocca & La Rocca (1975).

In our view, the following represents a reasonable summary of the present position:

1. Spore-forming bacteria of a heat-resistant nature are not infrequently found in commercially sound canned foods. Provided that cool storage conditions obtain, these organisms do not cause spoilage and their presence is no reflection on the heat processing conditions to which the food has been subjected
2. In fat-free canned products such as fruits and vegetables the presence of viable non-spore forming rods or cocci is an indication of under-processing or leakage of the container
3. Experimentally at least, it has been demonstrated that cocci heated in fat may survive temperatures sufficient to destroy resistant spore-forming spoilage organisms heated in aqueous media. Whether or not fat-entrapped cocci survive the pressure-processing of canned foods, the available evidence indicates that they have little or no practical significance since they remain trapped in the fat and are unable to develop
4. When viable cocci or other relatively non-heat resistant forms are found in large numbers in the aqueous phase and throughout any pressure-processed canned product, their presence is the result of under-processing or container leakage.

Canned cured meats

In connection with the microbiology of sound canned foods, cured meat products such as hams and chopped luncheon meats, call for special consideration. The pH of these products is invariably above 5.0 but the standard of heat processing universally accepted for other non-acid canned foods is not applied because this would result in excessive liberation of juices and general loss of quality. The level of heat processing for canned cured meats, therefore, is relatively low and preservation is attained through the contributory action of curing salts and in the case of large packs, also by refrigeration.

Canned cured meats of less than 1.35 kg (3 lb) weight generally receive a heat process in the range $F_0 = 0.2$ to $F_0 = 0.6$ measured at the centre of the pack. Since the heat transfer is almost entirely due to conduction, other regions of the can receive a heat treatment considerably in excess of the centre, leading to a significant increase in the average heating effect. The process is sufficient to render the packs shelf stable, i.e. whilst small numbers of viable spores of *Clostridium* and *Bacillus spp* may remain in some cans they are sufficiently heat injured for the curing salts, particularly sodium chloride and sodium nitrate, to inhibit their germination and growth. A low incidence of contaminating spores is a necessary prerequisite for good results. Although as pointed out in the section concerned with the curing of meat, all the factors — extent of heating, level of curing salts and extent of contamination — are interrelated. If spoilage does occur with these products it is generally due to the growth of spores of the genus *Bacillus*. Putrefactive spoilage is rare and should be regarded as an indication of a serious failure in the application of correct procedures.

Large packs of canned cured meat products, such as canned hams, which must be kept under refrigeration, often receive no more than a pasteurisation process and the temperatures reached at the centre of the can are of the order of 70°C maintained for a period of a few minutes. There is no legally required minimum centre temperature in the UK although the Department of Health and Social Security accept a minimum of 65.5°C for 30 minutes or 69°C for a shorter unspecified period provided the products are kept under adequate refrigeration throughout the distribution chain. According to the Meat Regulations of the USA a minimum temperature of 66°C is required. These minima are generally exceeded in order to lessen the spoilage risks and in most cases only spores of *Clostridium* or *Bacillus* remain. If vegetative cells survive they are generally enterococci of faecal origin which have an unusually high heat resistance (Zivanovic et al, 1965). According to Gardner & Patton (1975) the heat resistance of *Streptococcus faecalis* var *liquefaciens*, the causative organism of 'soft core' hams, depends on the salt content. Faecal enterococci are able to grow at refrigeration temperatures and if present in sufficient numbers this growth gives rise to mild off flavours and instability of the meat colour. The cans are not blown since gas is not produced and there is not thought to be a health risk if the contents are consumed.

Since pasteurised canned meats such as canned hams are, more often than not, a non-sterile product relying partly on correct storage conditions for maintaining an acceptable pack, it is difficult to differentiate between the bacteria associated with sound and unsound products. Potential spoilage organisms may be present in sound cans and differentiation between acceptable and non-acceptable produce is sometimes dependent on numbers rather than on types of organisms.

One obvious qualitative factor of value is the presence or absence of non-sporing organisms. It is common experience that some canned hams of European origin are consistently free from non-sporing organisms and it is Niven's (1955) experience that this applies also to the majority of USA canned hams. Jensen (1954), referring to canned cured meats that have received the standard process, states: 'We have not isolated the coliform group of bacteria, *Staphylococci, Serratia, Salmonella, Achromobacter, Pseudomonas, Flavobacterium, Chromobacterium, Lactobacillus, Proteus* and the miscellany of non-sporing rods'.

The possibility of faecal streptococci surviving the heat process has already been mentioned and the presence or absence of these organisms has been studied by Kelch & Stehl (1960) who examined the carcase surface of hogs before and after killing and from the 146 swabs taken they isolated *Streptococcus faecium* 48 times and *Strep. faecalis* var. *liquefaciens* 63 times. From 175 canned hams after incubation, 45 per cent contained faecal streptococci. *Strep. faecium* was present in nearly 90 per cent of all cases, presumably due to the fact that *Strep. faecium* is rather more heat resistant than *Strep. faecalis* (Leistner, 1960).

Kendereski (1971) in a survey of 2552 samples of semi-preserved meats such as canned hams, chops and frankfurters found 11.9 per cent contained streptococci (8.7 per cent *Streptococcus faecalis*). Buttiaux (1951) stated that 20 per cent of the canned hams of good quality he examined were sterile. Although the majority of the products were unsterile the viable organisms were relatively few (1 to 10 per g), the organisms usually present being spores of the bacillus group and *Lactobacillaceae* (*Strep. faecalis* and *Strep. thermophilus*); *Clostridium* spores were rarely present in sound samples. Ingram & Hobbs (1954), reporting a survey of canned hams, found that the great majority of apparently sound cans contain a few (less than 10 per g) spores of *Bacillus* species. In a small number of cans, counts of aerobic organisms exceeded 10^4, the organisms being largely faecal streptococci or *Pseudomonas fluorescens*; the latter organism was possibly derived from contaminated cooling water. In one of 175 apparently sound cans was found faecal *Escherichia coli* and in another *Clostridium perfringens*. Hobbs (1955) has later reported on 15 apparently sound cans and stated that 7 were sterile, 1 contained coliform Gram-negative organisms, 8 contained sporing bacilli and 2 contained streptococci. Clayson's (1955) experience is that some brands have bacterial counts of less that 100 per g surface and centre (aerobic spore-formers only) whilst others give counts up to 100 000 per g at the centre, mainly of coccus type. The survey by Kendereski (1971) showed that 47.5 per cent were sterile but that 3.2 per cent of contaminated cans contained more than 50 000 bacteria per g. Of the contaminated cans, spore-forming aerobes were present in 36.17 per cent, micrococci in 32.6 per cent, streptococci in 22.7 per cent, spore-forming anaerobes in 3.5 per cent, haemolytic bacteria in 7.24 per cent and gram-negative bacteria in 0.82 per cent. In a survey of refrigerated canned hams imported into the USA Surkiewicz et al (1977) found 179 out of 180 packs to have aerobic plate counts of 2000 or fewer per g and of these 174 had counts of less than 1000 per g.

In view of the fact that *Bacillus cereus* is implicated as a cause of food poisoning, the identification of aerobic spore-forming bacilli isolated from canned ham is of importance. Kelch (1961) investigated the various species of aerobic spore-forming bacilli isolated from canned ham. Fifty strains were identified, the majority of which were shown to be *B. licheniformis*. Seven were classified as *B. coagulans*, 7 were similar to *B. pantothenticus* but did not produce acetyl methyl carbinol, 5 as *B. stearothermophilus*, 5 as *B. laterosporus* and 3 as *B. subtilis*. Petrova (1975) isolated 50 strains of aerobic spore formers from ham. These were classified as *B. cereus* 2, *B. pumilis* 10, *B. subtilis* 13, and *B. licheniformis* 25.

Like other canned foods, cured meats are liable to leaker infection if the can seams are unsatisfactory. Bacteria entering the product by this route may not be susceptible to inhibition in the same way as survivors from the heat process. An outbreak of food poisoning in the UK caused by *Staphylococcus aureus* was attributed to leakage in canned chopped pork with faulty seams imported from Poland (Anon, 1976). Hobbs' (1955) opinion is that canned ham should be free from non-sporing organisms and the viable count should be less than 1000 per g (sporing organisms only). Goldenberg et al (1955) have proposed that bacteriologically satisfactory canned hams should meet the following specification:

Satisfactory

Esch. coli (faecal)	Absent from 1 g
C. welchii	Absent from 1 g
Faecal Strep.	Absent from 1 g
Total count at 37°C	Less than 10 per g
Total count at 22°C	Less than 10 per g
Spores	Less than 10 per g
Pathogenic bacteria	Not found

Reasonably satisfactory

As above, but total count 10 to 100 per g.

At an International Symposium on Semi-Preserved Meats, a Scientific Committee (1955) proposed the following standards:

(a) Pathogenic bacteria

1. *C. botulinum* should be absent. Or, if present, the conditions in the ham should be sufficient to prevent toxin formation
2. Toxigenic clostridia such as *C. perfringens* should be absent
3. *Salmonella* and *shigella* should be absent
4. Toxin-forming *Staphylococcus* and/or its toxin should be absent

(b) Bacteria indicative of inadequate processing

1. All members of the *Enterobacteriacae* should be absent
2. All members of the genera *Achromobacter*, *Flavobacter*, and *Pseudomonas* should be absent
3. Fungi should be absent

(c) Bacteria related to the stability of the pack

1. Gas-forming clostridia should be absent, except those which cannot develop in canned ham
2. Gas-forming denitrifying bacilli may be present. If present the cans may blow up if they are not kept in cool storage
3. *Micrococcus denitrificans* and *Cornyebacterium denitrificans* should be absent
4. Lactobacilli which can alter the colour, taste, or smell should be absent
5. Faecal streptococci, of Lancefield group D, should be absent

(d) Bacteria likely to be present, but not known to affect the quality of the pack if present in reasonable numbers.

The spores of inert clostridia, and of inert bacilli may be present, besides other inert thermoduric bacteria. The viable count of all these bacteria should not exceed 10^4 per gram. (In this context, inert means non-pathogenic and not able to change the quality of canned hams at temperatures below 5°C.)

In a pack such as canned ham the possibility of the survival of non-sporing organisms through fat protection must be considered. The fact, however, that some packers consistently produce canned hams free from non-sporing organisms without obvious loss of quality from over-processing suggests that the standard proposed by Hobbs (1955) is not unrealistic.

REFERENCES

Anon 1944 Food 13: 232
Anon 1966 Federal Std No 209a, Aug 10. Clean room and work station requirements, controlled environment
Anon 1968a ASTM, F 25 — 1968
Anon 1968b Millipore Corp Detection and analysis of particulate contaminants. Bedford, Mass
Anonymous 1976 Brit Med J 2: 1456
Bashford T E 1942 J Roy San Inst 62: 13
Buttiaux R 1951 2nd Int Congress on Canned Food (CIPC), Paris
Cameron E J, Esty J R 1940 Food Res 5: 549
Candy M R, Nichols A A 1956 J Dairy Res 23: 329
Cheyney E W 1919 J Med Res 40: 177
Clayson D H F 1955 Int Symp on Food Bacteriology, Annales de LInstitut Pasteur de Lille 7: 28
Crossley E L 1938 J Hyg 38: 205
Crossley E L 1944 Private Communication
Deichstetter J 1901 Zeitschr f Untersuch d Nahr w Genussmit 6: 1115
Fellers C R 1926 J Bact 12: 181
Foord D C, Crane C L, Clark B S 1943 Food Res 8: 489
Gardner G A, Patton J 1975 Proc of the European Meeting of Meat Res Workers No 21: 52. Ulster Curers' Assoc, 2 Greenwood Ave, Belfast
Goldenberg N, Sheppey C G, Robson J N 1955 Int Symp on Food Bacteriology, Annales de L'Institut Pasteur de Lille 7: 240
Golicz K 1969 Medycyna Weterynaryjna 25 (1): 50
Grosskopf J C, Harper W J 1974 Milchwissenschaft 29 (8): 467
Hirst F 1944 Private Communication
Hobbs B C 1955 Int Symp on Food Bacteriology, Annales de L'Institut Pasteur de Lille 7: 190
Ingram M 1955 Int Symp on Food Bacteriology, Annales de L'Institut Pasteur de Lille 7: 146
Ingram M, Hobbs B C 1954 J Roy San Inst 74: 1151
Jensen L B 1954 Microbiology of meats. Garrard Press, Champaign, Illinois
Johansson K R, Ferris D H 1946 J Infect Dis 78: 238
Kelch F 1961 Fleischwirtschaft 13: 294
Kelch F, Stehl E 1960 Fleischwirtschaft 12: 92
Kendereski S 1971 Rev des Fermentations et des Ind Alim 26 (1): 32
Leistner L 1960 Proc of 12th Res Conf of Am Meat Inst Found, Univ of Chicago, Circular No 61
Martin J H 1974 J Milk Food Technol 37 (2): 94
Mudd S, Mudd E 1924 J Exp Med 40: 633
Mudge C S, Foord D C 1940 Amer J publ Hlth 30: 273
Nichols A A 1939 J Dairy Res 10: 231
Niven C F 1955 Private Communication
Petrova L 1975 Veterinarnomeditskinski Nauki 12 (8): 82
Prescott S C, Underwood L 1897 Technology Quarterly 10: 183
Prescott S C, Underwood L 1898 Technology Quarterly 11: 6
Ridgway J D 1958 J appl Bact 21: 118
La Rocca P T, La Rocca R 1975 Dev Ind Microbiol 16: 339
Russell H L 1895 Ann Rept Wis Exp Sta p 227
Ruyle E H, Tanner F W 1935 Zbl Bakt Section 2, 92: 436
Savage W G 1923 Canned foods in relation to health. CUP, London
Savage W G, Hunwicke R F 1923 Studies in sweetened and unsweetened (evaporated) condensed milk. Fd Invn Bd Spec Rept No 13. HMSO, London
Scientific Committee 1955 Int Symp on Food Bacteriology, Annales de L'Institut Pasteur de Lille 7: 254

Senhaji A F 1977 J Food Technol 12: 217
Senhaji A F, Loncin M 1977 J Food Technol 12: 203
Surkiewicz B F, Harris M E, Carosella J M 1977 J Food Protection 40 (2): 109
Tanner F W 1944 Microbiology of foods 2nd edn. Garrard Press, Champaign, Illiois
Vaillard L 1900 Rept to 10th Congress of Hyg and Dem, Paris
Weinzirl J 1919 J Med Res 39: 349
Williams O B, Clark E D 1942 Food Res 7: 178
Yesair J, Bohrer C W, Cameron E J 1946 Food Res 11: 327
Zivanovic R, Oluski A, Tadic Z 1965 Technologija Mesa 6 (7-8): 198

14

Bacterial food-poisoning

Before the true role of bacteria in food-poisoning was properly appreciated, illness resulting from the consumption of food was attributed to the presence of ptomaines or toxic amines which were believed to have been formed as the result of the bacterial decomposition of protein. This theory is untenable since it has been shown that before these products are formed the food reaches such an advanced state of putrefaction that it would be abhorrent to normal persons. Furthermore, it has been demonstrated by Savage (1921) and others that these protein degradation products are only toxic when tested in quantities far greater than those which would be ingested under normal circumstances. There is no real evidence that such products in food material are in themselves harmful, and Tanner & Tanner (1952) have pointed out that the 'putrid poison' of the earlier workers may be known today as the toxins of *Clostridium botulinum* or *Staphylococcus* or *Salmonella* species.

Food-poisoning of bacterial origin arises through the consumption of food infected with certain types of living organisms or of food in which bacteria have developed and produced toxin. In other words, food-poisoning may be an infection or an intoxication. Six types of organisms are now recognised as the chief causal agents in bacterial food-poisoning. These are *Clostridium botulinum*, the toxin of which gives rise to the highly characteristic disease botulism, *Clostridium perfringens*, certain organisms of the *Salmonella* and *Staphylococcus* groups, *Bacillus cereus* and *Vibrio parahaemolyticus*. In addition to these, various other organisms have been held to be responsible agents in food-poisoning outbreaks. It is possible that some strains of *Streptococcus faecalis* may cause illness in man and streptococci of the Lancefield Groups A and D and the viridans type have, on some occasions, been found to be present in large numbers in the suspect food. Enteropathogenic strains of *Escherichia coli* may cause infant enteritis with fatalities in maternity wards, and some serotypes can cause diarrhoea in adults if food heavily contaminated with these strains is consumed. Outbreaks involving camembert cheese have been reported by Marier et al (1973). *Yersinia enterocolitica* and some strains of the genus *Campylobacter* have been isolated on a number of occasions from patients suffering from gastroenteritis; in the case of the latter organism the symptoms may be quite severe. Both organisms are known to be associated with animals and meat products but little is known about the spread of the infection or the minimum dosage required to promote illness. It may be that the food is involved only as a vector and the true status of these two organisms as food-poisoning agents remains doubtful. *Yersinia enterocolitica* can grow at temperatures as low as 4°C and may be of some significance to workers in chill stores, while *Campylobacter* infections have been reported among workers in poultry processing plants.

Botulism

The term botulism is derived from the word 'botulus' (a sausage), the illness originally being associated in Germany with the consumption of spoiled sausage. Van Ermengem (1897) first isolated the spore-forming anaerobe, now designated as *Clostridium botulinum*, responsible for this type of food-poisoning. *C. botulinum* is a Gram-positive anaerobic rod measuring 2 to 10 μm in length and from 0.5 to 2.0 μm in width. It produces oval-shaped, sub-terminal spores and is motile by means of peritrichous flagella. Seven toxicologically distinct types, designated A, B, C, D, E, F and G are recognised, although some workers sub-divide type C into C *alpha* and C *beta*. All these types can be divided into two groups depending on whether they are proteolytic or non-proteolytic in action. Smith (1977) does not regard *C. botulinum* as a single species but rather as a conglomerate of several culturally distinct groups of clostridia united by the fact that they all produce a powerful neurotoxin with similar pharmacological action. He recognises four groups based on cultural and serological characteristics as follows:

Group I	All strains of type A and proteolytic strains of types B and F
Group II	Non-proteolytic strains of types B and F and all strains of type E
Group III	All strains of types C and D
Group IV	Proteolytic but non-saccharolytic type G.

All strains are mesophilic but some from Group II will grow and produce toxin at temperatures as low as 3.0°C (Roberts & Hobbs, 1968). The toxin produced by Group II strains requires activation by trypsin and the spores produced by this group have a much lower heat resistance than those of Group I. Group III strains have an optimum growth temperature between 25° and 37°C and will not grow below 10°C; they are associated with botulism in animals and birds. Type G produces very small colonies, is very sensitive to oxygen and has an optimum growth temperature between 30° and 37°C. Little is known about this strain which was isolated from soil in Argentina by Gimenez & Ciccarelli (1970). Because *Clostridium sporogenes* is culturally identical to Group I strains and *C. novyi* is culturally identical to Group III strains that produce lecithinase, it is essential to include toxin tests in identification procedures (Smith, 1977).

Clostridium botulinum is widely distributed in nature having been isolated from virgin and cultivated soils in all the continents. Types A, B and E occur mainly in the northern hemisphere although their distribution is very uneven, while types B, C and D predominate in Australia, New Zealand and South Africa where they cause botulism in cattle. Type F isolations are infrequent and so far are confined mainly to northern latitudes. Meyer & Dubovsky (1922) found *C. botulinum* to be present in 9 out of 64 samples of soil collected from various English counties, and Leighton & Buxton (1928) located the organism in four samples of Scottish soil. At least 5 per cent and possibly 14 per cent of the English soils examined by Haines (1942) contained *C. botulinum*, some of the positive cultures being obtained from grassland. More recently Smith & Moryson (1975) carried out a survey of mud samples in the London area and found *C. botulinum* type B in 39 per cent of the samples, type C in 14.5 per cent and type E in 9 per cent of the samples, 3 per cent contained both B and E and 3 per cent B, C and E. Type D was present in 1.5 per cent samples and 3 per cent were untypable. A similar survey around Edinburgh gave an incidence of 57 per cent for type B and 14 per cent for type C. It appears therefore that *C. botulinum* occurs more frequently in the United Kingdom than the earlier surveys indicated. In America, Meyer & Dubovsky (1922) found type B to be the most common in soil from the eastern part of the country, while Parry (1946) found 11 per cent of soil samples from central New York to contain *C. botulinum* with type A predominating. He also found cultivated soils slightly more positive for the organism than other soils and the occurrence of *C. botulinum* was considerably higher in soils which had been sown with beans than with other crops. In the central United States *C. botulinum* is found less frequently, type B strains being most common, but Meyer & Dubovsky (1922) found 23 per cent of soil samples from the Pacific Coast States to contain *C. botulinum*, with 84 per cent of typable strains being type A. Kravchenko & Shishulina (1967) examined over 4000 soil samples in Russia for *C. botulinum* and *C. tetani*; they found a marked regional distribution with the frequency of both organisms increasing from the north to the south of the country. Type C strains were only found in the south. Type E was isolated only from soils rich in organic matter at the edges of streams and ponds. They concluded that soil was the actual habitat of type E (and probably other types) and that growth occurred in the soil. Type E was formerly thought to be associated chiefly with the marine environment but is now known to be fairly common in fresh water deposits in the northern hemisphere although its distribution is very uneven. Thus Eklund & Poysky (1970) found either type A or E in all the sediment samples they took from the Columbia river in North West America and in sediment samples from Lake Washington 18 to 35 type E cells per gram were isolated. In Lake Michigan, type E was found to be much more numerous in Green Bay than elsewhere and it has been suggested that this is associated with a high organic content resulting from drainage of fertilised fields in the area. Bott et al (1968), Sugiyama et al (1970) and Sugiyama et al (1972) found type E to be more prevalent in bottom sediment than on the shore, the inference being that the organism was growing in the sediment. Further evidence of the possible growth of *C. botulinum* type E in sediment was provided by Ward et al (1967). Investigating sediments from the Gulf of Mexico they found that type E isolations were associated with periods of warm weather. Cann et al (1965) could not demonstrate type E in the North Sea although it was frequently found in marine deposits off the Swedish coast. In a survey of British coastal waters Cann et al (1968) found type B in 3.5 per cent of samples examined. More recently a survey of trout farms in Britain has shown that the incidence of *C. botulinum* can be high in some farms, non-proteolytic strains of types B and F as well as type E being found (Hobbs, 1976).

Mainly types A, B, E and F are associated with botulism in man, type G has not as yet been implicated in a case and type F is known to have been responsible for only two outbreaks, one involving home-made liver paste described by Møller & Scheibel (1960) and one in America from venison jerky reported by Midura et al (1972).

In recent years much interest has been shown in type E which is now known to be more common than was formerly believed. In a review of botulism in America, Osheroff et al (1964) state that *C. botulinum* type E has been the major cause of cases of botulism from commercially prepared foods in the United States in the past two decades. One outbreak of type E botulism in 1963 which occurred in Detroit and has been reported by Johnston et al (1963) involved commercially canned tuna fish. The fish was used to prepare salad sandwiches which were eaten by three women. All three contracted botulism and two died; the third woman who had eaten less of the salad developed only minor symptoms and made a fairly rapid recovery. *C. botulinum* type E was isolated from the empty can and from autopsy specimens of the victims. Extensive investigations on stocks from the same manufacturer revealed a further 21 cans to be contaminated with *C. botulinum* type E. Most of these cans had defective seams and since the organism was also isolated from post-process handling equipment it was assumed that leakage had occurred. This theory is further strengthened by the very low heat resistance of type E spores. Other outbreaks of type E botulism were reported from America in 1963 involving commercially prepared smoked white fish from the Great Lakes. Recently an outbreak occurred in Britain involving four people who consumed a freshly opened can of Alaskan salmon. All four persons were admitted to hospital with severe symptoms of botulism and subsequently two of the patients died. Neurological symptoms occurred some 9 hours after consumption of the meal. Type E toxin was found in samples of sera from the patients and the organism was isolated from a small portion of the salmon (Gilbert, 1978). The infection was attributed to post-process leakage.

Outbreaks of botulism usually arise from the consumption of food which has been incorrectly preserved or stored and in which many of the competing micro-organisms have been destroyed. Over the years commercially canned foods have established an excellent record for safety, most of the recorded outbreaks of botulism originating from home-canned products. However, in 1971, commerically canned Vichyssoise soup containing type A toxin was responsible for two cases, including one death, in America. In 1973 a further outbreak occurred involving canned mild hot peppers packed in West Virginia. There were eight cases but no deaths, while in the same year a case occurred in Canada which was traced to marinated mushrooms packed in oil by a small cannery near Boston. Both products were found to contain type B toxin. A further outbreak in 1974, involving two cases with one death, was traced to a can of beef stew produced by another canner, type A toxin being responsible. During this period inspections of other canners' produce which had not been involved in any outbreaks revealed swollen No. 10 cans of mushrooms containing type B toxin, swollen cans of chicken and vegetable soup containing both A and B toxins and a leaking can of tuna fish containing type C toxin. In 1973 and 1974 a more disconcerting situation was revealed involving commercially canned mushrooms containing either pre-formed toxin or viable spores of *Clostridium botulinum*. The first case was detected by a can-making company investigating a spoilage problem. Type B toxin was confirmed by the FDA who eventually found 19 cans, from 7 codes belonging to this canner, to contain the same toxin. Later, during an inspection of a different cannery, the FDA discovered a further batch of canned mushrooms containing viable spores of *C. botulinum* type B but no toxin. Following these two incidents with mushrooms a complete survey was made of all stocks of canned mushrooms held in the United States from both domestic and imported sources. The results revealed that 30 cans were toxic and a further 11 contained viable spores but no toxin. One of the toxic cans contained both viable spores and toxin of *C. tetani* but no *C. botulinum*. The cans involved originated from seven American producers and two foreign canners (Ecuador and France). Half of the contaminated cans were the large No 10 size, 20 per cent were 454 g (16 oz) cans and the remaining 30 per cent were smaller than 284 g (10 oz). The products included whole mushrooms, sliced mushrooms and mushroom pieces, all contained *C. botulinum* type B and in addition 2 cans also contained type A. These outbreaks which occurred in the United States from 1970 to 1974 are reviewed by Lynt et al (1975). Why these problems should suddenly arise in the canning industry after many years of operation virtually free of contamination with *C. botulinum* is not clear, but intensive inspections by the FDA of the American canneries involved revealed a number of irregularities. In some cases faulty heat process control equipment was found, while in others the correct operating procedures were not being followed. The need to maintain all equipment in good working order and to adhere strictly to the scheduled operating instructions cannot be too strongly emphasised.

During growth in favourable food materials *Clostridium botulinum* synthesises within the cell an extremely potent neurotoxin which is released into the food when the cell dies and lyses. The seven types A-G are antigenically distinct, and antitoxin prepared against one is inactive against the toxin of the other types. These toxins are specific proteins and although some strains synthesise fully active toxins, the majority produce prototoxins which require activation in order to achieve full potency. Activation may result from the action of their own cellular enzymes, or the action of exogenous enzymes such as trypsin. The toxin attacks the nervous system, often with fatal results. The symptoms, which are highly characteristic, usually appear within 24 hours of consumption of the food, but the onset may be delayed for 72 hours or even longer. Similar symptoms are produced by all the types and include vomiting, constipation, double vision, thirst and difficulty in swallowing. Temperature is usually sub-normal and death is often the result of paralysis of the respiratory centres. Patients surviving for more than 8 days usually undergo a slow convalescence. Antitoxin, if administered at the very early onset of the intoxication, can save lives and the fact that it is now more available, coupled with more efficient communication systems, is probably responsible for a decline in the mortality rate. Smith (1977) gives the following mortality rates for various countries: USA 31 per cent; USSR 29 per cent; Japan 27 per cent; Germany 13 per cent; Poland 10 per cent and France 3 per cent. Although in Poland and France botulism is not uncommon the mortality rates are considerably lower. This may in part be due to the types involved. It is known that type B is less toxic to man that types A and E.

The purified toxins of *Clostridium botulinum* are destroyed by heating at 80°C for 10 minutes. However they may gain some protection when mixed with various food components such as protein. Although most of the toxin is produced by the vegetative cells, small amounts may be present in the spores when it will have the same heat resistance as the spore. The toxin content of type A spores has been calculated; 10^6 to 10^7 spores contain one mouse lethal dose when administered parenterally (Fernandez et al, 1969).

A very wide range of foods has been involved in outbreaks of botulism, including meat, fish, fruit, vegetable and milk products. Usually, the intoxicated food shows obvious signs of spoilage such as gas evolution, digestion and rancid odour, but there is abundant evidence that toxin may be formed without the food becoming noticeably spoiled. In canned foods, growth of *Clostridium botulinum* does not invariably result in the formation of 'swells'. Schoenholz et al (1923), investigating artificially infected canned products, found that asparagus and beets became toxic although the cans did not bulge; furthermore, the degree of spoilage of the contents was not sufficient to constitute a safeguard for the consumer. In canned spinach and string beans, results were variable, some cans becoming bulged while others remained flat. Townsend et al (1954) have confirmed that growth and toxin formation without gas production can occur in a variety of foods.

Scott & Stewart's (1944) findings on the subject of growth of *Clostridium botulinum* in canned vegetables are of great interest. They report that in internally lacquered cans of carrots and beetroot, growth and toxin production occurred regularly but not when unlacquered cans were used. The inhibition of growth in plain cans was found to be due to dissolved tin. The concentration of tin required to prevent growth varied with the vegetable, being about 150 p.p.m. in beetroot and about 30 to 60 p.p.m. in carrots. In a later report, Scott & Stewart (1945) showed that the inhibitory concentration of tin varied widely with different vegetables. This variation appeared to be related to the protein content of the vegetables, the higher the protein content the higher the concentration of tin required to inhibit growth.

As stated earlier (p. 181), the spores of *Clostridium botulinum* possess considerable resistance to heat, and the importance which attaches to the organism in relation to its growth in canned food has led to the adoption of its thermal death-time as a minimum standard of processing for foods with pH values higher than 4.5. This is based on the assumption that the organism is incapable of growth in environment of higher acidity than pH 4.5. On rare occasions, however, growth and toxin production have been suspected in food products with pH values lower than 4.5. Nine such cases have been reviewed by Slocum et al (1941). The foods concerned in the outbreaks were canned pears, canned apricots, tomato catsup, tomato-onion chilli sauce, and green tomatoes, all except the catsup being home-canned. The outbreak investigated by these workers was caused by home-canned tomatoes, the processing of which was inadequate and in addition to *C. botulinum*, a *Bacillus* sp. and a coccus were present. There appeared to be some evidence that the concomitant growth of other organisms permits the development of *C. botulinum* under conditions in which it fails to grow when present in pure culture. It is conceivable that the growth of other organisms in acid products could raise the pH to a level at which dormant spores of *C. botulinum* could germinate and grow. Odlaug & Pflug (1977) stored the spores of *C. botulinum* types A and B in tomato juice (pH 4.2) and citric acid phosphate buffer (pH 4.2) at 4°C, 22°C and 32°C for 180 days. The spore count was determined at 0, 12, 20, 47, 79, 99, 125 and 180 days when it was found that type A spores showed little decrease in either the tomato juice or the buffer solution at the three temperatures. Type B spores showed no decrease in numbers at 4°C but a decrease of approximately 30 per cent occurred after 180 days' storage at 22°C and 32°C. They concluded that *C. botulinum* spores could survive for long periods in an acid medium and that although acid foods represented no risk to adults they could be a source of spores in infant botulism. Dolman & Iida (1963) have recorded the growth, with toxin production, of *C. botulinum* type E in pickled herring at pH 4.0 to 4.2.

In a review of the literature on botulism in relation to acid foods, Ingram & Robinson (1951) concluded that the evidence available does not warrant unreserved acceptance of the view that *Clostridium botulinum* can grow below pH 4.5, whether accompanied by other micro-organisms or not. They point out that the probable increase in acidity of the product before the observations were made, due to slow diffusion of acid from imperfectly cooked fruit and the growth of *C. botulinum* and the accompanying organisms, has hitherto been ignored in the investigation of outbreaks involving acid packs.

Townsend et al (1954) have studied the pH levels at which

Clostridium botulinum spores germinated and produced toxin in a variety of foodstuffs acidified with citric acid. The lowest pH at which growth and toxin occurred was 4.8; it was concluded that the pH level of 4.5, which is commercially recognised as the maximum permissible for those foods where acidity is relied on to inhibit growth of *C. botulinum*, provides ample measure of safety.

Unlike its spores (except those of type E) the exotoxin of *Clostridium botulinum* is relatively heat-labile, 15 to 30 minutes at 80°C being adequate for the inactivation of types A, B and C (Meyer, 1928).

The effect of salt concentration and temperature on the growth and toxin production of *Clostridium botulinum* is discussed in Chapter 3.

For further details concerning botulism the reader is referred to Dolman & Iida (1963), Lewis & Cassel (1964), Riemann (1973) and Smith (1977).

Clostridium perfringens (C. welchii)

Knox & MacDonald (1943) in England and McClung (1945) in America, were among the first workers to demonstrate clearly that certain outbreaks of food-poisoning were caused by *C. perfringens*. Since then, many outbreaks of mild illnesses have been attributed to this organism and it is now second only to *Salmonella* as a cause of bacterial food-poisoning in England and Wales, being responsible for approximately one quarter of all reported cases (Hobbs & Gilbert, 1978). The illness is characterised by abdominal cramps and diarrhoea; usually without fever and vomiting, commencing 12 to 20 hours after eating contaminated food (Hobbs, 1960).

Clostridium perfringens is an anaerobic, non-motile, short, thick bacillus with blunt ends, occurring singly, in pairs or, less frequently, in short chains. Young cells stain Gram-positive but they may appear Gram-negative with age; subterminal ovoid spores are formed. It can be divided into five toxicological types (A to E) on the basis of four major toxins which may be produced. Of these five types only types A and C have been implicated in cases of human gastroenteritis, type A being the most common. Type C causes a more serious disease called enteritis necroticans which may have a high mortality rate: fortunately this condition is rare but outbreaks have been reported from Germany (Zeissler & Rassfeld-Sternberg, 1949) and New Guinea (Murrell et al, 1966). Type A may also cause gas gangrene.

Clostridium perfringens is widely distributed in soil and the faeces of man and animals but only type A strains form part of the microflora of both soil and intestinal tracts, type B, C, D, and E occurring as animal parasites (Smith, 1975). The earlier work of Hobbs et al (1953) appeared to indicate that only heat resistant strains of type A were involved in food-poisoning outbreaks. However, Hauschild & Thatcher (1967) working with a heat susceptible strain of type A known as S-79 found that if the vegetative cells were ingested by man, symptoms of gastroenteritis including abdominal pain and diarrhoea were produced. It is now generally accepted that both heat resistant and heat susceptible strains of type A *C. perfringens* can give rise to food-poisoning outbreaks. The symptoms are caused by the consumption of food containing large numbers of vegetative cells which undergo sporulation on reaching the intestine. The enterotoxin responsible for the illness is a heat sensitive protein and is produced during the process of sporulation (Hauschild et al, 1970; Duncan, 1973). Because of the heat resistance of its spores, and the low Eh required for its growth, food-poisoning from *C. perfringens* is frequently associated with cooked meat dishes and large joints of meat in which the heat-shocked spores have survived the cooking process, germinated, and then rapidly multiplied if the food is allowed to cool slowly. It tends to occur in commercial kitchens and institutions where large quantities of food are prepared. Useful reviews of *C. perfringens* food-poisoning are given by Hobbs (1965) and Bryan (1969) and additional information by Smith (1975).

Salmonella food-poisoning

Organisms of the genus *Salmonella* have for long been associated with food-poisoning in man and recent figures show that they are the most common and widespread cause of food-poisoning reported in England and Wales, accounting for approximately 75 per cent of all cases.

As a group, the members of the genus *Salmonella* are very closely related, in many cases their cultural and biological reactions being so similar that differentiation is possible only with the aid of complicated serological methods. By the use of such methods over 1700 serotypes are now recognised, all of which are capable of causing illness in man or animals, or both. These facultative anaerobic organisms are motile, Gram-negative, non-sporing rods of low heat resistance. A characteristic of this group of organisms is their inability to ferment lactose or saccharose although dextrose and other carbohydrates are attacked, usually with the production of acid and gas; they are not proteolytic.

Salmonella organisms are frequently associated with disease in cattle, pigs and poultry; rats and mice are also subject to epidemics and both man and animals may become carriers of the infection, suffering no clinical symptoms but continuing to excrete the organisms in their faeces. Although, as stated earlier, all serotypes are capable of causing disease in man and animals, some give rise to a more severe form of illness involving septicaemia and a higher mortality rate and serious losses can result in farm animals. Because these serotypes usually cause severe infection in specific hosts they have been termed 'host-adapted' serotypes, examples being *Salmonella choleraesuis* in pigs, *S. gallinarum* and *S. pullorum* in poultry and *S. typhi* and *S. paratyphi* B in man. Normally in host-adapted infections a small proportion of the cases which recover become carriers, the organisms being excreted in the faeces, thus ensuring the continued relationship between host and serotype. Host-adapted serotypes rarely infect other host species.

Most *Salmonella* serotypes cause acute intestinal infection with no septicaemia and the illness is usually of relatively short duration with few long-term carriers resulting. Over a period of time, however, certain serotypes may become associated with particular animal hosts due to the methods used in farming.

The foods most often concerned in *Salmonella* outbreaks are meat, milk, poultry and eggs. Especially involved have been those foods such as prepared meat, sausages and meat pies which are lightly cooked and subject to much manipulation. The appearance and taste of infected food may not be noticeably altered.

In England and Wales the *Salmonella* type most frequently encountered in human infections is *S. typhimurium*, this serotype was responsible for 70 per cent of salmonella infections in man in 1960 but only 37 per cent in 1970. The incidence of *S. typhimurium* decreased between 1960 and 1966 and then remained fairly constant. However the incidence of other serotypes greatly increased from 1966 to 1970. Several explanations have been advanced for this change, including the increase in tourism and the importation of food products. However, although these factors may have resulted in the introduction of some different serotypes, it appears that the main cause for the increase in serotypes other than *S. typhimurium* is the importation of animal feedstuffs. It has been demonstrated that if animals are given feed contaminated with low numbers of *Salmonella* they may show no symptoms of infection but will excrete the organisms and in the case of poultry the eggs will also be infected. Animals subject to intensive farming such as poultry and pigs receive most of their protein from imported feedstuffs and consequently are mainly responsible for the establishment of these 'new' serotypes. Human food-poisoning results from the contamination of carcass meat during slaughtering with salmonellas present in the intestines of healthy animals (McCoy, 1975). The serotypes which have been responsible for the increase in human salmonella infections in England and Wales from 1966 together with the main vehicle of infection are as follows: *S. panama* (pigs); *S. stanley* (pigs); *S. virchow* (poultry); *S. enteritidis* (poultry); *S.*

4, 12:d (poultry and pigs); *S. agona* (poultry and pigs) and *S. indiana* (poultry and pigs). In England and Wales cattle are the main source of *S. typhimurium* infections in man.

Imported commodities such as egg products, coconut, raw meat and fertilisers, in addition to the animal feedstuffs have all been incriminated as carriers of salmonellas. Various methods suggested for treating these materials include pasteurisation, storage at elevated temperatures and fumigation. Some success has been achieved in this direction: for example, desiccated coconut, once a major source of *Salmonella* contamination is now relatively free of this organism due to pasteurisation and improved manufacturing procedures. Since 1964 all liquid whole egg in this country (including imported supplies) has been required by law to undergo a pasteurisation before distribution. This has been particularly successful, outbreaks of human salmonellosis from eggs and egg products have since decreased from about 30 per cent to approximately 4 per cent. In addition the Zoonoses Order of 1975 requires that salmonella excreters in flocks and herds must be notified, while the proposed Protein Processing Order will seek to reduce the contamination of animal feedstuffs.

One strain of *Salm. senftenberg* known as strain 775W has been shown to possess exceptional heat resistance, a suspension of 2×10^7 organisms in heart infusion broth having a thermal death-time of 37 minutes at 60°C (Davison et al, 1966). These authors conclude, however, that in their opinion, heat-resistant strains of this serotype have played little part in the past in the occurrence of *Salmonella* infections. Reinke & Baker (1966) have demonstrated that a heat treatment of 2.5 minutes at 64.4°C is sufficient to guarantee liquid whole egg to be *Salmonella* free.

The *Salmonella* organisms do not elaborate soluble exotoxin but produce endotoxin, and when living organisms of the food-poisoning type are ingested, acute gastroenteritis results, in the majority of cases after a 12 to 24 hour incubation period. Mortality is low — 1 to 2 per cent (Dible, 1943). According to Dack (1956), it is a characteristic feature of *Salmonella* gastro-intestinal outbreaks that the incubation period varies with the individual but normally it is at least 3 hours before the onset of symptoms. He further states that in the past, cases of alleged *Salmonella* poisoning in which the incubation period was 2 to 3 hours were in all probability caused by staphylococci. *Salmonella* food-poisoning is not often fatal, but the illness, which is characterised by nausea, vomiting, diarrhoea, abdominal pain and fever, may persist for several days.

Useful reviews of the problem of *Salmonella* infection in foods are provided by Hinshaw & MacNeil (1951), Bowmer (1965), Lee (1974) and McCoy (1975).

Staphylococcal food-poisoning

Although reported as early as 1894, general recognition of the part played by staphylococci in food-poisoning outbreaks is a comparatively recent development. These free-growing organisms are facultative anaerobes; they are Gram-positive, non-motile and do not form spores and hence are relatively susceptible to heat destruction. Acid but not gas is produced from a number of carbohydrates and gelatin is liquefied by some strains. Staphylococci are extremely widespread being found in air, water, milk and sewage, but their natural habitat is the anterior nares and skin of man and animals. They are also present in the intestinal tract. When they penetrate the skin through small lesions, pimples, boils and carbuncles may be produced. Hair follicles are common sites of entry and growth. Approximately 40 per cent of the human population are estimated to be carriers of *Staphylococcus aureus* and it is believed that about half of these are carrying enterotoxigenic strains.

During their growth in food, certain strains of *Staphylococcus aureus* elaborate a filterable enterotoxic substance which gives rise to acute intestinal disturbance. A characteristic of staphylococcal food-poisoning is the rapid onset of the illness following ingestion of the enterotoxin, symptoms in the majority of cases appearing within 2 to 4 hours. The illness is characterised by nausea, vomiting, diarrhoea and abdominal cramp, and while these may be severe, recovery is rapid following elimination of the irritant from the body.

As with other types of food-poisoning, there may be no change in the appearance or flavour of intoxicated food. Manufactured or prepared foods which are normally eaten cold, for example, hams, pressed meats, pies, brawns, custard-filled pastries and milk products are those most often involved. Canned foods, usually considered as beyond suspicion as vehicles in the transmission of staphylococcal food-poisoning, are known to be involved also. Undoubtedly in many instances canned foods are wrongly incriminated due to contamination of the product after the can is opened. However, according to Cockburn (1960) there were 68 outbreaks in England and Wales during the period 1949 to 1960, where it is recorded that the contents were eaten within one hour after the can was opened. Products mainly involved were corned beef canned in East Africa and processed peas (canned reconstituted dried peas). It seems clear from the report by Bashford et al (1960), in the case of the canned peas, that post-process infection, due to leakage through the seams, occurred and that the source of the contaminating organism (*Staphylococcus aureus*) was the personnel handling damp cans after the cooling process.

Canned peas support the growth of staphylococci with toxin production. However, canned garden peas (made from fresh peas) usually show gas production resulting in blown cans and this acts as a safeguard for the consumer. With processed peas the infection may not be apparent. In some foods growth of the organism may occur without production of enterotoxin. Davidson & Dack (1942) found that enterotoxin was produced in experimentally inoculated canned corn and oysters but not in canned salmon, although growth occurred equally well in all three foods. Segalove et al (1943) have studied the growth of a food-poisoning staphylococcus in various canned foods, including asparagus, spinach, string beans, peaches, tomato juice, corn, peas, shrimps and salmon. Maximum growth was found in peas, corn and the fish products, but growth did not occur in the acid packs (tomato juice and peaches). In nearly all cases growth was better at 22°C than at 37°C. Although normally unable to ferment sugars with gas formation, the organism was found to produce blowing in canned peas and corn. Growth and toxin production are also influenced by other micro-organisms which may be present in the food.

Foods suspected of having caused staphylococcal food poisoning will normally contain large numbers of *Staphylococcus aureus* per g and the organism may also be isolated from the faeces of the patients. However, not all strains of *Staph. aureus* are enterotoxigenic and Haynes & Hucker (1944), who studied 114 strains of staphylococci, concluded that enterotoxin production is a characteristic which can be assumed by certain strains. The positive identification of food-poisoning types of staphylococci is important, therefore, when outbreaks are being investigated. Over the years a considerable number of selective media have been developed for the isolation of *Staph. aureus* including the gelatin medium of Stone (1935), mannitol salt agar by Chapman (1945), Baird-Parker agar (1962) and the phenolphthalein diphosphate agar with polymyxin of Hobbs et al (1968). Most enterotoxigenic strains of *Staph. aureus* produce coagulase and a tube coagulase test is considered necessary for the confirmation of all strains isolated (Gilbert, 1974). By the use of phage-typing, it is possible to identify specific strains of staphylococci involved in food-poisoning outbreaks and the technique is useful in tracing sources of contamination. A useful review of phage-typing is given by Parker (1972).

Situations sometimes occur in which food-poisoning strains of staphylococci have grown in a food, produced enterotoxin, and then died off leaving active toxin in the food; also because the toxin is heat stable, lightly cooked foods may contain toxin but no viable organisms. In such circumstances the ability to demonstrate the presence of enterotoxin is essential in any subsequent investigation of these foods. Progress in this direction has been slow, however, due to the absence of suitable laboratory techniques. The usual laboratory animals are not susceptible when fed enterotoxin unless the amount is large. The injection of animals has been criticised on the grounds that the method is unreliable and in this connection Rigdon (1938) has reported that typical symptoms were produced when kittens and puppies were injected with sterile media containing no toxin. Jones & Lochhead (1939) found that a pipette-feeding method, using kittens, possessed advantages over other methods. Hopkins & Poland (1942) have demonstrated the susceptibility of young pigs to intraperitoneal injection of the enterotoxin, but for many years the use of human volunteers was the most reliable test available.

During the 1960's workers in America achieved a significant breakthrough by isolating and purifying the enterotoxins produced by *Staphylococcus aureus* (Bergdoll et al, 1959; Casman, 1960; Bergdoll et al, 1965; Casman et al, 1967 and Bergdoll et al, 1971). Five enterotoxins were isolated and designated A, B, C, D and E. Once the antigenicity of the enterotoxins had been established the serological detection of enterotoxin in foods and culture filtrates became possible. A number of methods have been developed for this purpose but the one most favoured is the slide gel double diffusion technique of Wadsworth (1957) or one of the various modifications (Casman et al, 1967; Gilbert & Wieneke, 1973).

All five enterotoxin types are known to cause staphylococcal food-poisoning, but strains producing enterotoxin A and A and D have been implicated more frequently than the others in the UK (Gilbert & Wieneke, 1973).

The fact that the enterotoxins of *Staphylococcus aureus* exhibit considerable heat stability has been recognised for many years but until the development of serological methods for the quantitative detection of the enterotoxin no reliable data could be established. From the point of view of canned foods it is important to know the heat resistance of staphylococcal enterotoxin and the effect that various factors may have on its thermal inactivation in order to assess the probability of its survival in various canned products. Although as yet information on the subject is limited, some interesting work has already been reported, notably that of Denny and his colleagues at the NFPA in America. Denny et al (1971) studied the effects of toxin concentration on the heat inactivation of enterotoxin A in beef bouillon (pH 6.2) and in phosphate buffer. They found that in the beef bouillon enterotoxin A had a *z* value of approximately 27.8°C with three different concentrations of toxin (60 μg per ml, 20 μg per ml, and

5 μg per ml) and that heat inactivation was dependent on the initial concentration of enterotoxin. The inactivation times at 121.1°C were 40 minutes for a concentration of 60 μg per ml, 37 minutes for a concentration of 20 μg per ml and 27 minutes for the 5 μg per ml concentration. In 0.15 M phosphate buffer (pH 7.2) enterotoxin A was from 3 to 5 times less heat resistant although again the highest concentration of toxin had the greatest heat resistance. At the lowest concentration of enterotoxin used (5 μg per ml) they found heat inactivation times at 121.1°C of 27 minutes in the beef bouillon and 5 minutes in the phosphate buffer. Reiser & Weiss (1969) have reported that the maximum concentration of enterotoxin A that *Staph. aureus* could produce in ideal conditions is approximately 5 to 6 μg per ml but Denny and his co-workers were able to demonstrate only 0.2 μg per ml when the beef bouillon was inoculated and incubated for 14 days. They state that the low level of toxin produced in beef bouillon indicates that the heat resistance of enterotoxin A in this product would be considerably less than the 27 minutes they obtained for a concentration of 5 μg per ml. Humber et al (1975) investigated the effect of pH on the heat inactivation of enterotoxin A. The toxin was destroyed more rapidly in beef bouillon at pH 5.3 than at pH 6.2, the z values being 27° and 30°C respectively. They found the serological detection method to be more sensitive than monkey feeding tests and stated that although man is probably more susceptible to enterotoxin A than are monkeys, a positive serological test may not be a valid indication of the toxic level of enterotoxin present in a heated food.

Some of the earlier studies on the heat inactivation of staphylococcal enterotoxins have been criticised on the grounds that the concentration of enterotoxin used in the experiments was too high and unrealistic. Denny et al (1971) suggested that lower levels of toxin should be used and later Denny (1973) reported a heat inactivation time, for enterotoxin A in beef bouillon at pH 6.2, of 16 minutes at 121.1°C using a toxin concentration of 1 μg per ml. This figure was obtained using a serological assay for the enterotoxin. If a monkey feeding test was used the inactivation time at 121.1°C (F value) was 13 minutes. Soo et al (1973) using enterotoxin A at a concentration of 1 μg per ml in whole milk obtained a heat inactivation time of 18 minutes at 121.1°C. Tatini (1976) summarises the results on this subject obtained by a number of workers.

Various factors which affect the heat resistance of micro-organisms also affect the thermal stability of staphylococcal enterotoxin and the situations are closely paralleled, thus the heat resistance of both are influenced by the nature of the food, pH, presence of salt or sugar, initial concentration and the sensitivity of the detection method used.

However, despite the considerable advances achieved in recent years our knowledge of the thermal stability of staphylococcal enterotoxins is still inadequate, particularly at low concentrations of toxin. Meanwhile their significance in canned foods remains uncertain although it appears highly probable that enterotoxin, if present in sufficient concentration, will survive pasteurisation processes and may have some significance in certain HTST processes. In our opinion low acid canned foods adequately processed for *Clostridium botulinum* and with sufficient safety margin to prevent the growth of spoilage organisms, will present no hazard with respect to staphylococcal enterotoxin, bearing in mind the low concentration of toxin likely to be present, provided that sound raw materials are used.

The present methods used in the serological assay of foods for enterotoxin are complex and require a high degree of skill on the part of the analyst. For these reasons the use of the technique is normally restricted to specialist laboratories. It is clear, however, that raw materials of dubious microbiological quality should not be used in the production of canned foods.

Bacillus cereus

Hauge (1950, 1955) attributed four large food-poisoning outbreaks in Norway to the presence of large numbers of Gram-positive bacilli in the food. These were isolated and identified as *B. cereus*. The foods involved were starchy products, such as vanilla sauce, prepared in advance and stored under conditions permitting the growth of this organism. Christiansen et al (1951) described a similar outbreak in Sweden involving yellow pudding which had been stored in cups for 24 hours at a temperature suitable for bacterial growth. It appears from both investigations that a large number of cells must be present in the food before a toxic level is reached.

Hauge (1950, 1955) gives the symptoms as abdominal pain, diarrhoea, rectal tenesmus and nausea which seldom gives rise to vomiting. The incubation period is usually 12 to 13 hours.

Dack et al (1954) fed cultures of *Bacillus cereus* to human volunteers but were unable to produce the symptoms reported by the Scandinavian workers. The strains of *B. cereus* used by Dack and his colleagues, however, were isolated from cheese spreads and had not been involved in food-poisoning outbreaks; the significance of these negative results, therefore, is doubtful.

In recent years numerous outbreaks of food-poisoning caused by the consumption of foods heavily contaminated with *Bacillus cereus* have been reported. Most of these outbreaks have occurred in European countries including the USSR, with relatively few incidents being recorded from America and Canada where the extent of the problem has not yet been fully investigated. In England and Wales, during the years 1973–1975 inclusive, a total of 199 cases was reported which accounted for approximately 0.7 per cent of all recorded cases of food-poisoning. Many of these cases originated from contaminated rice which had been boiled and then held overnight, at ambient temperatures, to drain before receiving a short frying process the following day. The spores of *B. cereus* are capable of surviving this treatment after which they germinate and grow rapidly in the warm moist rice, producing a toxin which appears to survive the brief frying. The illness in the UK is often characterised by somewhat different symptoms than have been reported elsewhere. Severe vomiting can occur from 1 to 6 hours after the meal which may be followed later by diarrhoea. It is now known that two toxins are involved and a serological typing scheme has been developed which enables the vomiting and diarrhoeal strains to be distinguished. A number of serotypes have not been associated with illness (Hobbs & Gilbert, 1978).

Bacillus cereus is a Gram-positive, aerobic, spore-forming rod which may exhibit considerable variation in its biochemical reactions. The optimum temperature for growth is between 30° and 37°C but some strains may grow at temperatures as low as 10° to 12°C. It is widely distributed in cereals and other dried foods and is a common spoilage organism in milk and milk products. It is a causal agent of 'bitty' cream. The foods involved in *B. cereus* food-poisoning include rice dishes, various cereal products, puddings, sauces, soups, mashed potato, vegetables and minced meat. Hobbs (1974) gives a table comparing the characteristics of *B. cereus* food-poisoning with that produced by *Clostridium perfringens*; there are many similarities between them. She further states that foods containing less than 10^6 *B. cereus* cells per g are probably unlikely to cause illness, but cooked foods containing counts above 10^6 per gram, stored at ambient temperatures, may be suspect. A useful review of *B. cereus* food-poisoning is given by Goepfert et al (1972).

Vibrio parahaemolyticus

This organism was first reported as a causal agent of food-poisoning by Fujino et al (1951) after investigating an outbreak of food-poisoning in Japan involving 272 people with 20 fatal cases. It is now known to be responsible for over 50 per cent of all food-poisoning cases in Japan and outbreaks have been recorded from Australia, the Far East, America and Europe — including Britain.

The organism is a straight rod, rarely curved, with rounded ends. It occurs singly or in short chains. Pleomorphic cells are present. It is Gram-negative and motile by means of a single flagellum at one pole of the cell. The optimum temperature for growth is between 35° and 37°C, with a maximum of 44°C and a minimum temperature of about 10°C. At the optimum temperature growth is extremely rapid with generation times as low as 8 to 9 minutes having been recorded. It grows over a wide pH range (5 to 11) with an optimum pH of 7.5. It is moderately halophilic, tolerating salt concentrations up to about 7 per cent with an optimum of 3 per cent and it will not grow in culture media without the addition of salt. Many different strains exist but not all of them cause illness and a serological typing scheme has been developed by Japanese workers. A characteristic known as the Kanagawa phenomenon has proved very useful in distinguishing enteropathogenic strains. When grown for 24 hours at 37°C on the medium of Wagatsuma (1968) some strains produce a clear halo of haemolysis at the periphery of the colonies and are termed Kanagawa positive. Strains showing no haemolysis are termed Kanagawa negative. It has been shown by Sakazaki et al (1968) and Miyamoto et al (1969) that over 95 per cent of the strains isolated from the faeces of patients are Kanagawa positive, while of those isolated from sea water and marine animals, 1 per cent or less are Kanagawa positive.

The illness, which is an infection following the ingestion of live organisms in the food, has an incubation time of about 10 hours which may be extended to 2 days. The symptoms, which may be severe, are diarrhoea, abdominal pains, nausea, fever and cold shivers but occasionally dehydration and shock predominate the clinical symptoms, causing the illness to be confused with cholera (Peffers et al, 1973).

The natural habitat of the organism appeared at first to be the inshore waters, marine sediments and brackish waters around the coasts of Japan but it has since been demonstrated in similar situations in the Far East, America, Canada and Europe. In 1973 the first outbreak known to be caused by local produce occurred in Britain at a holiday camp on the south coast. The vehicle of infection was dressed crab-meat prepared from locally caught crabs (Hooper et al, 1974). Outbreaks occur most frequently during the summer months when the inshore waters become warmer allowing a more rapid proliferation of the organism. Sea-foods such as crabs, prawns and shellfish may become contaminated and if these are only lightly cooked, or eaten raw or fermented as in Japan, food-poisoning may result. In Britain the greatest risk of *Vibrio parahaemolyticus* food-poisoning would appear to be from the importation of frozen cooked sea food from the East, used to make up dishes such as prawn cocktail. A detailed review of

Vibrio parahaemolyticus food-poisoning is given by Cabassi & Mori (1976).

Relation of food-poisoning to canned foods

Considering all the evidence available and having due regard to the efficiency of modern processing conditions, commercially canned and preserved foods are of minor significance as vehicles for food-poisoning. Many of the cases in which canned foods have been concerned appear to have been based on clinical rather than bacteriological findings. In the past, considerable prejudice has existed concerning the public health aspects of canned foods and to some extent this prejudice continues among a very small section of the population, but it is an incontestable fact that canned foods are less liable than fresh foods to give rise to food-poisoning. Savage (1939) concludes: 'As regards our food supply generally, canned foods can be accepted as a safe and convenient source and one without any special hazards, nutritional, chemical or bacterial'.

In Britain the occurrence of botulism is extremely rare, only seven outbreaks having been recorded. Two of these involved glass packs (Savage, 1939), but in the most recent outbreak which occurred during 1978, canned salmon packed in Alaska was implicated. The freshly opened can of salmon was eaten with a salad. Type E toxin was found in samples of sera from the patients while the organism was isolated from a small portion of the salmon (Gilbert, 1978). Because of the low heat resistance of the spores of type E, the can probably became infected by leakage. The very low incidence of botulism in Britain is remarkable in view of the fact that *Clostridium botulinum* is not uncommon in the environment (Smith & Moryson, 1975). This is in part due to the high standards of canning maintained in the industry and the fact that home canning is not a popular method of preserving food in this country.

In the USA the majority of cases recorded since 1899 have resulted from home canning (Meyer & Eddie, 1965; Anon, 1974). In most cases the food involved was vegetables processed in boiling water. During the past 30 years, however, some imported commercially canned foods have been involved. They were antipasta and shallots processed in Italy, clams and crabs from Japan, sprats from Germany (type E) and smoked salmon from Labrador (type E). In 1963, tuna fish, commercially canned in California, was responsible for an outbreak of type E botulism in Michigan. In this instance the cans had defective seams and since type E organisms were isolated from post-process handling equipment in the factory concerned, it was assumed that leakage had occurred. The spores of *Clostridium botulinum* type E have a very low heat resistance and would certainly have been destroyed by the heat process the cans received. In the same year another outbreak was traced to liver paste which had been canned in Montreal, the contamination being attributed to under-processing. Some cans contained type A toxin and others type B toxin. This was the position in America until 1971 when two incidents occurred, the first concerned a can of Vichyssoise soup when two people became ill and one died; the soup contained type A toxin. The second incident was the discovery of types A and B toxin in cans of chicken and vegetable soup. In 1973 two outbreaks occurred involving mild peppers and marinated mushrooms and these were followed in 1974 by the discovery of type C toxin in a leaking can of tuna fish. An outbreak involving two people, one of whom died, was attributed to a can of beef stew containing type A toxin. In addition during 1973-1974 the canned mushrooms incidents occurred in which 41 cans were found to contain toxin or viable spores or both, mainly type B. Seven US canners were involved and two foreign producers. The FDA had reacted to some of the earlier incidents and in 1973 the final order of 'Good Manufacturing Practice Regulations for Thermally Processed Low-Acid Canned Foods in Hermetically Sealed Containers' was published. This regulation (Part 128b) gives recommended and required practices for canning operations to ensure the production of safe and wholesome foods. It covers raw materials, hygiene, processing methods, quality control, record keeping and training of operators. In 1974 Part 90 was published requiring the registration of all commercial processors, the filing of processing conditions and the data used in the calculation of F_0 values, the reporting of any spoilage or process deviation likely to cause a health hazard and the organisation of an efficient recall system. In the UK the DHSS, the industry and associated research establishments and other interested parties have since co-operated to produce a Code of Practice for the canners in this country although no cases of botulism have originated from commercial processers in Britain for over 50 years. From 1926 to 1971 approximately 775 billion cans were produced in the USA during which time only 3 outbreaks involving deaths occurred from commercially processed foods (Lynt et al, 1975), figures which perhaps justify the positive reaction of the FDA between 1971 and 1974. The proteolytic strains of *C. botulinum*, that is all strains of type A and some strains of types B and F, are most likely to be found in canned foods because of their greater heat resistance. Such strains usually, but not always, produce foul odours by protein decomposition and there is therefore a very good probability that the contents of the can would not be consumed. The non-proteolytic strains such as type E have a much lower heat resistance and usually only gain access to canned food through accidental leakage of the container. The incidence of non-proteolytic strains in canned foods is likely therefore to be much lower. However, these strains are unlikely to cause any noticeable deterioration of the food and the chances of it being consumed are greater. Heating the food thoroughly before consumption will destroy the toxin but some packs such as fish may be eaten cold. Nevertheless, Tanner & Tanner's (1952) statement that 'the spectacular symptoms, the suddenness with which the symptoms appear and the high death-rate combine to give botulism a place quite out of proportion to its significance as a cause of death', would appear to have a special point where commercially canned foods are concerned.

Cockburn (1960) has summarised data concerning food-poisoning outbreaks, including those attributable to canned foods, in England and Wales during the period 1949 to 1960. The total number of outbreaks due to canned foods was 321, which represented 10 per cent of the total recorded. One hundred and ninety outbreaks were associated with canned meat, 76 with canned fish, 48 with canned vegetables and 7 with canned fruit. Most were due to staphylococci, which accounted for two-thirds of the canned meat outbreaks, three-quarters of the canned fish outbreaks and four-fifths of the canned vegetable outbreaks. *Salmonella* organisms were involved in only two instances.

The low heat resistance of staphylococci, coupled with the destructive effect of modern processing conditions, makes it unlikely that staphylococci or their enterotoxins survive the canning process. Thus, allowing for the fact that some of the recorded outbreaks may have been due to infection after the can was opened, it seems clear that many must have been due to post-process leakage. According to Davidson & Dack (1942), commercial experience is that contamination after processing involves gaseous spoilage. While it is true that in the normal way the majority of leaky cans are eliminated as 'swells', it cannot be assumed that gas-forming bacteria must of necessity be present among the invading bacteria, or that the food is conducive to gas formation. If staphylococcal species effect entrance unaccompanied by gas-forming bacteria, the can may not 'swell' and on opening the contents may appear normal in spite of extensive development of the organisms. In view of the ability of these organisms to form enterotoxic substances, it seems highly probable that some of the earlier cases of canned-food poisoning, where the causal agent was considered to be heat-stable toxin, may have been due to post-process contamination with staphylococci.

An incident occurred in the UK in 1976 in which a man and his wife both became ill after eating sandwiches made from the contents of an imported can of chopped pork. *Staphylococcus aureus* was isolated from both the faeces of the patients and the remains of the pork. Apparently sound cans from the same batch were collected from the retailer and shown to contain *Staph. aureus* producing enterotoxins A and C. Examination of the can seams revealed excessive free space and insufficient overlap indicating that leakage had occurred.

There have been several reports in the literature of outbreaks of *Salmonella* food-poisoning and some of typhoid fever in which imported canned meats have been incriminated. The most spectacular of these was the Aberdeen typhoid outbreak which occurred in the summer of 1964. Over 400 cases were recorded and the source of the infection was attributed to a 6 lb can of corned beef canned in Argentina. The large number of cases involved was due to the fact that ample opportunity existed for cross-infection and growth in the retail shop. It was later established that the can in question was cooled with unchlorinated river water drawn from a point downstream of the town's sewage outfall. Investigations after the outbreak showed conclusively that small numbers of *Salm. typhi* inoculated into cans of corned beef would multiply extensively without producing gas or materially altering the condition of the meat (Lynne & Tomlinson, 1964; Meers & Goode, 1965; Monteverde et al, 1966). It was also shown that mixed cultures of *Salm typhi* and *Escherichia coli* would grow in corned beef without producing gas. Therefore, although all the cans produced are subjected to a factory incubation test it might be possible for a very small number of contaminated cans to remain undetected. Thus the committee set up to investigate the outbreak concluded that the can of corned beef probably became contaminated by leakage due to being cooled in unchlorinated river water.

In connection with post-process leakage, the importance of using cooling water of good bacteriological quality, and of the procedures used for handling cans after cooling cannot be over-emphasised. This subject is discussed in Chapter 6.

Occasionally, canned foods are incriminated in food-poisoning cases which arise through ignorance or negligence in handling the food after opening the can. Where food-poisoning involving canned food is concerned, the possibility of contamination and delayed usage of the contents after opening the container should not be overlooked. Dorling (1942) has reported a fatal case of staphylococcal poisoning caused by canned soup which had been opened and held for one week before consumption. Hobb's (1953) conclusion is that 'canned food is rarely a cause of food-poisoning unless the contents of the tin are contaminated with food-poisoning bacteria after the tin has been opened. The general standard of canning is high and it is unusual to find that either under-processed cans or those which suffer from structural defects have found their way to a kitchen . . . Occasionally, however, cans are badly manufactured, so that the seams leak, while tins stored for a long time may become rusty and develop pinholes through which contaminating organisms pass.'

Food idiosyncrasy

The consumption of certain foods induces illness in persons who are hypersensitive to protein contained in the food which in normal persons is without effect. The symptoms produced may be of a severe nature and are often similar to those occurring in bacterial food-poisoning with the result that illness following a meal is sometimes mistakenly attributed to unsoundness of the food. Among the foods which give rise to illness in hypersensitive persons are fish, especially shell-fish, eggs, cheese, tomatoes and strawberries, but almost any protein-containing food may be responsible. The reactions whereby the symptoms are produced in susceptible subjects are related to immunological reactions involving antigen and antibody, and are fully discussed by Wilson & Miles (1975).

REFERENCES

Anonymous 1974 Botulism in the US, 1899-1973. US Dept Health, Education and Welfare, Public Health Service, DHEW Pub No (CDC) 74-8279
Baird-Parker A C 1962 J appl Bact 25: 12
Bashford T E, Gillespy T G, Tomlinson A J H 1960 Staphylococcal food poisoning associated with processed peas. Fruit and Veg Canning and Quick Freezing Res Ass, Campden
Bergdoll M S, Borja C R, Avena R M 1965 J Bact 90: 1481
Bergdoll M S, Surgalla M J, Dack G M 1959 J Immunology 83: 334
Bergdoll M S, Borja C R, Robbins R, Weiss K F 1971 Infection and Immunity 4: 593
Bott T L, Johnson J, Foster E M, Sugiyama H 1968 J Bact 95: 1542
Bowmer E J 1965 J Milk Fd Technol 28 (3): 74
Bryan F L 1969 J Milk Food Technol 32 (10): 381
Cabassi E, Mori L 1976 Folio Veterinaria Latina 6: 335
Cann D C, Wilson B B, Hobbs, G 1968 J appl Bact 31: 511
Cann D C, Wilson B B, Hobbs G, Shewan J M, Johansen A 1965 J appl Bact 28: 426
Casman E P 1960 J Bact 79: 849
Casman E P, Bennett R W, Dorsey A E, Issa J A 1967 J Bact 94: 1875
Chapman G H 1945 J Bact 50: 201
Christiansen O, Koch S O, Madelung P 1951 Nord Med 3: 194
Cockburn W C 1960 J Roy Soc Hlth 80: 249
Dack G M 1956 Food Poisoning. Univ Chicago Press
Dack G M, Sugiyama H, Owens F J, Kirsner J B 1954 J Infect Dis 94: 34
Davison C M, Boothroyd M, Georgala D K 1966 Nature 212: 1060
Davidson E, Dack G M 1942 Food Res 7: 80
Denny C B 1973 Data presented at ASM Seminar on staphylococcal enterotoxins. Ann Meeting of American Society for Microbiology. Reproduced by Tatini S R in J Milk Food Technol 39 (6): 432

Denny C B, Humber J Y, Bohrer C W 1971 Appl Microbiol 21: 1064
Dible J H 1943 Practitioner 151: 103
Dolman C E, Iida H 1963 Can J Publ Hlth 54: 293
Dorling G C 1942 Lancet 242 (6187): 382
Duncan C L 1973 J Bact 113: 932
Eklund M W, Poysky F T 1970 In: Herzberg M (ed) Toxic micro-organisms. US Dept Interior, Washington DC
van Ermengem E 1897 Ztschr F Hyg u Infektionskr 26: 1
Fernandez E, Tang T, Grecz N 1969 J Gen Microbiol 56: 15
Fujino T, Okuno Y, Nakada D, Aoyama A, Fukai K, Mukai T, Ueho T 1951 J Japan Ass infect Dis 35: 11
Gilbert R J 1974 Postgraduate Medical J 50: 603
Gilbert R J 1978 Food Hyg Lab, London. Personal Communication
Gilbert R J, Wieneke A A 1973 In: Hobbs B C, Christain J H B (eds) The microbiological safety of food. Academic Press, London, p 273
Gimenez D F, Ciccarelli A S 1970 Zbl Bakt 1 Abt Orig 215: 221
Goepfert J M, Spira W M, Kim H U 1972 J Milk Food Technol 35 (4): 213
Gordon R A, Murrell W G 1967 Food Manuf 42 (8): 34
Haines R B 1942 J Hyg 42: 323
Hauge S 1950 Nord hyg tidskr 31: 189
Hauge S 1955 J appl Bact 18: 591
Hauschild A H W, Thatcher F S 1967 J Fd Sci 32: 467
Hauschild A H W, Niild L, Dorward W J 1970 Can J Microbiol 16: 331
Haynes W C, Hucker G J 1944 J Bact 48: 262 (Abs)
Hinshaw W R, McNeil E 1951 Advances in Food Res 3: 209
Hobbs B C 1953 Food foisoning and food hygiene. Edward Arnold, London
Hobbs B C 1954 The bacterial contamination of foods. Food Sc Abs 26: 601
Hobbs B C 1960 J Roy Soc Hlth 80: 267
Hobbs B C 1965 J appl Bact 28: 74
Hobbs B C 1974 Postgraduate Medical J 50: 597
Hobbs G 1976 Adv Food Res 22: 135
Hobbs B C, Gilbert R J 1978 Food poisoning and food hygiene, 4th edn. Arnold, London
Hobbs B C, Kendall M, Gilbert R J 1968 Appl Microbiol 16: 535
Hobbs B C, Smith M E, Oakley C L, Warrack G H, Cruikshank J C 1953 J Hyg 51: 75
Hooper W L, Barrow G I, McNab D J N 1974 Lancet, i: 1100
Hopkins E W, Poland E F 1942 Food Res 7: 414
Humber J Y, Denny C B, Bohrer C W 1975 Appl Microbiol 30: 755
Ingram M, Roberts T A (eds) 1967 Botulism, 1966. Proc of the 5th Int Symp on Food Microbiology, Moscow. Chapman and Hall, London
Ingram M, Robinson R H M 1951 Proc Soc Appl Bact 14: 73
Johnston R W, Feldman J, Sullivan R 1963 Publ Hlth Rep Wash 78 (7): 561
Jones A H, Lochhead A G 1939 Food Res 4: 203
Knox R, MacDonald E K 1943 Med Officer 69: 21
Kravchenko A T, Shishulina L M 1967 In: Ingram M, Roberts T A (eds) Botulism 1966. Chapman and Hall, London
Lee J A 1974 J Hyg Camb 72: 185
Leighton G, Buxton J B 1928 J Hyg 28: 79
Lewis K H, Cassell K (eds) 1964 Botulism. US Department of Health, Education and Welfare. Public Health Service Publication No 999-Fp-1
Lynne P M, Tomlinson A J H 1964 Mon Bull Min Hlth Lab Serv, 23: 147
Lynt R K, Kautter D A, Read R B Jr 1975 J Milk Food Technol 38 (9): 546
Marier R, Wells J G, Swanson R C, Callahan W, Mehlman I J 1973 Lancet ii: 1376
McClung L S 1945 J Bact 50: 229
McCoy J H 1975 J Hyg Camb 74: 271
Meers P D, Goode D 1965 Lancet 1: 426
Meyer K F 1928 Handb d path Mikroorg 4: 1269
Meyer K F, Dubovsky B J 1922 J Inf Dis 31: 600

Meyer K F, Eddie B 1965 Sixty-five years of human botulism in the US and Canada. Epidemiology and tabulation of reported cases 1899-1964. Hooper Foundation, Univ Calif, San Francisco Medical Centre
Midura T F, Nygaard G S, Wood R M, Bodily H L 1972 Apply Microbiol 24: 165
Miyamoto Y, Kato T, Obara Y, Akiyama S, Takizawa K, Yamai S 1969 J Bact 100: 1147
Møller V, Scheibel I 1960 Acta Path Microbiol Scand 48: 80
Monteverde J J, Simeone D H, Moran N, Chialve E J 1966 Revista de Medicina 48 (1): 3
Murrell T G C, Egerton J R, Rampling A, Samels J, Walker P D 1966 J Hyg Camb 64: 375
Odlaug T E, Pflug I J 1977 Appl and Environmental Microbiol 34 (1): 30
Osheroff B J, Slocum G G, Decker D V M 1964 Public Hlth Rep Wash 79 (10): 871
Parker M T 1972 In: Norris J R, Robbins D W (eds) Methods in microbiology, 7B. Academic Press, New York, p 1
Parry E W 1946 Food Res 11: 203
Peffers A S R, Bailey J, Barrow G I, Hobbs B C 1973 Lancet i: 143
Reinke W C, Baker R C 1966 Poultry Sci 45: 1231
Reiser R F, Weiss K F 1969 Appl Microbiol 18: 1041
Riemann H 1973 Can Inst Food Sci Technol J 6 (2): 111
Rigdon R H 1938 Proc Soc Exp Biol and Med 38: 82
Roberts T A, Hobbs G 1968 J appl Bact 31: 75
Sakazaki R, Tamura K, Kato T, Obara Y, Yamai S, Hobo W 1968 Jap J med Sci Biol 21: 325
Savage W G 1921 J Hyg 20: 69
Savage W G 1939 Lancet 237: 991
Schoenholz P, Esty J R, Meyer K F 1923 J Inf Dis 33: 289
Scott W J, Stewart D F 1944 Aust J Council Sci and Ind Res 17: 16
Scott W J, Stewart D F 1945 Aust J Council Sci and Ind Res 18: 173
Segalove M, Davison E, Dack G M 1943 Food Res 8: 54
Slocum G G, Welsh H, Hunter A C 1941 Food Res 6: 179
Smith L DS 1975 The pathogenic anaerobic bacteria, 2nd edn. Chas C Thomas, Springfield
Smith L DS 1977 Botulism, the organism, its toxins, the disease. Thomas, Springfield, Ill
Smith G R, Moryson C J 1975 J Hyg Camb 75: 371
Soo H M, Tatini S R, Bennet R W 1973 Abstr Ann Meeting American Society for Microbiology, p 1
Stone R V 1935 Proc Soc Exp Biol and Med 33: 185
Sugiyama H, Bott T L, Foster E M 1970 In: Herzberg M (ed) Toxic micro-organisms. US Dept Interior, Washington DC
Sugiyama H, Bott T L, Soucheck J A 1972 In: Halvorson H O, Hanson R, Campbell L L (eds) Spores V. American Society for Microbiology
Tanner F W, Tanner L P 1952 Food-borne infections and intoxications. Garrard Press, Champaign, Illinois
Tatini S R 1976 J Milk Food Technol 39 (6): 432
Townsend C T, Yee L, Mercer W A 1954 Food Res 19: 536
Wadsworth C 1957 Int Archives of Allergy and Appl Immunology 10: 355
Wagatsuma S 1968 Media Circle 13: 159
Ward B Q, Carroll B J, Garrett E S, Reese G B 1967 Appl Microbiol 15: 629
Wilson G S, Miles A A 1975 Principles of bacteriology and immunity (Topley and Wilson). Arnold, London
Zeissler J, Rassfeld-Sternberg L 1949 Brit Med J i: 267

15

Laboratory examination of canned foods

Culture media

In the past a great variety of culture media has been used for the bacteriological examination of canned foods. A common practice has been the employment of special media prepared from the food material to be examined but many authorities contend that the nutritional requirements of the micro-organisms concerned with the spoilage of canned foods are adequately served by a relatively few, easily prepared media. This is true once vegetative growth has been established but the recovery of dormant spores, as in bacteriological sterility tests, would seem to impose restrictions on the type of media employed for this purpose. The media discussed here are intended only for the primary isolation of micro-organisms. An extensive study of a particular group of organisms may call for special media, whilst the complete identification involves the use of standard media generally employed for pure culture studies. Details of these are to be found in the many textbooks dealing with practical bacteriology.

Culture media for the primary isolation of micro-organisms from canned foods can be grouped into those used for low-acid and medium-acid foods and those for acid foods. Under these headings, the media commonly used for routine purposes are listed below. Details of other media which have been found useful in the bacteriological examination and investigation of canned products are given under the heading 'Additional Media for Routine and Special Purposes'.

Routine media for low-acid and medium-acid foods (above pH 4.5)

Aerobic or facultative anaerobic bacteria

Dextrose-tryptone medium. Comparative tests by Williams (1936) demonstrated that dextrose-tryptone agar is the most suitable medium for the cultivation of the 'flat-sour' group of thermophilic bacteria. The medium is also satisfactory for the isolation of mesophilic aerobes or facultative anaerobic bacteria such as may be encountered in the examination of low and medium-acid canned products, and its use for routine culture purposes is recommended by Cameron (1936). Bashford (1948) has reported that the addition of ½ to 1 per cent of meat extract greatly improves the dextrose-tryptone medium; the addition of 0.1 per cent of yeast extract is also widely advocated (*see* Tryptone yeast-extract broth). The standard formula and method of preparation is as follows:

Bacto-tryptone	10 g
Dextrose	5 g
Bromocresol purple	0.04 g
Distilled water	1000 ml.

Mix the ingredients and steam until dissolved. Adjust the reaction to pH 6.8 to 7.0, filter and sterilise at 1 bar pressure for 20 minutes. If a solid medium is required 15 g of agar may be added. In view of the findings of Wynne & Foster (1948), the addition of 1 g of soluble starch to each litre of medium is advocated by some workers.

It has been reported by Townsend et al (1954) that some lots of Bromocresol purple are more inhibitory to bacteria than others; this indicator should, therefore, be checked against a spore suspension of known count before use.

In the liquid medium the production of acid from the dextrose by 'flat-sour' organisms causes the colour to change from purple to yellow. On the agar medium, surface colonies of typical 'flat-sour' organisms are about 2 to 5 mm in diameter, have a characteristic opaque centre and are surrounded by a yellow zone. With those types which are weak acid-producers the yellow zone may not be apparent on agar while the liquid medium shows turbidity with little or no colour change. Corpe & Frazier (1950) state that not all 'flat-sour' colonies produce the yellow zone on tryptone agar. Reduction of the tryptone content from 1 to 0.5 per cent is recommended because doing so reduces the buffering capacity of the medium and a larger proportion of the organisms in a 'flat-sour' culture show the typical yellow zone.

Tryptone glucose yeast agar (plate count agar). This is a useful medium for the enumeration of aerobic bacteria and general culture work. It supports a good growth of *Bacillus stearothermophilus* when incubated at 55°C. We use it for the first stage in the preparation of spore suspensions of this organism. (See Additional Media for Routine and Special Purposes.) The formula and preparation are as follows:

Tryptone	5 g
Yeast extract powder	2.5 g
Dextrose	1 g
Agar (Oxoid No. 1)	9 g
Distilled water	1000 ml.

The ingredients are added to the water which is heated to boiling and stirred to obtain thorough mixing. After dispensing into containers the medium is sterilised by autoclaving in steam at 1 bar for 15 minutes. The pH should be approximately 7.0. If a firmer set is required for preparing streak cultures or for washing off, additional agar may be added.

Liver broth medium. It is clearly an advantage in the routine examination of canned foods for the presence of anaerobic bacteria to employ a medium suitable for both mesophilic and thermophilic types. Of several media which have been recommended one which is consistently reliable is liver broth.

To prepare this medium boil 500 g of minced beef liver in 1000 ml of water for one hour. Adjust the reaction of the mixture to pH 7.0 and boil for a further 10 minutes. Strain through muslin and make up the volume to 1000 ml. Add 10 g of peptone, 1 g of dipotassium phosphate, 1 g of soluble starch and adjust the reaction to pH 7.0. When the medium is filled off, place about 1 to 2 cm of the liver particles in the bottom of each tube and cover with the broth to a depth of an inch. Sterilise at 1 bar pressure for 20 minutes. Before inoculation, this medium must be steamed for 20 minutes to remove dissolved air. After cooling and inoculation the surface of the medium is layered with a sterile petroleum jelly-paraffin wax mixture or sterile 3 per cent plain agar (cooled to 50°C). Amaha (1953) has reported advantages in the use of agar containing 0.2 per cent sodium thioglycollate for stratification of liver broth cultures.

Reinforced clostridial medium. This medium was designed by Hirsch & Grinstead (1954) for the cultivation of *Clostridium butyricum* but is equally suitable for other *Clostridium* spp. According to Gibbs & Hirsch (1956) the solidified medium compares favourably with Pork-pea Infusion Agar for counting anaerobes.

Yeast extract	3 g
Beef extract ('Lab-Lemco')	10 g
Peptone	10 g
Soluble starch	1 g
Dextrose	5 g
Cysteine hydrochloride	0.5 g
Sodium acetate, anhyd.	3 g

Sodium chloride	5 g
Agar	0.5 g (15 g)
Distilled water	1000 ml.

The medium is steamed to dissolve the ingredients, filtered through paper pulp and the pH adjusted to 7.4. It is sterilised by autoclaving at 1 bar pressure for 15 minutes.

Gibbs & Freame (1965) have modified this medium and to make it more selective have added sodium sulphite 0.04 per cent, and ferric citrate 0.07 per cent, the result being termed Differential Reinforced Clostridial Medium (DRCM). It is claimed by these authors that the fluid medium gives very good recoveries of clostridial spores from food samples and that the spores may be enumerated using dilution counts by the most probable number (MPN) technique. We recommend this medium for the detection of anaerobic bacteria.

Tryptone yeast-extract broth. This medium is suitable for the detection of anaerobes in the routine examination of canned foods.

Tryptone	10 g
Dextrose	5 g
Yeast extract	1 g
Dipotassium phosphate	1.25 g
Distilled water	1000ml.

Adjust the reaction to approximately pH 6.8, tube in 8 to 10 ml amounts and sterilise at 1 bar pressure for 20 minutes. Before use the tubes should be exhausted in flowing steam, and, after inoculation, they should be layered with a medium made from 0.5 per cent tryptone and 2 per cent agar. When used for thermophilic anaerobes the stratifying layer is first solidified by cooling and the tubes reheated in a water bath to 55°C before incubation.

Yeast-extract thioglycollate starch agar. For recovery and counting of clostridial spores Wheaton & Pratt (1961) and Wheaton et al (1959) demonstrated that, in general, sub-culture media prepared from fresh-meat infusions are preferred. Nevertheless, these media are troublesome to prepare and the possibility of using simpler media for routine purposes is attractive. One such medium which has proved popular in Europe is that proposed by Wynne et al (1955).

Yeast extract (Difco)	10 g
Thioglycollate supplement (BBL)	5 g
Soluble starch (Difco)	1 g
Dipotassium phosphate	2 g
Agar	15 g
Distilled water	1000 ml.

The medium is prepared in the normal manner, the pH adjusted to 7.4 and then tubed in 15 ml amounts. 0.3 ml of a Seitz-filtered solution (5 per cent) of sodium bicarbonate may be added to each of the tubes prior to use.

The medium is not specific for *Clostridia* since it also permits the growth of a few facultative anaerobes. However, it is pointed out by Mossel & Drion (1954) that this is not important in the sterility testing of canned foods because, in most cases, the surviving bacteria are of equal significance.

Routine media for acid foods (below pH 4.5)

Bacteria

The significant spore-forming organisms which fall under this heading are those of the *Clostridium pasteurianum* type and the 'flat-sour' thermophile — *Bacillus coagulans.*

Proteose peptone-acid medium. This medium known as Thermoacidurans Agar was used by Stern et al (1942). They found that the aciduric 'flat-sour' thermophile, *Bacillus thermoacidurans (B. coagulans)* grew poorly on most media. For the isolation and cultivation of this organism, they developed a medium which has the following composition:

Proteose peptone (Difco)	5 g
Yeast extract (Difco)	5 g
Glucose	5 g
Dipotassium phosphate	4 g
Distilled water	500 ml.

The ingredients are dissolved in the water and the pH adjusted to 5.0 with dilute hydrochloric acid. In a separate flask 20 g of agar are dissolved in 500 ml of water and then both solutions are sterilised at 1 bar pressure for 30 minutes. When the medium is ready to be used, the melted agar is mixed with the sterilised nutrients at 45° to 55°C. Incubate the inoculated medium at 55°C.

Liver broth with crushed tomatoes. This medium is suitable for the isolation of butyric anaerobes and other acid-tolerant bacteria in canned foods. It is prepared by mixing equal amounts of liver broth and crushed tomatoes. For the production of spores the addition of 10 per cent soil is recommended. Sterilisation is for 20 minutes at 1 bar pressure.

Orange-serum medium. This medium appears to have been developed independently by Hays (1951) and by Troy & Beisel, as reported by Murdock et al (1952). It is recommended by the American

Public Health Association (1976) as a routine medium for isolating acid tolerant organisms from canned foods. It gives good growth with *Bacillus coagulans*, lactobacilli and butyric anaerobes. The formula and method of preparation is as follows:

Tryptone	10 g
Yeast extract	3 g
Dextrose	4 g
Dipotassium phosphate	3 g
Orange serum	200 ml
(Agar)	17 g
Distilled water	800 ml.

Prepare the orange serum by heating 1 litre of freshly extracted orange juice to about 93°C. Add 30 g of filter aid, mix and filter through a Buchner funnel using suction.

Mix the ingredients and steam until dissolved. Sterilise at 1 bar pressure for 15 minutes. The pH should be about 5.5.

MRS broth. This medium was developed by de *M*an, *R*ogosa and *S*harpe (1960) for the cultivation of lactobacilli. It has the following composition:

Peptone (Oxoid L34)	10 g
'Lab-Lemco' powder (Oxoid L29)	8 g
Yeast extract	4 g
Dextrose	20 g
'Tween' 80	1 ml
Dipotassium hydrogen phosphate	2 g
Sodium acetate $3H_2O$	5 g
Triammonium citrate	2 g
Magnesium sulphate $7H_2O$	0.2 g
Manganese sulphate $4H_2O$	0.05 g
Distilled water	1000 ml.

Add the above ingredients to the distilled water maintained at 60°C and mix until completely dissolved. Adjust the pH to 6.2, dispense into tubes and sterilise in steam at 1 bar pressure for 15 minutes. To prepare the solid medium add 1 per cent agar, steam until dissolved, dispense and sterilise in the same manner.

Moulds and yeasts

Malt-extract medium. Linden (1936) recommends a malt-extract medium for yeasts and moulds. It is prepared as follows:

Malt extract (Difco)	100 g
Distilled water	1000 ml.

Dissolve the powdered malt extract by steaming. Adjust the reaction to pH 4.7 and cool to 50°C. Add slowly 100 ml of a 5 per cent suspension of Bentonite and mix vigorously. Hold at 50° to 75°C for 30 minutes, then filter through paper till clear. Heat for 10 minutes at 1 bar steam pressure and filter through paper to remove any precipitate. Sterilise at 0.7 bar pressure for 15 minutes and cool rapidly. For plating purposes add 2 per cent agar.

Potato dextrose agar. This medium is recommended by the APHA (1976) for the isolation of heat resistant moulds such as *Byssochlamys fulva*. The formula and method of preparation are as follows:

Potato extract	4 g
Dextrose	20 g
Agar	15 g
Distilled water	1000 ml.

Add the ingredients to the water and boil to dissolve. Dispense into tubes or bottles and sterilise at 1 bar steam pressure for 15 minutes. When required for use the medium is steamed until melted, cooled to 50°C and then acidified to pH 3.5 by adding 1 ml of sterile 10 per cent tartaric acid solution to each 100 ml of molten medium prior to pouring the plates.

Additional media for routine and special purposes

Iron-sulphite agar. For the detection of thermophilic anaerobes producing hydrogen sulphide a modification of the medium recommended by Cameron (1938) is used:

Tryptone	10 g
Sodium sulphite, anhyd.	0.5 g
Iron citrate	0.5 g
Agar	15 g
Distilled water	1000 ml.

Since it has been demonstrated that 0.1 per cent sulphite is inhibitory to certain bacteria the concentration of sodium sulphite has been reduced to 50 per cent of the level used in the original NCA medium.

The medium is tubed and used for deep-shake cultures. The type species, *Desulphotomaculum nigrificans*, produces distinct black spherical growths in the agar. Alternatively the medium may be used in Miller-Prickitt tubes. These tubes are oval in cross-section and facilitate the counting of colonies.

The problem of sulphite concentration in media for clostridia has been studied by Put (1961), who demonstrated that it should not exceed 0.05 per cent and that 0.03 per cent was probably better. She worked with a Wilson and Blair medium enriched with an equal volume of tomato juice or liver broth. By using two sulphite levels, 0.1 per cent and 0.3 per cent, she was able to divide the clostridia into two groups, sulphite resistant and sulphite sensitive respectively.

Pork-pea infusion medium. Andersen (1951) describes a rapid plate method of counting spores of *Clostridium botulinum* using an anaerobic Petri dish and a medium designed to give reliable germination of the spores. A feature of this medium is the incorporation of bicarbonate ions which were shown to be essential for the prompt and uniform development of colonies of *C. botulinum.* The basic medium without the addition of sodium bicarbonate is a good general anaerobic enrichment medium. The formula and method of preparation is as follows:

Pork infusion	800 ml
Pea infusion	200 ml
Peptone	5 g
Tryptone	1.6 g
K_2HPO_4	1.25 g
Soluble starch	1.0 g
Sodium thioglycollate	0.5 g
Agar	16.0 g.

(Adjust pH to 7.2. Omit agar for broth or reduce to 0.75 per cent for-semi-solid medium.)

Autoclave for 5 minutes at 1 bar pressure and clarify by adding 25 g of diatomaceous earth and filtering. The tubed media is sterilised for 15 minutes at 1 bar pressure.

Bacillus sporulation medium. We have found the following media to be very useful when preparing spore crops of *Bacillus stearothermophilus* with a high heat resistance for use in spore bulbs or inoculation trials. The organism is first inoculated on to plate count agar (PCA) and incubated at 55°C. When a good growth is obtained on the PCA it is removed with sterile distilled water whilst gently rubbing with a sterile swab. The resultant suspension is then centrifuged at 3000 r.p.m. for 3 minutes and the deposit re-suspended in a small volume of sterile distilled water. The slurry thus produced is used to heavily inoculate plates which have been poured with a medium having the following composition:

Beef extract	1.5 g
Yeast extract	3 g
Peptone	6 g

Tryptone	3 g
Agar	25 g
Distilled water	1000 ml.

Steam to dissolve ingredients, adjust pH to 7.0 and sterilise at 1 bar steam pressure for 15 minutes.

The inoculated plates are incubated at 55°C for 2 days after which the growth is harvested as before, pasteurised, and used to inoculate a second series of plates poured with a medium which is identical to the one above but which contains an addition of 1 p.p.m. of Mn^{++} ions, added either as the sulphate or chloride. These plates are incubated at 60°C and checked for sporulation after 2 to 3 days.

Bacillus spore recovery medium. For the enumeration of *Bacillus stearothermophilus* spores which have survived a heat process, e.g. in spore bulbs, we have found the following medium to give good results.

Beef extract	1.5 g
Yeast extract	3 g
Casitone	4 g
Peptone	6 g
Dextrose	1 g
Soluble starch	1 g
Agar	20 g
Distilled water	1000 ml.

Steam to dissolve ingredients, adjust pH to 7.3 and sterilise at 1 bar steam pressure for 15 minutes.

Nitrate-sugar-ham medium. Jensen (1954) gives details of a medium for the detection of gas-forming organisms of the *Bacillus* group which are responsible for spoilage in canned cured meats. Tubes are approximately half filled with finely chopped or ground spiced ham (taken from a sound can) and then wetted with a solution containing 0.4 per cent sodium nitrate and 2 per cent sucrose. After autoclaving at 1 bar pressure for 15 minutes, the tubes should appear as follows:

No trapped bubbles of air should appear in the medium. Five mm of liquid should cover the top of the meat.

A layer of fat from the spiced ham should make a seal which will prevent evaporation on storage. The medium should not occupy more than half the volume of the tube.

The medium is inoculated by introducing the inoculum with pipettes to the bottom of the tubes. Care should be taken that no air bubbles are allowed to remain in the tubes after inoculation. Incubate at 37°C and 49°C for 48 hours. Gas production is more rapid at the latter temperature.

'Micro-ham' medium. Buttiaux & Beerens (1955) have described a medium (based on the composition of canned ham) designed for the study of bacteria isolated from semi-preserved meat packs. Gelatine is included in the medium since Beerens (1955) has shown that some species of clostridia (e.g., *C. histolyticum*) do not produce gas in normal media but will do so in the presence of gelatine.

The composition and preparation of the medium is as follows:

(i) Proteose peptone	10 g
Meat extract	4 g
Yeast extract	3 g
Dipotassium phosphate	5 g
Gelatine	200 g
Distilled water	1000 ml.

The ingredients are dissolved in a water bath at 50°C and sodium chloride added at half the concentration present in the ham (if this is unknown 2 per cent salt is added). Adjust pH to 6.5.

(ii) Fifteen-millimetre cubes of ham or luncheon meat are placed in 16-mm tubes with a short piece of iron wire; these are autoclaved for 25 minutes at 120°C. Ten millilitres of the above gelatine base are poured into each tube and the whole sterilised for 15 minutes at 120°C.

(iii) At the moment of use add to each tube enough sterile potassium nitrite solution to bring the final concentration to 0.05 per cent and add 1 ml of sterile tomato extract.

When testing *Clostridium* species, 0.1 per cent dextrose should be added to the medium. The medium should be 'regenerated' in a boiling water bath immediately before use and 2 per cent plain agar used to layer the gelatine.

Incubation for at least 7 days at 37°C will show gas formation with bacteria that cause swelling of cans of semi-preserved products.

Osmophilic yeast media. For the detection of osmophilic yeasts, Lochhead & Farrell (1930) used a honey medium which is prepared as follows:

Honey	800 g
Peptone	1 g
Dipotassium phosphate	1 g
Magnesium sulphate	0.5 g
Ammonium tartrate	0.5 g
Sodium chloride	0.1 g
Calcium chloride	0.1 g
Distilled water to	1000 ml.

Mix the ingredients, the honey being previously warmed. Sterilise at 1 bar pressure for 20 minutes. The reaction of this medium is approximately pH 4.2.

For the isolation of osmophilic yeasts in sugar products Scarr (1959) uses a medium made by dissolving Difco wort agar in a 45 per cent w/w syrup consisting of 35 parts of sucrose and 10 parts of glucose. The medium is sterilised by autoclaving at 0.33 bar pressure for 20 minutes. *Zygosaccharomyces* spp. are usually evident after 72 to 96 hours incubation and less common osmophiles within 5 days.

Note. Some of the above media can be obtained in dehydrated form from Oxoid Limited, Wade Road, Basingstoke, Hants., RG 24 OPW (Oxoid); Difco Laboratories Limited, Central Avenue, West Molesey, Surrey, KT8 OSE (Difco); Becton Dickinson (UK) Limited, York House, Empire Way, Wembley, Middlesex, HA9 OPS (BBL).

Outline of examination procedure

The culturing of processed cans to determine the types and numbers of organisms which may be surviving the heat treatment provides useful information when new products or different processing methods are being introduced into the factory. Investigations of spoilage problems will also require the preparation of cultures from both spoiled and sound cans taken from the same batch in order that comparisons can be made of their respective flora. However, apart from the occasional search for specific organisms, as for example in any food poisoning investigation, the above situations are the only times in which it is normally necessary, or in fact advisable, to culture processed cans. In addition to being very time consuming, complete microbiological analyses of relatively small numbers of cans will provide insufficient data for the routine control of factory processes.

For those situations in which it is either necessary or advantageous to prepare cultures the method of examination will depend on the nature of the canned food. For low- and medium-acid packs, that is those with a pH above 4.5, the heat process is usually theoretically adequate for the destruction of all micro-organisms except a few of extreme heat resistance and these are not significant from the point of view of public health. For this class of product, the presence or absence of potential spoilage organisms is the main issue, and the examination is usually (1) to test the adequacy of the heat-process or (2) to determine spoilage causes. There is, however, the type of product such as cured or pickled meats containing curing salts, which, because of their composition permit the use of a less stringent process. Such products are normally unsterile and the examination involves the isolation and classification of the organisms present. In this country most canned products come within the definition for the first group and it to this group that the methods of examination given here mainly apply.

Size and selection of sample

In connection with sterility tests on canned foods, a point which frequently receives insufficient attention is the selection of samples. A negative result on a single sample is without significance except for that particular sample, since sterile cans are frequently present in under-processed batches. Sufficient cans should be examined to give a reasonable chance (say 95 per cent) of detecting abnormalities in the population from which they have been drawn. Unfortunately, unless the proportion of unsterile containers in the batch is very high, this involves taking a large number of cans for examination. The number of cans, drawn at random, required to detect various levels of spoilage is shown in Table 12 published by the National Canners' Association (1960).

Table 12 Sampling rates to detect various levels of spoilage

%Spoilage in lot	Number of cans required Probability[1] 0.95	Probability[1] 0.99
0.5	597	919
1	298	458
2	148	228
3	99	151
4	74	113
5	58	90
6	48	74
7	41	64
8	36	55
9	32	49
10	28	44
11	26	40
12	24	36
13	22	33
14	20	31
15	18	28
16	17	26
17	16	25
18	15	23
19	14	22
20	13	21
25	10	16

[1] Probability that at least one can in the incubated sample will show spoilage

Thomas & Cheftel (1955) and Scott (1953) discuss the taking of samples for assessment of the microbiological quality of canned goods and give further tables concerned with the size of samples.

This subject is also discussed in '*Micro-organisms in Foods*' volume 2 (IAMS, 1974).

For satisfactory control it follows that sterility tests should be the concern of the factory laboratory where an adequate number of samples are available for testing. Furthermore, these should be supplemented by a knowledge of heat-penetration data, process time and temperature conditions and the results of experimental inoculation of cans with organisms of known resistance. Under these conditions regular testing of smaller samples from successive batches provides an adequate safeguard. When information on the history of the product is not available it is usually necessary to reach some form of compromise between adequate sampling and the dictates of economics. For such cases the methods of sampling have been reviewed by Mossel & Drion (1958).

The cans examined should be truly representative of the production. In batch processes samples should be drawn in a random manner with respect to disposition within the retort and from continuous sterilisers in a random manner with respect to time of sampling. Inadequate sampling because of size or a non-random nature may lead to erroneous conclusions not only in respect of the extent of abnormalities but also of the cause of spoilage. For example, in a batch of cans which have spoiled as the result of underprocessing there may be a few cans which have become infected through leakage, and if such cans are taken as representative of the batch, the presence of typical leakage organisms (cocci, non-sporing rods) may give rise to confusion.

When taking samples of blown cans these should be collected, if possible, during the early stages of spoilage. The development of high internal pressures is liable to affect seam characteristics as well as falsifying the results of any air leakage tests subsequently made on the opened containers. It is important also to collect samples of sound cans to provide reference samples on the general quality level of the product and the container.

Preliminary external examination

After recording details of code, can size and net weight the container should be examined for obvious mechanical defects in the seams, perforations, rust, damage, 'peaks' or other abnormalities which may be usefully correlated with the bacteriological findings.

It is advisable also at this juncture to carry out measurements of external seam dimensions (*see* Ch. 17) since these may be useful later in the diagnosis of leaker spoilage.

Preliminary incubation of samples

Except in the case of swelled cans, all samples should be incubated before cultures are made. The organisms which survive heat-processing may be extremely few in number and the purpose of preliminary incubation is to stimulate multiplication of the survivors and thus increase the chance of securing them in the inoculum. In view of the fact that sub-lethal exposure to heat may cause dormancy of spores, the time allowed for preliminary incubation is important. Various periods, ranging from one to several weeks, have been recommended, but it frequently occurs that the urgency of the examination is the controlling factor. Unless further samples are available for prolonged incubation tests, a minimum period of one week at 37°C is desirable for low- and medium-acid products. Some workers e.g., Mossel & van Diepen (1956), have suggested the use of 30° to 32°C as a pre-incubation temperature for canned foods and this has merits for certain classes of product. For the type of canned food considered here, that is those normally rendered free from mesophilic spores by the heat treatment, we continue to use 37°C as a pre-incubation temperature except when leaker spoilage is suspected. This is mainly in the interest of obtaining results more quickly. With the supplementary incubation tests, for which the times involved are longer, the use of 30° to 32°C may be advantageous in certain circumstances. In addition, samples should be examined for thermophilic bacteria, but in view of the tendency of these to undergo rapid auto-sterilisation the samples should not be incubated for longer than 4 or 5 days at 55°C. Acid products may usefully be given 10 days' preliminary incubation at 25°C; they should not be incubated at higher temperatures since the activity of the acid may be increased to a point where it offsets any stimulative effect which the increased temperature may have on the development of the organisms. A similar proviso is given for any product known to contain preservative substances such as sulphur dioxide.

Sampling can contents

In the preparation of cultures from canned products the main points at which contamination may occur are during the opening of the container and the removal and transfer of the inoculum to the medium. In order to reduce the risk of contamination to a minimum the use of a laminar air-flow cabinet is strongly recommended. Such a cabinet manufactured to Class 100 (*see* p. 268) will provide a working area free of particles of 0.5 μm or larger with an efficiency of 99.997 per cent. The interior of the cabinet, but not the filter screen, should be thoroughly cleaned and treated with a bactericidal solution and then the fans should be operated for at least half an hour before the cabinet is used. The opening of 'hard swells' in these cabinets is not advisable because of the danger of contaminating the filters. If a laminar air-flow cabinet is not available the operations should be conducted in a still, dust-free atmosphere, preferably in a small room reserved for this purpose and which can be easily cleaned and disinfected. The importance of contamination, already discussed in Chapter 13, cannot be over-emphasised, and all possible precautions should be observed.

The can should first be thoroughly washed with hot soapy water or a detergent solution, rinsed and then immersed in a solution of a good bactericide for 10 to 15 minutes. In the case of round sanitary cans, the opening is made at one end (the plain end without an embossed code if possible) but flat oval cans should be pierced on the body, there being a tendency for the contents to flood out of these cans when pierced at the top or bottom. Easy-open ring pull cans should be pierced at the plain end leaving the score mark intact for possible leakage examination later. The area to be pierced is flooded with 70 per cent alcohol which is allowed to act for a minute or two and the excess is then removed with a squeezed-out swab. The can is dried by burning off the residue and the area is then flamed gently by rotating the can slowly over a bunsen flame. The end of the can is then covered with a sterile half Petri dish.

Flexible pouches are supported vertically in a suitable rack, sterilised externally with a 50:50 alcohol/ether mixture, allowed to dry and then opened by cutting off one of the corners with a pair of sterilised scissors.

If knowledge of the vacuum in the can is desired, this may be obtained by means of a specially fitted vacuum gauge. A suitable piece of apparatus for this purpose has been described by Dilley & Everton (1966). Their design, although basically similar to earlier versions, allows both easier operation and preparation of pre-sterilised components. Essentially the equipment consists of an adapter block into which is sealed a vacuum gauge and a detachable piercing unit, a small valve is located in the side of the block. Non-absorbent cotton wool is used to plug the bore of the piercer which is sterilised by autoclaving before use. The adapter block with its vacuum gauge and piercer is held by a bridge which is supported between two adjustable uprights; a cam operated lift pad forms the base of the apparatus. To make a vacuum determination the uprights are adjusted for correct container height and a clean, prepared can is placed on the lift pad. By operating a lever the can is brought into contact with the piercer and the vacuum noted. The small bleed valve on the side of the adapter block is now opened and the vacuum neutralised by filtered air flowing into the can through the cotton wool plug in the piercer. The apparatus is shown in Figure 25.

Fig. 25 Vacuum gauge (courtesy of Metal Box Co Ltd)

The can may be opened for sampling in a number of ways and the following method is quite suitable:

1. For liquid products cut a hole about 1.27 cm in diameter with a tapered punch such as that described by Bashford (1940). The tool is sterilised by flaming before use and when the puncture is made, the air replacing the vacuum in the can (assuming the vacuum test involving perforation has not been made) is sterilised by directing a Bunsen flame to the point of the tool (Fig. 26).

2. For solid or semi-solid products make a small central puncture observing the precautions as above and then cut out a circular disc with a sterilised circular-type can opener.

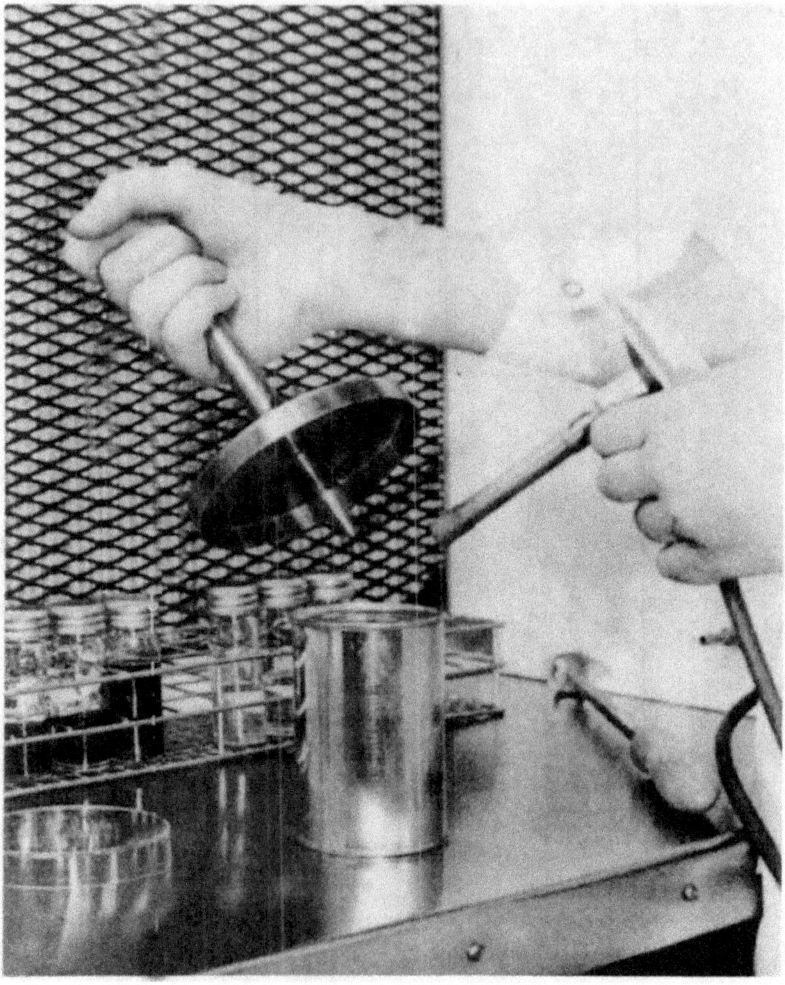

Fig. 26 Aseptic sampling of can contents

The method of removing the inoculum and the site from which it is taken depends on the nature of the product. With solid packs, a cylinder is removed with a sterile cork borer which is firmly plugged at the top with cotton wool. A rod to expel the sample from the borer is necessary. A fresh, sterile borer is required for each can, but the expelling rod may be used repeatedly, since it is prevented from making contact with the sample by the sterile cotton plug in the borer. In sterility tests on solid packs the inoculum is taken from the centre, since this portion receives least heat-treatment and viable organisms are, therefore, most likely to be found at this point. The sample is expelled from the borer into a wide-mouth bottle (100 ml capacity) containing about 50 ml of sterile 0.1 per cent peptone water and some chips of broken glass. Where the product contains fat or oil, the emulsifying diluent used by Williams & Clark (1942) may be employed. This consists of a sterile solution of a mixture of gum tragacanth and gum arabic (2 and 1 per cent respectively). At least 25 g of the sample is broken up by vigorous shaking to provide a suitable suspension for inoculating, by means of a sterile pipette, the appropriate media. For preparing suspensions, screw capped bottles have been found most useful; they are much superior to flasks, since they can be shaken more vigorously.

When examining solid packs for spoilage, it is important that the sample taken should include material from the exterior of the can contents; whereas in sterility tests the organisms are most likely to be found in the centre of the product, they will probably be confined to the surface if the can has leaked for any reason. In such cases, therefore, sampling should be carried out by pushing the borer vertically and obliquely in several directions to make contact with the container, thus ensuring that material is obtained from the surfaces of the contents adjacent to both end seams and the side seam. When score lines are present, for example in certain tear open corned beef cans, it is also important to obtain surface samples from the region of the score marks.

With liquid products or those in which there are solid and liquid constituents, the liquid portion should be sampled, using large (20 ml) pipettes and inoculated direct into the media. The bore at the tips of the pipettes should not be too narrow. For semi-solid or viscous products lengths of untapered glass tubing (4 mm bore) are convenient for sampling.

When a large number of samples have to be examined it may be necessary to modify slightly the above methods which in these circumstances would necessitate the use of a large number of sterile Petri dishes and borers. As substitutes for Petri dishes, sterilised circles of tinplate may be used for protecting the can ends after sterilisation. With solid products, the can is punctured in the manner stated; the can is then opened with a flame-sterilised can opener of the blade cutter type, care being taken to avoid damaging the seam. The top portion of the contents is removed with a flamed knife and a sample from the centre is taken with a flamed spatula, dispersed in sterile 0.1 per cent peptone water and inoculated into the various media. Liquid products do not present any problem since it is usually possible to maintain an ample stock of sterile glass pipettes or tubing.

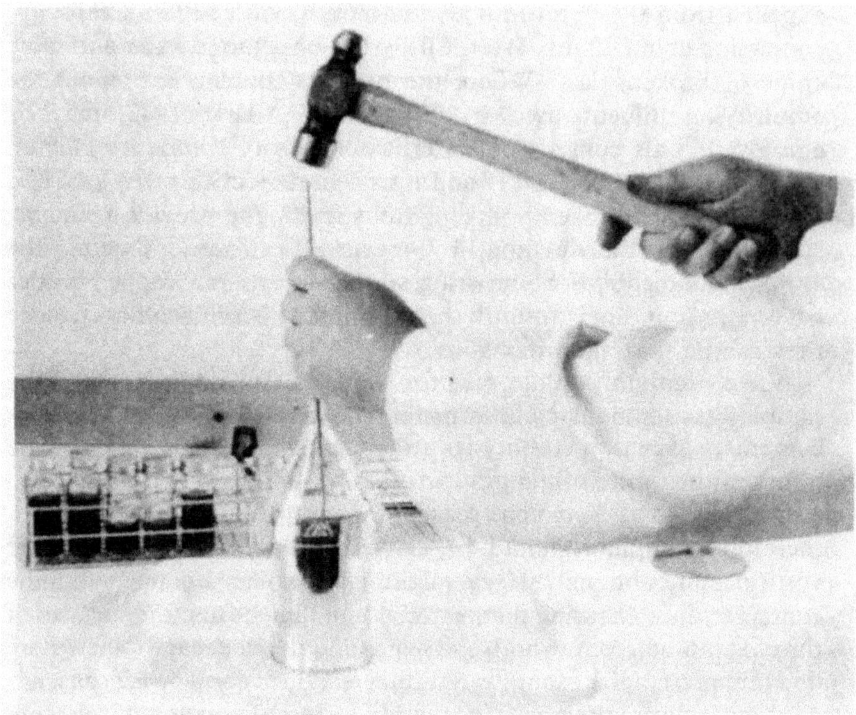

Fig. 27 Sampling the contents of a blown can

Technique for sampling swelled cans

When sampling swelled cans, some precautions must be taken to minimise the spurting of the contents. An apparatus, originated by Cheftel (1935), giving excellent protection to the operator, consists of a large glass filter funnel, firmly plugged in the stem with cotton wool through which is passed a thin steel rod with a sharpened tip. The whole apparatus is wrapped in stout brown paper and sterilised under steam pressure. In use, the funnel is inverted over the previously cleaned and sterilised can, the sharp tip of the rod resting on the can end (*see* Fig. 27). A blow on the blunt end of the rod forces the tip through the can, the funnel preventing the spurting contents from contaminating the operator. The can should stand in a shallow tray of dilute hypochlorite to receive any material which may drip from the funnel. When the pressure has subsided the inoculum is withdrawn in the usual way.

Preparation of cultures

The cultures prepared and the choice of media will depend on the objects of the investigation and the type of product concerned. In certain circumstances it is advisable to prepare a number of each type of culture, as, for example, when the preliminary incubation period has been unavoidably shortened and the possibility exists that organisms may not be secured in a single culture.

The usual spoilage types of organisms are not particularly fastidious in their growth requirements and elaborate sub-culture media are not generally used. The amount of inoculum will depend on whether the can contents are already spoiled or whether the search is for viable but dormant organisms as in sterility tests. In the former instance 1 g of inoculum per 10 ml of medium is adequate, but, for sterility tests the amount should be increased.

The system of cultures shown in Table 13 may serve as a guide.

Table 13 System for culturing canned foods

Type of product	Media	Incubation conditions
Low- and medium-acid foods	Dextrose-tryptone broth	Aerobic 37°C 48 to 72 hr
	Dextrose-tryptone broth	Aerobic 55°C 48 to 72 hr
	Differential reinforced clostridial medium or liver broth or tryptone-yeast extract broth	Anaerobic 37°C 48 to 72 hr
	Differential reinforced clostridial medium or liver broth or tryptone-yeast extract broth	Anaerobic 55°C 48 to 72 hr
Acid foods	Orange-serum broth	Aerobic 37°C 72 hr
	Liver broth with crushed tomatoes	Anaerobic 30°C 72 to 96 hr
	Malt-extract broth or wort broth	Aerobic 25° to 30°C, 7 days
	Proteose-peptone yeast agar	Aerobic 37°C 48 to 72 hr
	Proteose-peptone yeast agar	Aerobic 55°C 48 to 72 hr

Micro-examination of contents

After preparation of the cultures, smears of the contents stained with gentian violet stain or methylene blue stain are examined microscopically, or alternatively a wet preparation may be examined under phase contrast. This examination may serve as a check on the results obtained from the culture tests, but it should be noted that correct interpretation of the microscopic evidence is sometimes difficult. It has already been mentioned that normal unspoiled cans may contain large numbers of non-viable organisms and that this condition may be due to the poor hygienic quality of the raw materials or to development of organisms during the period between the preparation of the food and heat-processing. Occasionally, cultures prepared from spoiled cans are sterile, although microscopically large numbers of organisms may be visible in the contents. Such cases are due to auto-sterilisation of organisms, which usually stain poorly or unevenly and show other signs of degeneration. Positive cultures are not often obtained from incubated cans which show no organisms in direct smears unless the product contains substances inhibitory to growth, as, for example, curing salts.

Hydrogen-ion concentration

The reaction of the can contents should be determined, particularly where the presence of 'flat-sour' organisms is suspected. For comparison, it is necessary that the normal pH of the food be known. Although reasonably accurate results are obtainable for most products by colorimetric methods, pH determinations are most satisfactorily made with a glass electrode.

Condition of contents

The odour, colour and texture of the can contents and the presence of any breakdown of gel structure should be noted. Extra firm vegetables and hard fish bones are two conditions which may indicate gross under-processing or possibly non-processing, whilst the internal surfaces of can bodies, when unlacquered, will usually remain bright and shiny in non-processed cans.

Detailed examination of container

Finally a complete examination of the container should be made. An outline of the procedure for stripping can seams is given later.

Examination of glass packs

The bacteriological examination of heat-preserved foods packed in glass is made on lines similar to that for canned foods. Sampling of these products is simplified as the metal cap is easily removed, but precautions must be taken in neutralising the vacuum. Caps should be sterilised before piercing with a sterile tool through a Bunsen flame. This procedure is essential for vacuum-closed packs and is desirable even for packs with non-vacuum type caps closed at atmospheric pressure, since a moderate vacuum is frequently formed automatically with such closures. To avoid disturbing the gasket, the position of which may have an important bearing on the results of the examination, the caps should be removed as carefully as possible. In the examination of the container attention is directed to the following points which are the main causes of leakage:

1. Defects or chipping of the glass rim resulting in imperfect seating of the rubber gasket or flowed-in compound
2. Incorrect placing of the paper liner so that it laps over the rim of the glass, thus preventing proper seating of the gasket
3. Faulty or uneven distribution of the lining compound resulting in uneven seating on the lip of the jar
4. Twisting of the gasket
5. Perforation of the metal cap.

Incubation tests (cans and glass packs).
It has already been mentioned earlier in this chapter that culturing a relatively small number of cans will provide insufficient information for the routine monitoring of sterilisation processes. Furthermore, such 'end product' examination should never be regarded as a substitute for an 'on-line' process control system. An effectively organised control system will ensure that all parameters capable of influencing the sterilisation process such as filling temperatures, net weights, process temperatures and times and conditions of cooling, etc. are continuously checked. Such a control system will often enable minor deviations, which may occur in these critical parameters, to be corrected during the production run, thus preventing the manufacture of suspect cans.

Where absolute sterility has not been attained it is clearly a matter of importance that the keeping properties of the food be examined. Because the only effective means of checking 'commercial sterility' is to determine whether or not the organisms are capable of developing in the unopened can when held at a suitable temperature, incubation tests are recommended for the routine examination of finished stocks. The use of incubation tests eliminates the risk of contamination which is always present in culture preparation work and which could lead to difficulties over interpretation of the results. However, the great advantage of incubation tests is that they enable non-destructive examinations to be made on a statistically significant number of cans. For routine purposes incubation tests normally operate at two levels, a laboratory procedure involving a small number of cans which are opened for examination after the period of incubation and a factory test in which a larger number of cans are incubated, generally at a lower temperature, examined visually for any blown or leaking cans and then returned to stock.

One suggested laboratory procedure for low-acid and medium-acid products is to hold them at 37°C and 55°C for at least 30 and 20 days respectively; these represent minimum periods, and where possible they can with advantage be extended. Since some micro-organisms found in air and water grow poorly, or not at all, at 37°C such a system cannot be relied upon to indicate leaker spoilage. For the detection of spoilage due to leakage additional cans should be held at 25°C for 10 days. Mossel & van Deipen (1956) have demonstrated that an incubation temperature of 30° to 32°C is adequate for mesophilic spoilage bacteria and many workers favour the use of this temperature. Since leaker spoilage is normally also evident under these incubation conditions, a temperature of 30° to 32°C is recommended for the factory incubation procedure which involves a larger number of cans and where, in the interests of economy, a single incubation temperature may be preferred. The cans should be held at 30°C for a minimum period of 10 days.

Commercially sterile packs may contain viable spores of *Bacillus stearothermophilus*. However, provided the cans are properly and quickly cooled (below 40°C) before casing, it is our experience that the presence of these spores does not give rise to spoilage under storage conditions normally experienced in the UK. By selecting an incubation temperature of 40°C it is possible to detect thermophiles of commercial spoilage significance and exclude the growth of these non-significant strains of *B. stearothermophilus*. In laboratories without the services of a skilled bacteriologist, where the testing of canned foods by incubation procedure is carried out, selection of 40°C as an alternative incubation temperature to 55°C has certain advantages. Cans that behave satisfactorily at this temperature may be safely released onto the market but any spoilage which becomes apparent during incubation is likely to be of commercial significance. The use of this temperature for incubation purposes is not meant to exclude the use of 55°C which gives a more complete picture of the overall sterility of the pack, but the results obtained are capable of interpretation by less skilled personnel.

At the end of the incubation period cans from the laboratory incubators should be cooled to room temperature and vacuum and pH determinations made; it is also a useful procedure to carry out a direct microscopic examination of can contents, since bacterial growth may occur without appreciable changes in vacuum or hydrogen-ion concentration. Products susceptible to 'flat-souring' may also be stored at 25°C for 3 months. From the pH determinations of samples so stored the 'flat-souring' activity at normal storage temperatures of any organisms can be assessed.

Acid products should be stored at a temperature of 25°C for a minimum period of 30 days, after which the vacuum is determined. Canned or bottled fruits are examined for signs of spoilage by *Byssochlamys fulva* or other organisms. Acid sauces, ketchups, pickles, etc., should be examined for evidence of fermentation. Gaseous spoilage in glass packs of this type is frequently not apparent until the closures are released, due to the fact that the gas is held in solution under pressure.

Apart from the final examination, it is, of course, imperative that frequent inspection of samples be made throughout the incubation period and that samples be removed for detailed examination as soon as spoilage is apparent.

Incubation tests provide very useful information concerning the 'commercial sterility' of products but should not be used in isolation for the release of stocks onto the market. Stock release control should involve careful consideration of the results of incubation tests in conjunction with heat penetration data and all the relevant information from the 'on-line' process control system. For further information concerning the factors which may affect the sterility of canned foods including the establishment of safe heat processes and their control and manufacture under hygienic conditions, the reader is referred to Technical Manual No. 3 '*Guidelines for the Establishment of Scheduled Heat Processes for Low-Acid Canned Foods*' published by the Campden Food Preservation Research Association (1977) and the '*Code of Practice on the Canning of Low Acid Foods*' (DHSS, 1979).

Aseptic packaging

Aseptic packaging lines contain complex equipment and in order to ensure a safe product it is necessary to monitor more control points than with continuous conventional canning lines. The process can be divided into stages as follows: (1) sterilisation of equipment; (2) sterilisation of containers and closures; (3) sterilisation of product and (4) filling sterile product into sterile containers and applying sterile closures within a sterile environment.

Aseptically filled cans may be subjected to the same test procedures as conventionally processed cans, although the results may be more difficult to interpret. For example a can may spoil due to leakage through the can seams, or leakage into the aseptic part of the plant, similarly heat-resistant spores may be present due to understerilisation of either the product or the plant. In the latter case the importance of thorough cleaning cannot be over-emphasised. Samples for examination should be drawn at regular intervals throughout the production run paying particular attention to the intial and final periods of production. Aseptic lines are designed to run continuously and intermittent running can lead to sterility problems, particularly if the line pressure falls. Can jams, which sometimes occur on the seamer, may require breaking asepsis to free and clean the line. In this case the affected portion of the line must be re-sterilised. In the case of easily released can jams on Dole seamers the plant is usually re-run immediately, rejecting the cans immediately adjacent to the affected production. After all such stops, whether intermittent or complete, it is important to take samples for examination as soon as normal production recommences. In addition to the checks carried out as part of the 'on-line' process control a log book should be kept detailing all process deviations, the times at which they occur and the remedial action taken.

The bacteriological examination of canned cured meats

The level of heat processing applied to this class of product is relatively low and consequently the bacterial flora is different from that of the canned foods which have been processed to F_0 values in excess of 2½. Therefore the simple sterility tests described earlier are not sufficient and the examination should include incubation tests, microscopic examination, viable counts of both aerobic and anaerobic organisms and direct and enrichment cultures for the isolation and identification of pathogens. Examination of the 'semi-preserves', canned hams especially, has been described by numerous workers (Buttiaux, 1953; Goldenberg, et al, 1955; Hobbs, 1955; Rievel, 1955; Jepson, 1957) and the reader is referred to the original papers for details of the methods used. Seidel (1959), discussing the problems of standardising the methods of examination, adequately reviews the work of the leading bacteriologists in this field.

With regard to the type and number of organisms to be found, Buttiaux & Mossel (1957) propose standards for semi-preserved, low-acid, canned foods in which the contents after incubation for 5 days at 30° to 32°C shall show:

1. No appreciable change in taste, odour, colour, pH or consistency, in comparison with the contents of unincubated cans
2. Absence of pathogenic, toxinogenic and faecal microbes, as well as their toxins
3. Absence of illegally added antimicrobial substances
4. Aerobic and anaerobic plate-count not over 10^4 per g.

REFERENCES

Amaha M 1953 Food Res 8: 411
American Public Health Association 1976 Speck ML (ed) Compendium of methods for the microbiological examination of foods APHA Washington
Andersen A A 1951 J Bact 62: 425
Bashford T E 1940 Food Manuf 15: 181 276
Bashford T E 1948 Private Communication
Beerens H 1955 Int symp on food bacteriology. Annales de l'Institut Pasteur de Lille 7: 75
Buttiaux R 1953 Food Manuf 28: 113
Buttiaux R, Beerens H 1955 Lab Practice 4: 234
Buttiaux R, Mossel D A A 1957 Annales de l'Institut Pasteur de Lille 9: 138
Cameron E J 1936 J Ass Off Agric Chem 19: 433
Cameron E J 1938 J Ass Agric Chem 21: 452
Campden Food Preservation Research Association 1977 Guidelines for the establishment of scheduled heat processes for low-acid canned foods. Technical Manual No 3 Chipping Campden
Cheftel H 1935 Food 4: 301
Corpe W A, Frazier W C 1950 Food tech 4: No 11 A8
Department of health and social security 1979 Code of practice on the canning of low-acid foods. In press — HMSO
Dilley A E, Everton J R 1966 Lab Pract 15 (3): 318
Gibbs B M, Freame B 1965 J appl Bact 28 (1): 95
Gibbs B M, Hirsch A 1956 J appl Bact 19: 129
Goldenberg N, Sheppey C G, Robson J N 1955 Annales de l'Institut Pasteur de Lille 7: 240
Hays G L 1951 Proc Florida State Hort Soc 64
Hirsch A, Grinsted E 1954 J Dairy Res 21: 101
Hobbs B C 1955 Annales de l'Institut Pasteur de Lille 7: 190
International Association of Microbiological Societies 1974 Micro-organisms in foods. University of Toronto Press, Canada, vol 2
Jensen L B 1954 Microbiology of meats. Garrard, Champaign, Illinois
Jepson A 1957 Meat hygiene. World Health Organisation, Geneva, p 420
Linden B A 1936 J Ass Off Agric Chem 19: 440
Lochhead A G, Farrell L 1930 Canad J Res 3: 51
De Man J C, Rogosa M, Sharpe M E 1960 J appl Bact 23: 130
Mossel D A A, Drion E F 1954 Ned Melk-en Zuiveltijdschr 8: 106
Mossel D A A, Van Deipen H M J 1956 Annales de l'Institut Pasteur de Lille 8: 132
Mossel D A A, Drion E F 1958 Food 27: 333
Murdock D I, Folinazzo J F, Troy V S 1952 Food Tech 6: 181
National Canners' Association 1960 Research information No 40
Put H M C 1961 Annales de l'Institut Pasteur de Lille 12: 175
Rievel H 1955 Annales de l'Institut Pasteur de Lille 7: 203
Scarr M P 1959 J Sci Fd and Agric 10: 678
Scott W J 1953 Food Pres Quart 13: 21
Seidel G 1959 Fleischwirtschaft 11: 732, 911, 1006
Stern R M, Hegarty C P, Williams O B 1942 Food Res 7: 186
Thomas G, Cheftel H 1955 Ann Falsif 48: 223, 283
Townsend C T, Somers I I, Lamb F C, Olson N A 1954 Laboratory Manual for the Canning Industry. National Canners' Assoc Res Lab, Washington D C
Wheaton E, Pratt G B 1961 J Fd Sci 26: 261
Wheaton E, Pratt G B, Jackson J M 1959 Food Res 24: 134
Williams O B 1936 Food Res 1: 217
Williams O B, Clark E D 1942 Food Res 7: 178
Wynne E S, Foster J W 1948 J Bact 55: 61
Wynne E S, Schmieding W R, Daye G T 1955 Food Res 20: 9

16

Examination of raw materials, plant and miscellaneous methods

In the main, bacteriological control work in the cannery resolves itself into the examination of raw materials, plant, utensils, etc., in order to determine their suitability for canning purposes. The routine tests are principally quantitative and may concern either specific types of spoilage organisms or, where the sanitary condition of plant, etc., is being assessed, the general bacterial population.

At present, standard methods for the bacteriological examination of canning ingredients are mainly concerned with the estimation of thermophilic spores in sugar, starches and spices. Such methods, however, may be readily modified or adapted for the detection of other groups of organisms in most of the materials used as canning ingredients. The chief difficulty in the bacteriological examination of raw materials for canning purposes is that except in the case of thermophilic spores, no standards exist by which the suitability of a material for canning may be judged. There is a wide field here for investigation. With regard to the examination of plant, utensils and other canning equipment, official tests do not as yet appear to have been established, but methods devised for other purposes can be usefully applied.

Examination of raw materials for thermophilic spores

Sugar

The following procedure for the detection of thermophilic bacteria in sugar is that recommended by the National Canners' Association (1968). The latest method from the American Public Health Association (1976) is similar.

Sampling. From each delivery, 227 g (½ lb) sample is taken from each of five bags.

Preparation of sample. Place 20 g of sugar in a sterile 250 ml flask marked to indicate a volume of 100 ml. Add sterile water to the 100 ml mark. Bring rapidly to boiling and boil for 5 minutes, or heat in an autoclave at 101.7°C for 15 minutes to 20 minutes. Replace evaporation with sterile water.

Thermophilic 'Flat-sour' Spores. Into each of five Petri dishes, pipette 2 ml of heated sugar solution. Cover with dextrose tryptone agar and mix thoroughly. 'Spreaders' may be inhibited by covering the cooled dextrose-tryptone medium with a thin layer of 2 per cent plain agar. The plates are incubated at 55°C for 36 to 48 hours. The total number of typical acid-producing colonies for the five plates is determined and this number indicates the 'flat-sour' spores in 2 g of original sugar. This number is multiplied by five to give the spore content of 10 g of sugar. Non-acid types of colonies are not included in this count.

Standard. For the five samples examined there should be a maximum of 75 spores and an average of not more than 50 spores per 10 g.

Total thermophilic spores. This count includes the non-acid in addition to the acid-producing colonies.

Standard. For the five samples examined there should be a maximum of 150 spores and an average of not more than 125 spores per 10 g.

Thermophilic anaerobic spores (not producing H_2S). Divide 20 ml of the heated sugar solution approximately equally among six freshly exhausted liver broth tubes and stratify the liquid medium with plain agar. Solidify the agar rapidly by placing the tubes in cold water, preheat to 50° to 55°C and incubate at 55°C for 72 hours. This method does not provide an exact quantitative result but serves as a means for estimation. Results are expressed as the number of tubes positive or negative, e.g. +++ – – – .

Standard. Spores of this type are permissible in not more than three (60 per cent) of the five samples and in any one sample to the extent of not more than four (65 + per cent) of the six tubes inoculated.

Thermophilic sulphide spoilage spores (thermophilic anaerobes producing H_2S). Twenty ml of the heated sugar solution are divided in roughly equal amounts between six tubes of melted sulphite agar. After thorough mixing the medium is quickly solidified by placing the tubes in cold water. After pre-heating to 55°C they are incubated at 55°C for 48 hours. The typical black colonies are counted and the results expressed as the number of spores per 10 g.

Standard. Spores of this type are permissible in not more than 2 (40 per cent) of the 5 samples and in any one sample to the extent of not more than 5 spores per 10 g. This would be equivalent to 2 colonies in the 6 inoculated tubes.

Gillespy & Thorpe (1961) consider it desirable to modify the American standards for anaerobic spores as follows:

Thermophilic anaerobes. For single samples representing strikes, there should be not more than one positive tube in six.

For the five samples representing a heterogenous consignment, there should be a maximum of not more than two positive tubes in any one sample and an average of not more than one positive tube.

Sulphide spoilage spores. These should not be detected at all.

Knock & Baumgartner (1947) proposed a modification of the American methods for sugar and starches, namely, autoclaving the solutions for 30 minutes at 110°C before making the counts. Gillespy (1947) supported their view that this gives a better picture of the infection in sugar intended for canning purposes. In South Africa, standards for canners' sugar (Standard Specification for Canners' Sucrose, SABS 420; 1952) are based on the NCA method except that the sugar solutions are heated for 30 minutes at 110°C. The following standards then apply:

'Flat-sour' spores. Not more than 10 spores per 10 g of sugar.

Total thermophilic spores. Not more than 100 per 10 g of sugar.

Thermophilic gas-producing anaerobes. These shall not be detected at all in a set of six tubes.

Sulphide spoilage spores. Not more than 1 per 10 g of sugar.

Starches

The method and standards suggested by the National Canners' Association are as follows:

Sampling. As for sugar.

Preparation of samples. Place 20 g of starch in a dry, sterile, 250 ml flask and add sterile cold water to the 100 ml mark with intermittent shaking. This mixture is then thoroughly stirred and used for detecting the various groups of spoilage organisms.

'Flat sour' spores. Ten millilitres of the starch suspension are added to 100 ml of melted dextrose tryptone agar at a temperature of 55° to 60°C. The starch suspension is kept under constant agitation during the pipetting operation. After the starch has been added to the tryptone agar, shake the flask in boiling water for 3 minutes to thicken the starch and then heat in the autoclave at 0.33 bar pressure for 10 minutes. Cool with gentle agitation to prevent aeration and then distribute equally among five Petri dishes. When hardened, stratify with a thin layer of plain 2 per cent agar. Incubate at 55°C for 48 to 72 hours.

Anaerobic spores (not producing H_2S). Divide 20 ml of the cold starch suspension approximately equally among 6 liver broth tubes. These tubes should be spun before heating and during the 15 min heating period in a boiling water bath. In the liver broth tubes, it has been found desirable to use finely ground liver in very small amounts. When the starch thickens, the fine liver will remain in permanent suspension throughout the medium and this is considered an advantage. It appears to be unnecessary to stratify the liver broth tubes.

Sulphide spoilage spores. Divide 20 ml of the cold starch suspension approximately equally among six sulphite agar tubes. These tubes also should be spun before heating and during the heating period in a boiling-water bath.

In inoculating and heating the liver broth and sulphite agar tubes, it is essential to maintain an even suspension of the starch, and for this purpose it is necessary to agitate them at frequent intervals. The following is a suggested procedure:

1. The tubes are spun in the hands, three at a time, just after the addition of the unheated starch
2. After the tubes are placed in the water bath the individual tubes are spun with the fingers
3. After all tubes are in the bath they are spun three or four times during the first 5 minutes of heating. After this, the heating is continued for 10 minutes
4. After the liver and sulphite tubes have been heated they are placed in cold water to solidify. The tubes are pre-heated to 50° to 55°C after which the liver tubes are incubated at 55°C for 72 hours and the sulphite tubes at 55°C for 48 hours.

Standards. These are the same as for sugar.

Milk powder

Sorensen (1938) has recommended a procedure for the examination of skimmed milk powders for thermophilic 'flat-sour' spores. The method was evolved from that used for sugar, but certain modifications were found necessary. As a diluent, N/60 sodium or lithium hydroxide is used. This eliminates difficulty in counting plates caused by undissolved flakes of milk powder and heat-coagulable material. Sorensen found that no detrimental effects on the thermophilic counts followed the use of alkaline diluent and states that acid-producing colonies are more readily discernible, since some powders are sufficiently acid to affect the indicator in the medium when water is used as the diluent. Autoclaving the milk powder solution is recommended in place of boiling, since foaming occurs when the latter procedure is attempted. In the sugar method, 20 g are made up to 100 ml with water, but when milk powder solution is prepared at this concentration the resulting sample is unsatisfactory for plating purposes and the quantity recommended is 10 g per 93 ml of water. The method is as follows:

1. Weigh 10 g of powder into 93 ml of sterile sodium or lithium hydroxide (N/60)
2. Shake vigorously and autoclave at 0.33 bar pressure for 10 minutes
3. Add 1 ml of the autoclaved sample to each of ten Petri dishes and pour with dextrose-tryptone agar. Incubate at 55°C for 36 to 48 hours
4. Count colonies on 10 plates, recording total as the total thermophilic spore count per g. Acid-producing colonies are noted separately. No standards are given.

Spices

A procedure for the microbiological examination of spices has been suggested by the American Public Health Association (1976).

Sampling. It is recommended that a 200 g sample be taken aseptically from each of five containers representing the delivery.

Preparation of samples. Due to differences in bulk and character the following procedures are recommended:

1. Whole spices such as berries, seeds and herbs; aseptically weigh 10 g into a 4 or 8 oz (113 or 226 g) wide mouth bottle containing 90 ml sterile phosphate buffer. Shake on a mechanical shaker for 5 minutes at 200 strokes per minute
2. Whole spices of larger size, e.g. bark and roots; aseptically weigh 100 g into a sterile, dry, blender jar, blend at lowest setting for 30 seconds or until a moderately fine particle size is obtained. Aseptically weigh 10 g of the macerated material into a 4 or 8 oz (113 or 226 g) wide mouth bottle containing 90 ml sterile phosphate buffer. Shake at least 25 times in a 30 cm arc
3. Ground spices and seasonings; aseptically weigh 10 g into a 4 or 8 oz (113 or 226 g) wide mouth bottle containing 90 ml sterile phosphate buffer. Shake at least 25 times in a 30 cm arc.

It is recommended that inoculations into the appropriate media be made within 15 minutes of preparing the initial dilution.

Bacterial counts. Counts should be expressed as numbers of bacteria, spores, etc., per g of spice.

1. Total viable count. Standard plate count agar is used to determine the total viable count. Incubate at 30° to 32°C for 48 hours.
   It is important to prepare sufficiently high dilutions to overcome the natural inhibitory effects of some of the spice ingredients; low dilutions may give false low counts due to this effect

2. Aerobic mesophilic sporeformers. Use tryptone glucose extract agar in which the dilutions have been pasteurised. Incubate at 35°C for 48 hours
3. Aciduric flat sour sporeformers. Use thermo-acidurans agar and incubate at 55°C for 48 hours
4. Thermophilic flat sour sporeformers. Use dextrose-tryptone agar and incubate at 55°C for 48 to 72 hours
5. Putrefactive anaerobes. Use Anderson's pork-pea agar in Miller-Prickett tubes stratified with thioglycolate agar or in pour plates in a Gaspak system. Incubate for 48 to 72 hours at 32°C. The production of gas by anaerobes often splits the agar, making enumeration of the colonies difficult. A more accurate method which gives better reproducibility is to make an MPN count using liver broth stratified with agar, or PE-2 medium
6. Thermophilic anaerobes not producing H_2S. Use liver broth or PE-2 medium stratified with 2 per cent sterile agar. Preheat tubes to 55°C and incubate at that temperature for 72 hours
7. Thermophilic anaerobes producing H_2S. Use sulphite agar in screw-capped bottles or Miller-Prickett tubes. Preheat to 55°C and incubate at that temperature for 24 to 48 hours.

No standards for the various counts are given.

Bacteriological examination of plant equipment, cooling water and containers

Swab-rinse plate count

With this method, the surface to be examined is wiped with a swab which is then rinsed in a suitable diluent, the number of organisms being estimated by a plate count. The swabs should be made from good-quality white absorbent cotton-wool twisted on wooden applicators; they should be sterilised by packing into plugged glass tubes and autoclaving for 20 minutes at 1 bar. Alternatively, ready prepared and sterilised swabs, packed individually in plugged cardboard tubes may be obtained from Medical Wire and Equipment Co. (Bath) Ltd., Potley, Corsham, Wiltshire.

A procedure suitable for the examination of canning plant and utensils, including post-process handling equipment, is given below.

Apparatus and materials. Sterile Petri dishes, sterile 1 ml pipettes, standard plate-count agar, sterile frames (10 x 10 cm) of tinplate with 10 sq cm openings, sterile cotton swabs, sterile 9 and 10 ml dilution blanks in screw-capped bottles. The dilution blanks should be made up with 0.1 per cent peptone water. If the presence of bactericides is suspected on the surfaces to be examined, for example, post-processing equipment wetted with cooling water, then suitable inactivators should be added to the 10 ml dilution blanks. For chlorine compounds a small crystal of sodium thiosulphate may be added before sterilisation; if quaternary ammonium compounds are in use then 0.5 ml of a sterile Lecithin/Lubrol mixture should be added aseptically just before use (4 per cent Lecithin and 6 per cent Lubrol in distilled water).

Collecting samples. A swab is moistened in a 10-ml dilution blank, excess liquid being removed by pressing the swab against the side of the container. By placing a sterile tinplate frame against the surface being tested 10 sq cm are slowly and thoroughly rubbed with the moist swab. After completing the swabbing, the swab is replaced in the container.

Laboratory procedure. The swab stick is broken just above the swab by applying side pressure against the rim of the bottle. The bottle cap is replaced and the bottle shaken vigorously to disentangle the swab. Allow to stand for 20 minutes, shaking periodically. Using sterile 1-ml pipettes and 9-ml dilution blanks prepare serial dilutions. Pipette 1 ml of each dilution into a sterile Petri dish and add 15 ml of melted plate-count agar (avoid pouring the agar directly on to the 1 ml inoculum). Mix together the inoculum and agar by rotatory and to-and-fro movements, allow to set and incubate for 5 days at 25°C. Count the plates and express the results as the total number of organisms per 10 sq cm. The counts should not exceed 500 organisms per 10 sq cm.

By the use of appropriate media and incubation conditions the method can be used for the detection of specific spoilage types (e.g. flat-sour spores) in canning plant.

In connection with methods of estimating the bacterial contamination of surfaces the experiments of Angelotti et al (1958) have indicated that the swabbing of several separate areas with one swab gives greater precision than the examination of one small area only.

For swabbing larger areas the American Public Health Association (1976) recommend the use of a cellulose sponge technique. Cellulose sponges, which should measure 5 x 5 x 5 cm, are placed individually in Kraft paper bags and steam sterilised at one bar for 20 minutes. The sterile sponge is removed from the bag with sterile tongs, or by hand if sterile gloves are worn, and moistened with 10 ml of sterile 0.1 per cent peptone water. It is claimed that by using such a sponge an area of several square metres can be effectively swabbed. Because of the greater area that can be covered, the method is particularly useful in searches for specific organisms such as *Salmonella,* when, after swabbing, the sponge can be introduced directly into an enrichment medium.

Agar sausage technique

This method has been described by Ten Cate (1963, 1965) and has proved very useful for checking the efficiency of plant and utensil cleaning operations. It is simple, rapid and economical in use, and therefore particularly suited to small laboratories or firms wishing to extend their hygiene control. The agar medium is sterilised in a cylindrical plastic coating which, when the ends are sealed, and the agar has set, resembles a sausage. When required for use, one end of the 'sausage' is cut off with a sharp sterilised knife, gentle pressure is applied at the opposite end causing the agar to protrude slightly and the exposed agar surface is used to make an impression of the surface under examination. With a broad knife a slice of the agar bearing the impression is removed and placed, inoculated side uppermost, in a Petri dish for incubation. In this way a large number of impressions or samples can be quickly taken and it is possible to incubate 3 or 4 in each Petri dish. Another advantage of this technique is that different selective media may be used in the 'sausage' thus enabling the direct-enumeration of different bacterial types. The method is not suitable for irregular surfaces or for heavily contaminated surfaces when overcrowding on the agar surface makes colony counting impossible.

Generally lower counts are obtained with the 'sausage' technique than with conventional swabbing methods and it has been reported (Riddle, 1967) that this is due to the distribution of bacteria on the surface. Small clumps of bacteria produce only one colony on the agar impression whereas swabbing separates the individual cells thus giving higher results in the final count.

Examination of containers

The efficiency of bottle-washing procedures may be determined by the following method which is based on that proposed by the American Public Health Association (1972). Briefly, the method is as follows:

Twenty millilitres of sterile buffered rinse solution or 0.1 per cent peptone water is introduced into the bottle which is then closed with a sterile cap: The bottle is then held in the horizontal position and shaken vigorously 10 times over a distance of about 20 cm. This treatment is repeated a further 3 times, turning the bottle 90° between each shaking. With the bottle in the vertical position, it is then swirled vigorously 20 times in a small circle, inverted and swirled again. When bottles contain residual chlorine, 0.5 per cent 0.1N sodium thiosulphate should be added to the rinse solution. After shaking, 10 ml of the rinse solution is withdrawn and distributed between 3 sterile Petri dishes to which melted plate count agar is then added. If bottles are suspected of being heavily contaminated, two 1 ml and two 0.1 ml amounts are plated.

An alternative method of examining glass containers is the following:

As the bottle is collected from the washing machine it is plugged with a sterile cotton plug or cap. A small quantity of melted agar is poured into the bottle, which is then rotated in cold water so that the interior surface is evenly coated with a film of agar. When the medium is set, the bottle is incubated as for plates.

Cans may be examined by the APHA method. The rinse solution (sterile 0.1 per cent peptone water) is added to the can and a sterile cover is immediately placed in position and seamed. After agitation, the cover is sterilised with hypochlorite solution, pierced with a sterile tool and the rinse solution withdrawn for plating by means of a pipette.

Flexible containers should be placed on a smooth, firm, horizontal surface and with the hands or a roller, the rinse solution 'worked' back and forth 10 times so that all internal surfaces are completely contacted. Ten ml of the rinse solution are aseptically withdrawn and plated out as before.

Aseptic canning lines

These vary considerably in design and complexity and it is not possible, therefore, to give detailed descriptions of examination methods. The following general procedures have been used and found useful on various occasions.

A check of the overall sterility of the installation including the filling machine and the containers may be obtained by processing a bacteriological medium, such as dextrose-tryptone broth, in place of product. The medium after passing through the sterilising and cooling sections is filled into pre-sterilised containers in the normal manner. It is not necessary to fill the containers completely with the medium provided that they are well shaken in order to rinse the interior surfaces. The sealed containers can then be incubated at any desired temperature. After incubation the cans are opened and then examined for a change in colour or other signs of growth in the medium. With some types of filler it may be necessary to thicken the medium with starch in order to obtain a clean filling operation. This procedure demonstrates whether or not the system as a whole is sterile but gives no indication as to the location of any possible trouble spots.

Unsterility of the plant can occur due to: (1) excessive microbial contamination due to inadequate or inefficient plant cleaning or excessively contaminated raw materials; (2) leakage of air, water or other unsterile material into the sterile areas; (3) inadequate pre-sterilisation of the equipment or (4) selection of the wrong processing conditions, that is, time and/or temperature. If plant sterility is suspect the following procedure may be adopted:

Bacteriological medium thickened with 5 per cent starch and sufficient agar to give a reasonably firm set is prepared and processed in place of product. When the medium reaches the filling zone the plant is closed down and left under a positive pressure of sterile air for up to 48 hours. The object is to allow the medium to form a coating on the internal surfaces of the plant so that after incubation growth in the medium will indicate the site of the infection. Therefore after the necessary incubation period the plant is carefully stripped down, all pipework, joints and valves being examined for signs of growth in the medium. There is usually sufficient retained heat in the plant to provide an adequate incubation temperature.

The efficiency of the container sterilisation may be assessed by inoculating the cans and ends with a suspension of heat resistant spores of *Bacillus stearothermophilus*, allowing it to dry into an adherent film on the internal surfaces and passing the cans through the normal sterilisation process. Sterile dextrose-tryptone agar medium filled into the cans, which are then incubated at 55°C for 5 days, will show whether the can sterilisation process is satisfactory. Separate trial runs can be carried out with inoculated can ends and inoculated can bodies to determine which part of the container sterilisation process is in doubt.

In the event that the containers are not effectively sterilised, for example in Dole equipment, then further investigations may be carried out by suspending discs of tinplate inoculated with spores of *Bacillus stearothermophilus* in various parts of the equipment in order to locate more accurately the source of the problem.

Aseptically filled cans are seamed at high temperatures which results in the sealing compound being in a semi-fluid condition while the vacuum is forming. Potentially the cans are more susceptible to post-process leakage than cans from conventional processes, although, because no post-cooling operation is required one source of contamination, namely the cooling water, is eliminated. However, it is desirable to ensure that an adequate time is allowed for the sealing compound to set before any can washing operations are attempted and adequate conveyor length should be provided for this purpose. Needless to say a high standard of post-canning hygiene must be maintained if spoilage is to be avoided and special attention should be given to the section immediately after the seamer.

For those containers and their lidding material which are sterilised by hydrogen peroxide rather than by heat a suitable test organism with a high resistance to peroxide is required for carrying out plant checks. *Bacillus globigii* is normally used for this purpose. Ito et al (1973) found this organism to have a resistance six times greater than the heat resistant strain of *Clostridium botulinum* (169B) to 35 per cent hydrogen peroxide at a temperature of 87.8°C. Spore crops of *B. globigii* are prepared by inoculating an overnight culture in nutrient broth on to nutrient agar plates which are incubated at 30°C for 4 days. The spores are harvested, washed several times in cold distilled water and pasteurised at 80°C for 10 minutes. Pre-formed containers, usually plastic pots, may be inoculated by pipetting a small amount of suspension containing a known number of spores into the pot, swirling round and allowing the suspension to dry. Web material may be inoculated by wiping the internal surface with a pad moistened with spore suspension to give a known number of spores per unit area. Survival of the spores may be detected by techniques similar to those described above but in this case using plate count agar as the medium. Surviving spores of *B. globigii* will produce characteristic orange-brown colonies.

Examination of can cooling water

Unless the bacteriological condition of the water is above suspicion, as when drawn direct from the main supply of a reliable public undertaking, the examination of cooling water should be regarded as an essential routine procedure in the canning laboratory. Where water is re-circulated and chlorinated, regular estimations of both the residual chlorine and the bacterial contents are necessary. The frequency with which it is necessary to make these tests in order to secure adequate control will vary with the type of plant and method of operation and can only be established by trial.

Estimation of bacterial content. Under certain circumstances it may be desirable to examine the water for specific types of organism, but a simple plating test is all that is necessary for controlling the purity for can cooling purposes. In the examination of water for drinking purposes greater importance attaches to the count at 37°C that that at 22°C, since to the former group belong the parasitic types which serve as an index of pollution by sewage or excretal matter. The majority of the organisms concerned in leaker spoilage are water and soil types, growing best at about 25°C, so that a plate count at this temperature is most significant where can cooling is concerned.

The sample should be drawn from the spray in the retort or cooling tunnel or, where tank-cooling is in operation, at the point of inflow. In all cases, the water should be allowed to run for a few minutes before sampling into sterile flasks or bottles. If the water is chlorinated, it is advisable to add a small crystal of sodium thiosulphate to the sample bottle before its sterilisation in order to remove free chlorine from the sample of water. If the water is expected to be very heavily contaminated, dilutions ranging up to 1:1000 should be prepared in sterile water. When the water is known to be of good quality it is sufficient to prepare only a 1:10 dilution. After thorough mixing by shaking twenty-five times with an up-and-down movement, 1 ml of the sample and of each dilution prepared is pipetted in duplicate into sterile Petri dishes. Ten to 15 ml of sterile plate-count agar or dextrose-tryptone agar, cooled to 45°C, is poured on and carefully mixed with the inoculum by rapid to-and-fro and centrifugal movements while the plates are kept flat on the bench. When set, plates are incubated at 25°C for a minimum of 5 days. Plates containing not more than 300 colonies should be selected for counting. Plates with less than 30 colonies should not be counted unless the undiluted water plates contain less than this number. If there is a marked discrepancy between the two plates of any dilution, the count should be viewed with suspicion. No generally accepted standards for cooling water appear to have been established, but a count of less than 100 per ml can be regarded as satisfactory, and the use of water of this quality is unlikely to result in appreciable leaker spoilage unless can seams are of a poor standard.

It was previously thought that the yellow and orange colonies of '*Flavobacteria*' which frequently occur on cooling water plates were insignificant from a spoilage point of view. Recent work, however, has shown that when present in sufficient numbers they may gain entry into processed cans through the seams even when the latter are of sound construction. This leakage potential exists until the can seams are completely dry. *Flavobacteria* may constitute a particular hazard where continuous cooker-coolers are in use, the organisms being carried over from the cooling water and deposited on to the exit runways by the passage of cans through the steriliser. If the wet runways are not treated frequently with sanitising agents a considerable build up of organisms occurs. Bean & Everton (1969) have made detailed investigations of the problem and state that *Flavobacteria* cause spoilage of milk-based products such as creamed rice pudding, macaroni, cream-style soups and evaporated milk. The spoilage results in a breakdown and thinning of the product, there is no fall in pH and no gas is produced. They report that the spoilage may take from 2 to 3 months to occur at room temperature.

Estimation of chlorine content. The total chlorine content of cooling water may be estimated colorimetrically, using the 'Chlorotex' reagent supplied by British Drug Houses Ltd. A colour chart is also supplied by this firm for comparing the colour produced when 5 ml of the reagent are mixed with 50 ml of the water, the various tints on the chart corresponding to amounts of chlorine ranging from 0.1 to 1.0 parts per million.

The free chlorine content should be estimated using the Palin-DPD method. A Lovibond Comparator with the appropriate disc is used to equate the colours produced when the reagent (in tablet form) is added to a 10 ml sample of the water under test. The tablet should first be dissolved in a small amount of the sample water which is then made up to the 10 ml mark. Clearly defined results are obtained and the method is very simple in use. Comparator discs and the reagent together with instructions for use may be obtained from The Tintometer Ltd., Waterloo Road, Salisbury, Wiltshire, or from B.D.H. Ltd. Laboratory Chemicals Division, Poole, Dorset.

High readings for total chlorine together with very little or no free chlorine indicate a high chlorine demand by the cooling water. Such a situation is normally the result of organic contamination of the water, particularly in the case of re-circulated systems, and the cause of the contamination (usually burst cans or food matter gaining access to collecting sumps, etc.) should be sought and rectified. If the high chlorine demand persists the entire system should be drained, washed out and refilled with fresh water. Recirculated water cooling systems should be drained and thoroughly cleaned out at least once a week.

Chlorinated cooling water should have a total chlorine content of no less than 5 parts per million some of which must be in the form of free chlorine.

Miscellaneous methods

Dilution cultures

In addition to heat-preserved packs, many canners produce lines which are preserved with edible acids, salts, sugar or combinations of these. In some cases these products, which include salad cream, sauces, pickles, etc., may be given a moderate heat treatment, but often they are unsterile. Especially where heating is not applied, the contaminating organisms may include bacteria, yeasts and moulds. The main problem is to decide whether the organisms present will cause spoilage. Demonstration of growth in a nutrient medium will not be of much assistance in this respect except in so far as to indicate the presence of possible spoilage types. A procedure which has been found useful involves the graded dilution of the product with a suitable nutrient medium. For example, salad cream known to be contaminated with yeast or aciduric bacteria may be diluted 5, 10, 20 and 50 per cent with wort broth and held at 25°C. If fermentation occurs in the lowest dilution it may be deduced that there is little margin of safety with regard to the product's resistance to spoilage. If considerable dilution is necessary before fermentation occurs it may be inferred that the product is reasonably resistant to spoilage. An alternative method has also proved useful:

Suitable dilutions of the product are inoculated with a significant spoilage organism and the results of periodic counts of the viable organisms in each mixture are plotted. Useful information can be deduced from the curves so obtained.

REFERENCES

American Public Health Association 1972 Hausler W J (ed) Standard methods for the examination of dairy products, 13th edn. APHA, Washington

American Public Health Association 1976 Speck M L (ed) Compendium of methods for the microbiological examination of foods, APHA, Washington

Angelotti R, Foter M J, Busch K A, Lewis K H 1958 Food Res 23: 175

Bean P G, Everton J R 1969 J appl Bact 32: 51

Gillespy T G 1947 Ann Rept Fruit and Veg Preserv Res Sta, Chipping Campden, p 40

Gillespy T G, Thorpe R H 1961 Cannery hygiene. Tech Bull No 6, The Campden Food Preservation Research Association, Chipping Campden

Ito K A, Denny C B, Brown C K, Yao M, Seeger M L 1973 Food Tech, 27 (11): 58

Knock G G, Baumgartner J G 1947 Food Manuf 22: 11

National Canners' Association 1968 Laboratory manual for food canners and processors, 3rd edn, Vols 1 and 2. Avi Publishing Co, Westport, Connecticut

Riddle W 1967 The agar sausage technique and its use as a routine bacteriological control measure in food factories. Tech Circ No 372. The British Food Manufacturing Industries Research Association, Leatherhead

Sorensen C M 1938 Food Res 3: 421

Ten Cate L 1963 Fleischwirtschaft 15: 483

Ten Cate L 1965 J appl Bact 28: 221

17

Examination of containers

The mechanical condition of the containers may bear such an important relationship to the results that seam examination must be regarded as an essential part of any bacteriological investigation on canned foods. The bacteriological condition of the contents can be determined by cultural, microscopic and other tests, but these provide no indication as to whether the cans are likely to leak or, if spoilage has already occurred, whether this was the result of such defects as substandard seams, excessively contaminated cooling water, strained or damaged seams.

The criterion of a satisfactory can seam is that it should be gas-tight. This has naturally led to the use of vacuum or air pressure for the detection of leaks in containers. One method is to test the can by means of compressed air whilst it is held under water, leakage being indicated by the appearance of air bubbles. Another is to apply the vacuum method for leak detection which is recommended by Lefebvre (1952).

The difficulties involved in testing cans for leakages have been reviewed by Sanders (1949) who pointed out that cans do not necessarily behave in the same way when subjected to different types of leakage tests. In our opinion, negative leakage tests on filled, processed cans, especially those which have been spoiled, must always be viewed with caution. Cans which have been proved by bacteriological tests to have leaked may subsequently become airtight, the original leak being sealed by the can contents. Support for this view is derived from the fact that leakers very frequently become hard swells, the defect in the seam being so effectively sealed that the pressure in the can builds up and may eventually cause it to burst. Soaking or other treatment to soften and wash away the obstruction is not always successful, presumably because the material is inside the seam and is not reached by the soaking solution. According to the Metal Box Company Ltd. the vacuum leakage test is a little more sensitive than the air pressure leakage test, but both fail to reveal leakage in roughly 30 per cent of blown cans.

The 'Bios' test is a microbiological method for assessing seam tightness. It involves the filling of cans with a nutrient medium. The cans are then closed, processed and cooled in water inoculated with an organism capable of producing gas in the medium in the cans. After incubation the number of swelled cans is determined. Much useful information can be obtained in this way but the application of the process is largely limited to the testing of sealing compounds and new seaming techniques.

It is generally accepted in the industry that in the evaluation of can seam quality, inspection and measurement of seams and their components are essential. It should be appreciated that only the relatively gross defects are determinable by inspection and measurement and failure to establish the existence of mechanical defects must be anticipated in a proportion of spoiled cans when the bacteriological evidence indicates leakage. A tolerance of five to ten thousandths of an inch is usually permissible in seam dimensions and it is not always sufficiently appreciated that since the average bacterium is less than one five-thousandth of an inch in length, it is possible for infection to take place through apparently sound seams.

The incidence of leakage among cans with apparently sound seams may be raised by a number of factors, including heavily infected cooling water, straining of seams by excessive internal pressure in the can as the result of rapid reduction of steam pressure at the end of the retorting process and disturbance of lining compound caused by damage to the seam. It will generally be found that such leakers are most prevalent in thin or watery products, probably for the reason that in the more viscous packs such as thick soups, any microscopic channels in the seams tend to be sealed by the contents of the can.

A seam of normal commercial quality is expected to conform to certain standards which include the correct dimensions for its components. These dimensions are directly governed by the setting of the machine and the condition of its parts. They are to a very large extent interdependent, and when the dimension of one component is abnormal there is usually a complementary abnormality in another part of the seam or the can. To some degree the dimensions depend on the thickness and the mechanical qualities of tinplate being reasonably constant and on the amount and even application of the lining compound.

Significant seam dimensions are shown in Figure 28. The measurements are made with an ordinary micrometer screw gauge or with a seam micrometer which is a specialised form of this instrument.

Selection of cans for examination

The examination of can seams is normally carried out either in connection with the investigation of spoilage, or as a routine control operation concerned with correct adjustment and efficiency of seaming machines.

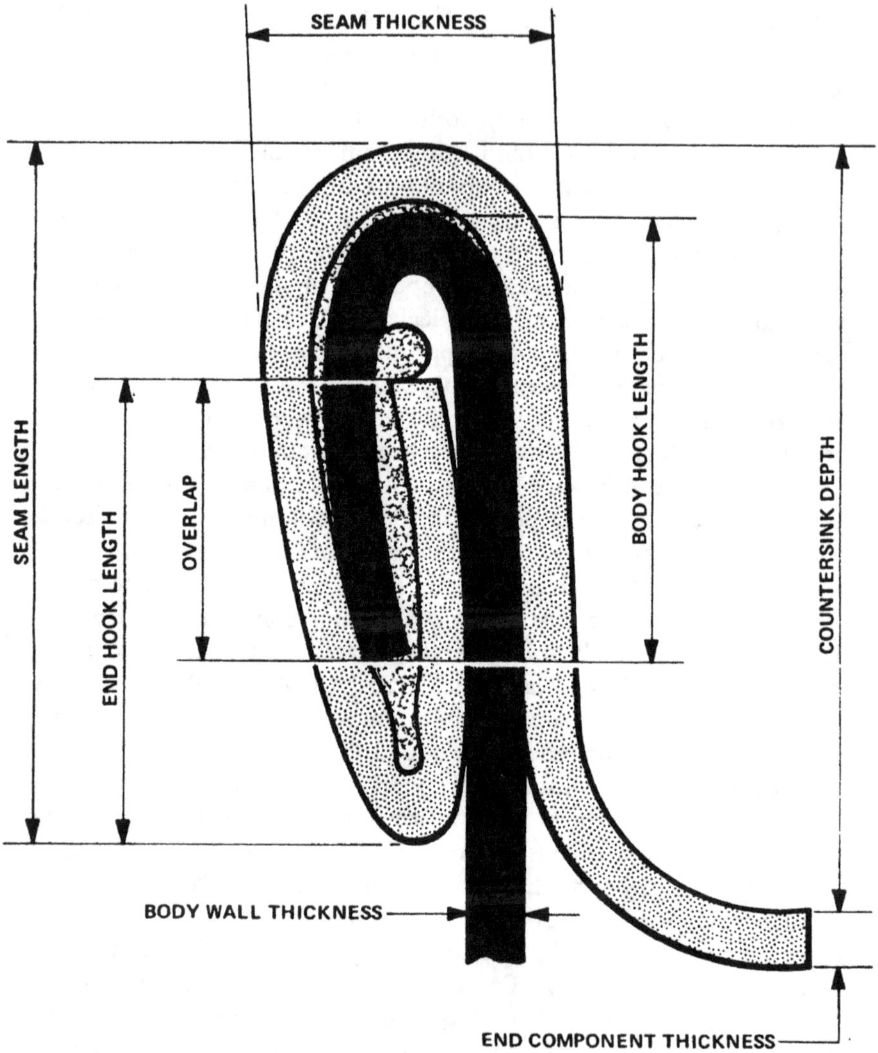

Fig. 28 Section of can seam

In the investigation of the cause of spoilage it is important that conclusions concerning a large number of blown cans are not based on the examination of one or two samples. In any batch of canned foods a very small proportion of leakers must be anticipated, the normal level being something less than 1 per 1000. These spoiled cans may arise from a multiplicity of causes which are more or less uncontrollable; undoubtedly a major factor is that it is impossible to manufacture every can seam to the ideal mechanical standard. When the proportion of spoiled cans shows a marked increase over the normal level it is usually due to some relatively gross defect or defects in the process of can fabrication which can be detected by seam examination on the lines indicated here. It follows, therefore, that haphazard selection of one or two samples from a batch of spoiled cans may give rise to misleading results if the cans chosen are of the 'non-apparent' leaker type. Unless the number is unduly large, at least 10 per cent of the samples involved should be examined in detail after having ascertained the general features shown by the majority of the cans. When a large number of cans are involved, the detection of the defect can often be facilitated by the careful selection of cans for detailed examination. Generally, the most easily diagnosed cans are those showing moderate pressure with leakage. An obvious defect will usually be found at the site of leakage in such cans, and if cans of this type form a fair proportion of the batch they should be selected for examination. It is, however, essential that any cans selected should show the same feature as those present in the majority; if the general feature is absence of apparent leakage, the cans selected should be of this type. An adequate sample of 'sound' cans should also be taken so that it is possible to establish the general level of quality.

For the quality control of seaming operations it is necessary to collect samples from the machine at regular time intervals during normal running, supplemented by special samples at start-up and after machine breakdowns. The results of the examination, giving details of important measurements, should be logged, so that the performance of the seaming machine is readily apparent. Because there are inherent dimensional variations in can seams produced by a machine functioning normally, some form of graphical control procedure is preferred in which the limits are based on the statistical interpretation of machine performance. The use of such a statistical chart will indicate clearly when a machine is drifting out of control and will avoid unnecessary adjustments being made for minor fluctuations.

The procedure to be followed in examining containers is as follows:

1. General preliminary examination of can for obvious defects such as perforation or fractures
2. Examination of rolled seams
3. Examination of soldered seam (lock seam and lap joints).

These points are amplified below.

General preliminary examination

A visual examination of all cans should first be made. A careful search should be made for pinholes, fractures and other defects in the body and ends of the can. Unless the defect is very small it can usually be detected by leakage of the can contents, but occasionally a minute fracture or hole may become so effectively blocked by the contents of the can that a 'swell' results. Minute fractures in tinplate are sometimes caused by embossing code marks too heavily. Pinholes frequently result from external rusting, and spoiled cans showing rust formation require very careful inspection. After thorough cleansing with water and fine abrasive washing powder, the corroded areas should be scrutinised with a hand lens. In examining rusted cans, the chief difficulty lies in deciding whether suspicious 'pits' in the tin coating are actually minute perforations which have become sealed by the contents. The practice of probing with a fine needle is highly questionable since the metal, which may not have corroded through completely, may be perforated by the exploratory needle.

An external examination of the seam should then be made. From this preliminary inspection, experienced persons may be able to detect abnormalities in the contours of seams so that it may only be necessary to strip seams in order to confirm opinions formed as the result of the external examination. For example, a long can seam with normal

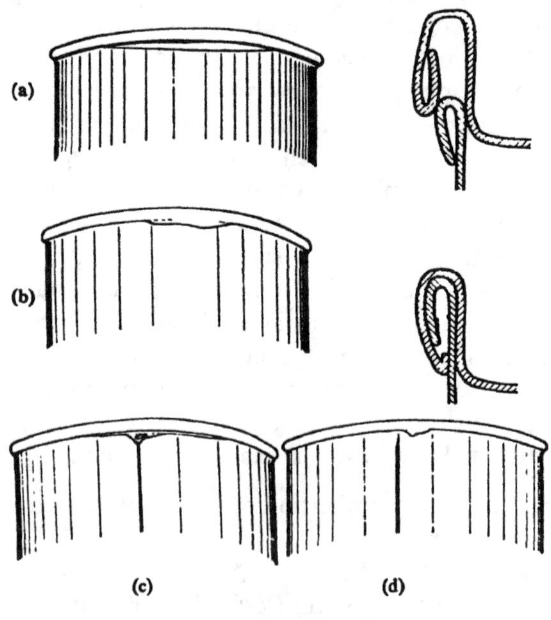

Fig. 29 Defects in can seams

countersink invariably indicates a short cover hook and a narrow rounded seam is an indication of looseness.

Gross defects in the can seam, easily seen during the external examination, are a mis-locked double seam caused by complete failure of the cover hook to engage with the body hook, a knocked down flange (*see* Fig. 29a), a knocked down curl (*see* Fig. 29b) and an incompletely soldered side seam. These faults invariably lead to leakage. Other faults which may give rise to leakage if the defect is sufficiently pronounced are 'skidders' caused by incomplete second operation rolling, spurs (*see* Fig. 29d), metal fracture at the droop (*see* Fig. 29c) and cut-overs.

To a certain extent, visible leakage of the contents through any seam can be of diagnostic importance. As a general guide, it may be taken that if the ends of a 'swell' can be 'sprung' with moderate thumb pressure, a leak indicates a defective seam. On the other hand, if the internal pressure is so great that the can is a 'hard swell', any leakage of the contents through the seam should be viewed with reserve, since with very high pressure the contents may be forced out, even though the seam is of normal quality.

Examination of rolled seams

It is generally acknowledged by most authorities that a seam should be judged as a whole and not solely by its appearance or its external dimensions or those of its separate parts. The examination procedure should follow a set pattern, first the examination and measurement of the seam externally, followed by stripping of the seam and measurement and examination of the component parts.

The important external features are:

Seam length. This measurement is required in order to calculate the percentage overlap. Provided the seam is not short due to looseness or long due to inadequate overlap some tolerance may be allowed on the absolute dimensions which are normally as follows:

Can diameter	*Seam length*
202	0.116 in ± 0.005 in
211—401	0.120 in ± 0.005 in
502—603	0.130 in ± 0.005 in

Seam thickness. Variation in the thickness of tinplate, not only from sheet to sheet but from one part of a sheet to another, makes it difficult to lay down standards for seam thickness which are generally acceptable. There should not, however, be any marked variation in the thickness of any given seam apart from the normal increased thickness at the juncture of the rolled seam and the side seam. This increased thickness is due to the two extra layers of metal at this point. Measurements should be taken at four equidistant points on the circumference, with the side seam juncture midway between two of the points.

Variation in seam thickness may be due to a worn chuck, grease or dirt on the base-plate causing the can to slip during the seaming process, second operation wheel too tight, worn machine bearings causing chuck to be off centre or uneven application of lining compound.

Countersink depth. This may be measured either by the special seam micrometer, which is fitted with a special attachment for the purpose, or by means of a dial depth gauge. Normally the countersunk depth should exceed the seam length by at least 0.005 in.

Droop. At the point where it passes over the side seam, the rolled seam is increased in length with a corresponding decrease in the length of the cover hook. This is caused by the extra thickness of metal in the side seam lap joint. The increased length of the rolled seam at this point is termed droop and should not exceed 20 per cent of the remainder of the seam.

Common causes of abnormal droop are excess solder in the side seam lap joint, excessive pressure on the seaming rolls and abnormally long can hook caused by heavy base-plate pressure. Where droop is excessive, there is a tendency for the metal at this point to be fractured or badly scored by the seaming rolls.

After measurement of the external features detailed above the can contents may be removed for bacteriological examination if desired. In order to separate the hooks for examination and measurement the seam must be stripped. In our view a satisfactory method of doing this is as follows:

With a circular cutter, remove the end of the can except for a rim about 0.6 mm wide. At this stage the air leakage test may be carried out if so desired. Continue the stripping operation by tearing off the narrow rim of tinplate with a pair of pincers, using the top of the can as a support for the tool. If the rim is removed by short, outward pulls, it will be found that the tear runs easily round the top of the seam. After cutting through the seam with the pincers and gently easing the remaining portion of the cover off the top of the body hook with the pincers, it will be found that the cover hook can easily be disengaged. Figures 30 and 31 illustrate this method of stripping.

Hook overlap. The end cover hook and body hook lengths are now measured, each at the same four positions used for the external seam measurement. The values obtained are used to calculate the percentage overlap which is given by the formula:

$$\frac{x+y+1.1te-L}{L-(2.2te+1.1tb)}\times 100$$

Where x = body hook length
y = cover hook length
L = seam length
te = end plate thickness
tb = body plate thickness

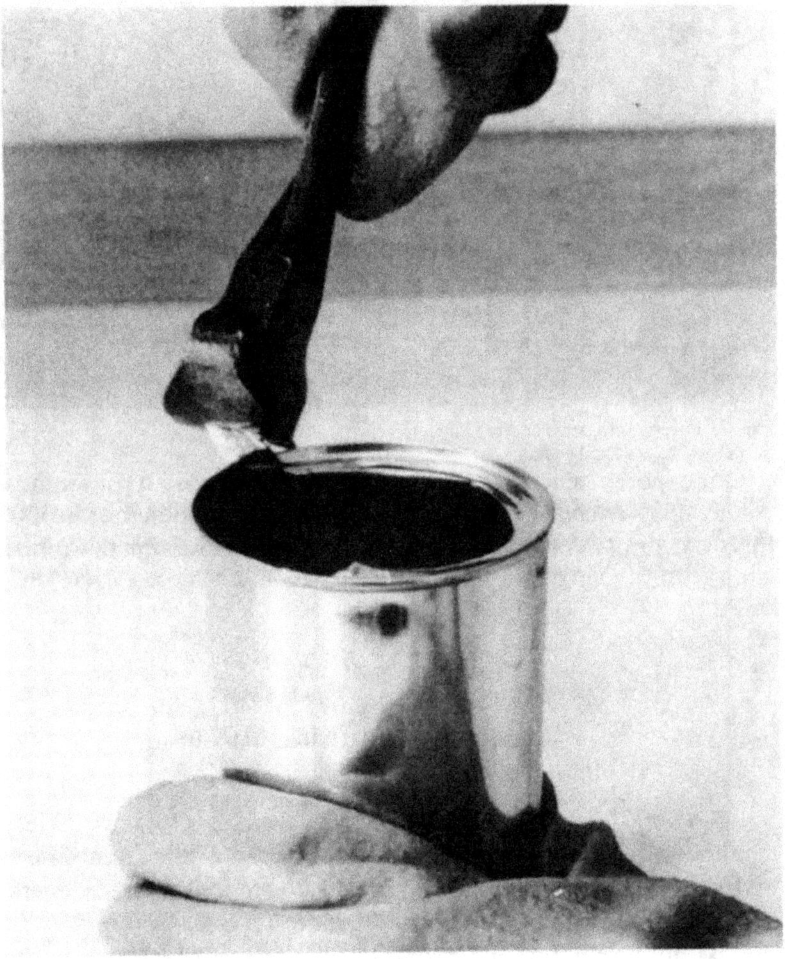

Fig. 30 Tearing off end rim

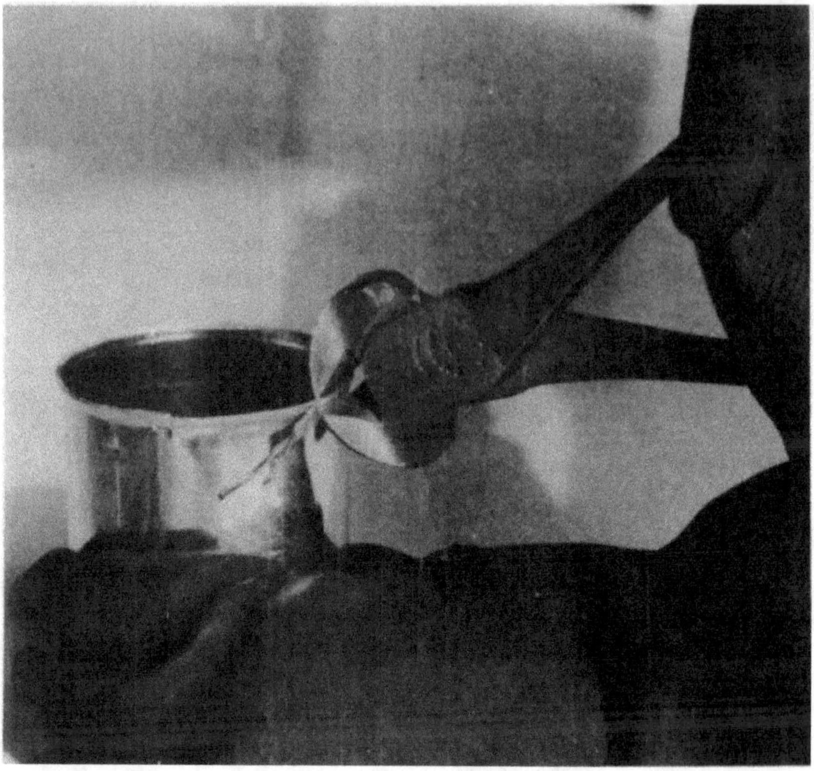

Fig. 31 Disengaging end hook

Our experience has shown that the overlap should have a minimum value of 55 per cent if micro-leakage is to be avoided in products with a high vacuum. Normally the hook lengths are within the following values:

Can diameter	*Body and cover hook length*
202	0.075 in $\pm$ 0.005in
211—401	0.080 in $\pm$ 0.005 in
502—603	0.085 in $\pm$ 0.005 in

Wrinkling. The hooks should be free from excessive wrinkles, the presence of which normally indicates looseness of the seam. The usual system of rating is to give a numerical value to the wrinkle, depending on the depth to which it extends down the cover hook (Fig. 32). Thus, if maximum wrinkling (i.e. the full length of the hook) is given a value of 5, the rating of a wrinkle extending three-fifths of the length of the hook would be 3. On this basis, wrinkle ratings of not higher than 2 and

1 respectively for the A1 and UT size cans should be permitted. An undesirable feature, similar to wrinkles in appearance, is caused by folds or pleats in the cover hook formed as a result of the metal folding back on itself during the first operation and being partly flattened in the second operation. The toroidal seaming panel ends developed by The Metal Box Co. Ltd. substantially reduced the tendency for wrinkle formation.

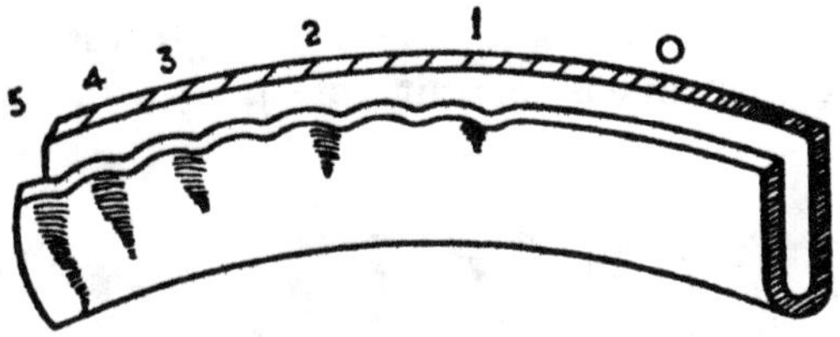

Fig. 32 System of rating wrinkles

The development of high speed can closing machines, a reduction in thickness of tinplate and other advances in can-making technology, have led The Metal Box Co. Ltd. (1973) to propose an up-dating of double seam assessment. In their Double Seam Manual they have summarised the requirements of a double seam. These include:

1. Correct tightness to ensure the lining compound is held under compression. Tightness is assessed by
   a. Degree of wrinkling, together with
   b. Free space
   (Free space = seam thickness − [2 (tb) + 3(te)])
2. Body hook correctly embedded in lining compound. This is assessed by the 'per cent body hook butting' which should have a minimum value of 70 per cent.

$$\text{Per cent body hook butting} = \frac{\text{BH} - 1.1\,tb}{\text{SL} - 1.1\,(2\,te + tb)} \times 100$$

Where BH = body hook
SL = seam length

3. Correct body and end hook overlap. This should be as large as possible without giving rise to wrinkling. This actual overlap can be measured with an optical seam projector or calculated from the formula

$$\text{Actual overlap} = \text{EH} + \text{BH} + 1.1\,te - \text{SL}$$

Where EH = end hook length

4. Freedom from obvious defects. There should be no gross local distortions in the seam, such as spurs, droops and cut-overs, etc.

Special slide rules are available for calculating free space, per cent body hook butting and actual overlap.

Defective soldering. Defective soldering of lock seams or of lap joints is almost always attended by leakage of the contents. Frequently, in cases of marked solder deficiency, the tinplate in the vicinity of the defect exhibits an area of sulphide staining, indicating that the contents have leaked during the retorting process. Occasionally, however, minute channels permit infection without leakage of

Fig. 33 Testing for defective soldering of seam

contents so that it is unsafe to rely on visible external leakage in judging soldered seams. In order to separate the lap joints for examination, the rolled seams at the ends of the can must first be taken apart. The body hook formed by the laps is then levered straight with a screwdriver or similar tool. By firmly gripping the body close to the edge of the inner lap with pliers, and pulling outwards, the lap should come apart (Fig. 33). If there is marked deficiency of solder the laps separate easily, but considerable force may be required if they are well soldered. A common defect is the presence of a small unsoldered channel running through an otherwise well-soldered lap joint into the end of the fold of the side seam.

Side seams may be opened for inspection with the aid of an apparatus which consists of two semi-cylindrical lengths of metal which when fitted together form a cylinder of slightly smaller diameter than the can. The can body to be opened is slid over this cylinder and, by operating a cam, the cylinder is expanded until the side seam of the can is forced open. Before opening the side seam it is necessary to remove the cover hooks at both ends of the can. The opened seam should be carefully examined for solder-free patches with a hand lens. Among causes of soldering defects, may be mentioned springy or greasy tinplate, poor fluxing, solder containing dirt or impurities, dirty solder roll and inefficient cooling fan.

Oval and rectangular cans

The foregoing remarks are concerned with round sanitary cans, but the methods of examination given can be equally applied to other cans, such as the drawn oval or rectangular types. Due to the rapid changes in the curvature of the seam, uniform seaming of rectangular and small oval cans is more difficult than with the round type. The most common causes of leakage are excessive wrinkling or short hooks at the corners or sharp bends of the seam. With regard to seam and hook dimensions of these cans, individual manufacturers set their own standards, taking into consideration the weight of tinplate, size of can, radius of corners or sharp bends, etc. For the larger-sized cans (i.e. 8 oz (226 g) and upwards) it appears to be generally agreed, however, that the hooks on the straight or gently curved portions of the seam should be equal to the standard for round cans; slightly shorter hooks are permissible at the corners or sharp curves where the flange width is reduced to minimise wrinkle formation. To compensate for these unavoidable mechanical deficiences, it is a common practice to use a greater amount of gasketing material for this type of can.

Some of the common defects which result in leaker spoilage have been briefly mentioned; these and others are summarised in Table 14.

Table 14 Summary of some common seam defects

Defects	Cause	Comments
Long seam	Insufficient 1st operation	Cover hook is short — whole seam is stretched
Short seam	Excessive 1st operation	Sharp bend or 'cut-over' at top inside edge of seam
	Insufficient 2nd operation	Seam loose — hooks wrinkled
Loose seam	Insufficient 2nd operation	Seam thick and below normal length: hooks bow-shaped and not tightly engaged — show wrinkling
Uneven seam	Worn rollers, roller pins on chuck spindle or other machine parts Excessive 1st or 2nd operation	Variation in seam tightness
Excessive droop (increased seam length at junction with side seam)	Excess solder in side seam laps Seaming rolls set too tightly Excessive base pressure (long can hook)	Cover hook at site of droop is short; base of droop often cut or scored by seaming rolls
Short body hook	Insufficient base pressure. Excessive sive clearance between rolls and chuck	Can is too long — body hook being short by an equal amount. Top of seam has rounded contour
Long body hook	Excessive base pressure	Always results in corresponding decrease in cover hook. 'Spurs' may be formed
Short cover hook	Insufficient 1st operation	Seam longer than normal — body hook normal. 'Spurs' may be formed
Long cover hook	Excessive 1st operation	'Cut-over' at top inside edge of seam.
Wrinkled hooks	Insufficient 2nd operation	Seam loose, hooks curved
Deep countersink	Insufficient base pressure Excessive clearance between rolls and chuck Chuck flange too thick	Deep, countersink is always accompanied panied by short cover hook
Shallow countersink	Worn chuck	Does not ensure tightness at correct point on seam
'Skidders' — term denoting cans which slip or skid during seaming operations	Insufficient base pressure Worn chuck Worn serrations on base plate Greasy base plate or chuck Excessive 1st or 2nd operation	A portion of the seam is unduly wide and very loose
'Cut-over' — sharp edge or cut on top inside edge of seam	Excessive base pressure Excessive 1st operation Too much clearance between rolls and chuck Worn chuck	'Cut-overs' are specially apparent near side seams
'Spur' — a 'V' shaped protrusion at base of seam	Insufficient 1st operation Excessive 2nd operation Too much solder at overlap Excessive base pressure Harsh plate	Frequently occur near side seam Deficient overlap at the site of the spur
Damaged can flanges	Can not centring on chuck. Damage in transit, rough handling etc.	Where flange is knocked down there may be complete failure to engage with cover hook

Flexible pouches

The formation of high quality seals is vital to retort pouch integrity and a system for the examination and control of seals is essential if unacceptable spoilage rates are to be avoided. As far as potential leakage is concerned there are two basic differences between can double seams and pouch seals. Firstly, pouch seals when correctly formed, are a continuous weld and as such provide a permanent barrier to the ingress of micro-organisms, whereas some 2 to 3 per cent of cans with good commercial seams may exhibit temporary leakage after processing (Thorpe & Everton, 1968). Secondly most cans possess a headspace and a vacuum but the flexible nature of the pouch will not support the same conditions and there is not, therefore, the same tendency for micro-organisms to be sucked into a pouch.

Bios test. As a test of the overall integrity of pouches the 'Bios' test, already described for cans, may be used. However, because of the lack of 'suction' mentioned above, it is necessary to flex the pouches during the test. This technique also serves as a check for pinholes and minute cracks which may occur in the laminate. The usual test organism is *Aerobacter aerogenes*.

The 'Bios' test, however, is more suitable for laboratory work and is of little use for the on-line control of seal quality, where ideally a 100 per cent non-destructive test would be preferable. To this end a number of techniques including ultrasonics, low current X-rays, polarography and infrared thermography have been investigated by various workers but found to be unsuitable, the main reasons being low sensitivity, low throughput or high cost. At the present time the usual method of controlling seal quality consists of taking samples from the line on a regular basis and examining their critical parameters, particular attention being given at start-up and after breakdowns.

Both the manufacturers' seals (if pouch stock is used) and the processors' closing seals should be examined and it is advisable to make checks on samples of the closing seal from unprocessed and processed pouches.

Burst test. The most widely employed test for seal strength is the burst test. This may be applied either as an inflation burst test or as a compression test. In the former, a flat empty pouch is inflated through a clamp fixed to the open side, while the thickness of the pouch is restricted by two rigid parallel plates. The pouch may be inflated to the point at which it bursts, the bursting pressure being noted on a pressure gauge, or it may be inflated to a pre-set level and then held for a pre-determined time. For the compression test the pouch is filled with water, sealed, and then placed between two parallel plates which are then gradually brought closer, thus squeezing the pouch and applying a strain on the seals. As with the inflation test, the pressure can be increased until the seal fails, or a preset maximum pressure can be held for a set time.

Tensile strength. The tensile strength of seals can be assessed by the use of instruments such as the Instron. A 1.5 cm strip of the seal is cut from the pouch and held between two jaws which are then moved apart to apply tension to the seal. A good heat seal will normally require a force of about 2.1 kg per 10 mm to cause failure which is usually the result of delamination, but the actual value will vary depending on the laminate used. This is a useful test for checking the suitability of laminates and the sealing conditions of the equipment.

Visual examination. Finally a visual inspection of the seals should be made. A proportion of those leaving the sealing machine should be examined for the presence of wrinkles and occluded particles. Pouches showing either of these defects should be rejected. Tsutsumi, quoted by Lampi (1977), reporting on Japanese experiences of 100 per cent visual inspections both before and after retorting, gives failure rates of 0.2 per cent and 0.02 per cent respectively. Lampi (1977) summarises the criteria which define a high performance seal for retort pouches.

REFERENCES

Lampi R A 1977 Adv in Food Res 23: 305
Lefebvre P H 1952 9th Congr Int Ind Agr, Rome
Metal Box Company Ltd 1973 Double seam manual. The Metal Box Co. Ltd, Reading
Sanders R K 1949 Chem Ind p 151
Thorpe R H, Everton J R 1968 Post-process sanitation in canneries. Tech Man No 1. The Campden Food Preservation Res Ass, Chipping Campden.

Appendix

Temperature of saturated steam

Steam pressure		Temperature	
lb/sq in	bar	°F	°C
0	0.000	212.0	100.0
1	0.069	215.4	101.9
2	0.138	218.5	103.6
3	0.207	221.5	105.3
4	0.276	224.4	106.9
5	0.345	227.1	108.4
6	0.414	229.6	109.8
7	0.483	232.3	111.3
8	0.552	234.7	112.6
9	0.621	237.0	113.9
10	0.689	239.4	115.2
11	0.758	241.5	116.4
12	0.827	243.7	117.6
13	0.896	245.8	118.8
14	0.965	247.8	119.9
15	1.034	249.8	121.0
16	1.103	251.6	122.0
17	1.172	253.4	123.0
18	1.241	255.4	124.1
19	1.310	257.0	125.0
20	1.379	258.8	126.0
21	1.448	260.5	126.9
22	1.517	262.3	127.9
23	1.586	263.7	128.7
24	1.655	265.3	129.6
25	1.724	266.8	130.4
26	1.793	268.3	131.2
27	1.862	269.8	132.1
28	1.931	271.4	133.0
29	1.999	272.6	133.6
30	2.068	274.0	134.4

Sizes of some common UK round open-top cans

Trade Name	Dimensions[1]		Capacity	
	Diameter	Height	Fl oz	ml
Baby Food	202	$213\frac{1}{4}$	4.5	128
5 oz	211	202	5.5	156
6 oz Evap	211	205	6.1	173
U8 Handy	300	207	7.9	224
Picnic	211	301	8.3	236
A1	211	400	11.1	315
UT/Tall	300	$408\frac{3}{4}$	15.7	446
1 lb Flat	401	212	16.5	469
A2	307	408	20.2	574
A2½	401	411	29.7	843
A6	603	600	92.5	2627
A10	603	700	109.0	3096

1 Dimensions are given in inches and sixteenths of an inch in accordance with accepted convention, for example, 202 is $2\frac{2}{16}$ in.

Example of calculation of process lethalities

Data

Temperature of the heating medium (retort)	= 115°C
Initial temperature of the product	= 68°C
Coming-up time of the retort	= 15 min.
Process time at 115°C	= 108 min.
$f_h = f_c$ (*see* p.199)	= 65
$j_h = j_c$ (*see* p.199)	= 1.6

Requirement

To calculate F value at the centre of the can.

Method

$I = 115 - 68 = 47$ (*see* p. 199)
$jI = 1.6 \times 47 = 75.2$ (*see* p. 199)
B = The process time corrected for the contribution of the coming-up time by adding 0.4 × coming-up time.
$B = 108 + 0.4(15) = 114.$

Since $B = f_h(\log jI - \log g)$ (*see* p. 199)
$114 = 65(\log 75.2 - \log g)$

$$g = 1.325$$

that is, the pack centre temperature reaches 113.67°C.

From tables giving the relationship between g and $\frac{f}{U}$ ($g = 1.325 \times 1.8$: *see* p. 201)

$$\frac{f_h}{U} = 2.00 = \frac{65}{U}$$

$$U = 32.5$$

Now $U = FF_i$ and $F_i = \log^{-1}\frac{121.1 - T}{z}$ (*see* p. 198).

Substituting the values obtained

$$F_i = \log^{-1}\frac{121.1 - 115}{10}$$
$$= \log^{-1} 0.611$$
$$= 4.07$$
$$32.5 = F \times 4.07$$
$$\mathbf{F = 8.0}$$

Lethal rates (F_{250} per minute)

$$\text{Lethal rate} = \frac{1}{\log^{-1}\dfrac{(250 - RT)}{z}}$$

Temp. °F	z value 12	14	15	16	17	18	19	20	22
200	0.000	0.000	0.000	0.001	0.001	0.002	0.002	0.003	0.005
201	0.000	0.000	0.001	0.001	0.001	0.002	0.003	0.004	0.006
202	0.000	0.000	0.001	0.001	0.002	0.002	0.003	0.004	0.007
203	0.000	0.000	0.001	0.001	0.002	0.003	0.003	0.005	0.007
204	0.000	0.001	0.001	0.001	0.002	0.003	0.004	0.005	0.008
205	0.000	0.001	0.001	0.002	0.002	0.003	0.004	0.006	0.009
206	0.000	0.001	0.001	0.002	0.003	0.004	0.005	0.006	0.010
207	0.000	0.001	0.001	0.002	0.003	0.004	0.005	0.007	0.011
208	0.000	0.001	0.002	0.002	0.003	0.005	0.006	0.008	0.012
209	0.000	0.001	0.002	0.003	0.004	0.005	0.007	0.009	0.014
210	0.001	0.001	0.002	0.003	0.004	0.006	0.008	0.010	0.015
211	0.001	0.002	0.003	0.004	0.005	0.007	0.009	0.011	0.017
212	0.001	0.002	0.003	0.004	0.006	0.008	0.010	0.013	0.019
213	0.001	0.002	0.003	0.005	0.007	0.009	0.011	0.014	0.021
214	0.001	0.003	0.004	0.006	0.008	0.010	0.013	0.016	0.023
215	0.001	0.003	0.005	0.006	0.009	0.011	0.014	0.018	0.026
216	0.001	0.004	0.005	0.007	0.010	0.013	0.016	0.020	0.029
217	0.002	0.004	0.006	0.009	0.011	0.015	0.018	0.022	0.032
218	0.002	0.005	0.007	0.010	0.013	0.017	0.021	0.025	0.035
219	0.002	0.006	0.009	0.011	0.015	0.019	0.023	0.028	0.039
220	0.003	0.007	0.010	0.013	0.017	0.022	0.026	0.032	0.044
221	0.004	0.008	0.012	0.015	0.020	0.024	0.030	0.035	0.048
222	0.004	0.010	0.013	0.017	0.022	0.028	0.034	0.040	0.054
223	0.005	0.011	0.016	0.020	0.026	0.032	0.038	0.042	0.059
224	0.006	0.014	0.018	0.023	0.029	0.036	0.042	0.050	0.066
225	0.008	0.016	0.022	0.027	0.033	0.041	0.048	0.056	0.072
226	0.009	0.019	0.025	0.031	0.038	0.046	0.054	0.062	0.081
227	0.012	0.022	0.029	0.036	0.044	0.052	0.061	0.070	0.090
228	0.014	0.022	0.034	0.041	0.050	0.060	0.069	0.079	0.100
229	0.017	0.031	0.040	0.048	0.058	0.068	0.078	0.089	0.110
230	0.022	0.037	0.046	0.056	0.066	0.077	0.088	0.100	0.123
230.5	0.024	0.040	0.050	0.060	0.071	0.083	0.094	0.106	0.129
231	0.026	0.044	0.054	0.065	0.076	0.088	0.100	0.112	0.136
231.5	0.029	0.048	0.058	0.070	0.081	0.094	0.106	0.119	0.144
232	0.032	0.052	0.063	0.075	0.087	0.100	0.112	0.126	0.152
232.5	0.035	0.056	0.068	0.081	0.093	0.107	0.120	0.133	0.160
233	0.038	0.061	0.074	0.087	0.100	0.114	0.128	0.141	0.169
233.5	0.042	0.066	0.079	0.093	0.107	0.121	0.136	0.150	0.178
234	0.046	0.072	0.086	0.100	0.115	0.129	0.144	0.158	0.187
234.5	0.051	0.078	0.093	0.107	0.123	0.138	0.153	0.168	0.197
235	0.056	0.085	0.100	0.116	0.131	0.147	0.162	0.178	0.208
235.5	0.062	0.092	0.108	0.124	0.140	0.156	0.172	0.188	0.219
236	0.068	0.100	0.117	0.133	0.150	0.167	0.183	0.200	0.230
236.5	0.074	0.108	0.126	0.143	0.160	0.178	0.195	0.211	0.243
237	0.083	0.118	0.136	0.155	0.172	0.190	0.207	0.224	0.256
237.5	0.091	0.128	0.147	0.165	0.184	0.202	0.220	0.237	0.269
238	0.100	0.139	0.159	0.177	0.197	0.215	0.234	0.251	0.284
238.5	0.111	0.151	0.171	0.191	0.209	0.230	0.249	0.266	0.300
239	0.121	0.164	0.184	0.205	0.225	0.246	0.264	0.282	0.316
239.5	0.133	0.178	0.200	0.220	0.241	0.261	0.280	0.299	0.332
240	0.147	0.193	0.216	0.237	0.258	0.278	0.298	0.316	0.351
240.25	0.154	0.201	0.224	0.245	0.267	0.287	0.307	0.326	0.364
240.5	0.161	0.209	0.233	0.255	0.276	0.297	0.316	0.335	0.369
240.75	0.168	0.219	0.242	0.264	0.286	0.306	0.327	0.345	0.380
241	0.177	0.228	0.251	0.273	0.296	0.316	0.337	0.355	0.390
241.25	0.186	0.237	0.261	0.284	0.306	0.327	0.347	0.365	0.400
241.5	0.195	0.247	0.272	0.294	0.315	0.337	0.357	0.376	0.411
241.75	0.205	0.258	0.282	0.304	0.326	0.348	0.370	0.387	0.422
242	0.216	0.268	0.292	0.316	0.339	0.359	0.380	0.398	0.433
242.25	0.226	0.279	0.305	0.327	0.350	0.371	0.392	0.410	0.445

Lethal rates (F_{250} per minute) *Cont.*

$$\text{Lethal rate} = \frac{1}{\log^{-1}\dfrac{(250-RT)}{z}}$$

Temp. °F	z value 12	14	15	16	17	18	19	20	22
242.5	0.238	0.292	0.316	0.340	0.362	0.384	0.404	0.421	0.457
242.75	0.249	0.304	0.329	0.352	0.370	0.396	0.415	0.435	0.469
243	0.261	0.316	0.341	0.364	0.388	0.408	0.429	0.448	0.481
243.25	0.273	0.330	0.356	0.378	0.402	0.421	0.440	0.460	0.493
243.5	0.287	0.344	0.369	0.392	0.415	0.435	0.457	0.473	0.508
243.75	0.301	0.358	0.383	0.406	0.429	0.450	0.469	0.487	0.520
244	0.316	0.373	0.398	0.421	0.444	0.464	0.485	0.501	0.533
244.25	0.331	0.379	0.415	0.436	0.459	0.479	0.500	0.516	0.548
244.5	0.348	0.405	0.431	0.452	0.474	0.495	0.515	0.531	0.563
244.75	0.365	0.422	0.448	0.469	0.490	0.511	0.532	0.546	0.577
245	0.384	0.440	0.464	0.487	0.509	0.527	0.545	0.562	0.593
245.25	0.401	0.459	0.483	0.505	0.526	0.545	0.565	0.578	0.608
245.5	0.421	0.477	0.501	0.523	0.544	0.562	0.581	0.596	0.624
245.75	0.442	0.497	0.521	0.543	0.562	0.581	0.599	0.612	0.641
246	0.462	0.519	0.541	0.563	0.580	0.600	0.616	0.631	0.658
246.25	0.486	0.540	0.565	0.584	0.602	0.619	0.637	0.649	0.675
246.5	0.511	0.562	0.584	0.606	0.621	0.639	0.658	0.668	0.695
246.75	0.536	0.585	0.609	0.625	0.645	0.660	0.676	0.688	0.711
247	0.561	0.610	0.631	0.650	0.666	0.681	0.695	0.707	0.730
247.25	0.590	0.637	0.658	0.674	0.690	0.703	0.719	0.728	0.750
247.5	0.618	0.662	0.681	0.699	0.714	0.726	0.741	0.750	0.770
247.75	0.652	0.690	0.709	0.724	0.735	0.749	0.763	0.772	0.790
248	0.683	0.720	0.736	0.750	0.763	0.774	0.785	0.794	0.811
248.25	0.712	0.750	0.769	0.776	0.789	0.799	0.813	0.818	0.833
248.5	0.748	0.781	0.794	0.806	0.815	0.825	0.834	0.841	0.855
248.75	0.788	0.814	0.825	0.834	0.844	0.854	0.859	0.865	0.877
249	0.824	0.849	0.857	0.865	0.872	0.880	0.885	0.891	0.900
249.25	0.865	0.884	0.891	0.897	0.902	0.909	0.914	0.917	0.925
249.5	0.907	0.920	0.925	0.930	0.934	0.938	0.943	0.944	0.950
249.75	0.952	0.960	0.962	0.965	0.967	0.969	0.971	0.972	0.975
250	1.000	1.000	1.000	1.000	1.000	1.000	1.000	1.000	1.000
250.5	1.10	1.09	1.08	1.08	1.07	1.07	1.06	1.06	1.05
251	1.21	1.18	1.15	1.16	1.14	1.14	1.13	1.12	1.11
251.5	1.33	1.28	1.26	1.24	1.22	1.21	1.20	1.19	1.17
252	1.47	1.39	1.37	1.33	1.31	1.30	1.28	1.26	1.23
252.5	1.62	1.51	1.47	1.43	1.40	1.38	1.35	1.33	1.30
253	1.78	1.64	1.59	1.54	1.50	1.47	1.44	1.41	1.37
253.5	1.96	1.78	1.71	1.66	1.60	1.56	1.53	1.50	1.44
254	2.16	1.93	1.85	1.78	1.72	1.67	1.63	1.59	1.52
254.5	2.37	2.09	2.00	1.91	1.85	1.78	1.73	1.68	1.60
255	2.61	2.27	2.15	2.06	1.97	1.90	1.84	1.78	1.69
255.5	2.87	2.47	2.33	2.21	2.11	2.02	1.95	1.88	1.78
256	3.16	2.68	2.51	2.37	2.26	2.16	2.08	2.00	1.88
256.5	3.49	2.91	2.71	2.55	2.41	2.30	2.20	2.11	1.98
257	3.83	3.17	2.93	2.74	2.58	2.45	2.34	2.24	2.08
257.5	4.22	3.44	3.17	2.94	2.76	2.61	2.48	2.37	2.19
258	4.65	3.73	3.42	3.16	2.96	2.79	2.64	2.51	2.31
258.5	5.10	4.05	3.69	3.39	3.16	2.97	2.80	2.66	2.43
259	5.62	4.38	3.99	3.65	3.38	3.16	2.98	2.82	2.56
259.5	6.18	4.76	4.30	3.92	3.62	3.37	3.16	2.99	2.70
260	6.80	5.18	4.65	4.22	3.88	3.60	3.37	3.16	2.86

Lethal rates ($F_{121.11}$ per minute)

$$\text{Lethal rate} = \frac{1}{\log^{-1}\dfrac{(121.11-RT)}{z}}$$

Temp. °C	z value 7	8	9	10	11	12
90.0	0.000	0.000	0.000	0.001	0.001	0.003
91.0	0.000	0.000	0.000	0.001	0.002	0.003
92.0	0.000	0.000	0.001	0.001	0.002	0.004
93.0	0.000	0.000	0.001	0.002	0.003	0.005
94.0	0.000	0.000	0.001	0.002	0.003	0.006
95.0	0.000	0.001	0.001	0.002	0.004	0.007
96.0	0.000	0.001	0.002	0.003	0.005	0.008
97.0	0.000	0.001	0.002	0.004	0.006	0.010
98.0	0.000	0.001	0.003	0.005	0.008	0.012
99.0	0.001	0.002	0.003	0.006	0.010	0.014
100.0	0.001	0.002	0.005	0.008	0.012	0.017
101.0	0.001	0.003	0.006	0.010	0.015	0.021
101.5	0.002	0.004	0.007	0.011	0.016	0.023
102.0	0.002	0.004	0.008	0.012	0.018	0.026
102.5	0.002	0.005	0.009	0.014	0.020	0.028
103.0	0.003	0.005	0.010	0.015	0.023	0.031
103.5	0.003	0.006	0.011	0.017	0.025	0.034
104.0	0.004	0.007	0.013	0.019	0.028	0.038
104.5	0.004	0.008	0.014	0.022	0.031	0.041
105.0	0.005	0.010	0.016	0.024	0.034	0.045
105.5	0.006	0.011	0.018	0.027	0.038	0.050
106.0	0.007	0.013	0.021	0.031	0.042	0.055
106.5	0.008	0.015	0.024	0.035	0.047	0.061
107.0	0.010	0.017	0.027	0.039	0.052	0.067
107.5	0.011	0.020	0.031	0.044	0.058	0.073
108.0	0.013	0.023	0.035	0.049	0.064	0.081
108.5	0.016	0.027	0.040	0.055	0.071	0.089
109.0	0.019	0.031	0.045	0.062	0.079	0.098
109.5	0.022	0.035	0.051	0.069	0.088	0.108
110.0	0.026	0.041	0.058	0.077	0.098	0.119
110.5	0.030	0.047	0.066	0.087	0.109	0.131
111.0	0.036	0.054	0.075	0.097	0.120	0.144
111.5	0.042	0.063	0.086	0.109	0.134	0.158
112.0	0.050	0.073	0.097	0.123	0.149	0.174
112.5	0.059	0.084	0.110	0.138	0.165	0.192
113.0	0.069	0.097	0.126	0.155	0.183	0.211
113.5	0.082	0.112	0.143	0.173	0.203	0.232
114.0	0.096	0.129	0.162	0.195	0.226	0.256
114.5	0.114	0.149	0.184	0.218	0.251	0.281
115.0	0.134	0.172	0.209	0.245	0.278	0.310
115.2	0.143	0.182	0.220	0.256	0.290	0.322
115.4	0.153	0.193	0.232	0.269	0.303	0.334
115.6	0.163	0.205	0.244	0.281	0.316	0.347
115.8	0.174	0.217	0.257	0.294	0.329	0.361
116.0	0.186	0.230	0.271	0.308	0.343	0.375
116.2	0.199	0.243	0.285	0.323	0.358	0.390
116.4	0.212	0.258	0.300	0.338	0.373	0.405
116.6	0.227	0.273	0.315	0.354	0.389	0.421
116.8	0.242	0.289	0.332	0.371	0.406	0.437
117.0	0.259	0.306	0.349	0.388	0.423	0.454
117.2	0.276	0.325	0.368	0.406	0.441	0.472
117.4	0.295	0.344	0.387	0.426	0.460	0.491
117.6	0.315	0.364	0.407	0.446	0.480	0.510
117.8	0.337	0.386	0.429	0.467	0.500	0.530
118.0	0.360	0.409	0.451	0.489	0.522	0.551
118.1	0.372	0.420	0.463	0.500	0.533	0.561
118.2	0.384	0.433	0.475	0.512	0.544	0.572
118.3	0.397	0.445	0.487	0.524	0.555	0.583
118.4	0.410	0.458	0.500	0.536	0.567	0.595
118.5	0.424	0.472	0.513	0.548	0.579	0.606
118.6	0.438	0.486	0.526	0.561	0.591	0.618
118.7	0.453	0.500	0.540	0.574	0.604	0.630
118.8	0.468	0.514	0.554	0.587	0.617	0.642
118.9	0.483	0.529	0.568	0.601	0.630	0.654

Lethal rates ($F_{121.11}$ per minute) *Cont.*

$$\text{Lethal rate} = \frac{1}{\log^{-1}\frac{(121.11 - RT)}{z}}$$

Temp. °C	z value 7	8	9	10	11	12
119.0	0.500	0.545	0.583	0.615	0.643	0.667
119.1	0.516	0.561	0.598	0.630	0.657	0.680
119.2	0.534	0.577	0.613	0.644	0.670	0.693
119.3	0.551	0.594	0.629	0.659	0.685	0.707
119.4	0.570	0.611	0.646	0.675	0.699	0.720
119.5	0.589	0.629	0.662	0.690	0.714	0.734
119.6	0.609	0.648	0.680	0.706	0.729	0.748
119.7	0.629	0.666	0.697	0.723	0.744	0.763
119.8	0.650	0.686	0.715	0.740	0.760	0.778
119.9	0.672	0.706	0.734	0.757	0.776	0.793
120.0	0.694	0.727	0.753	0.774	0.793	0.808
120.1	0.717	0.748	0.772	0.792	0.809	0.824
120.2	0.741	0.770	0.792	0.811	0.827	0.840
120.3	0.766	0.792	0.813	0.830	0.844	0.856
120.4	0.792	0.815	0.834	0.849	0.862	0.873
120.5	0.818	0.839	0.856	0.869	0.880	0.890
120.6	0.846	0.863	0.878	0.889	0.899	0.907
120.7	0.874	0.889	0.900	0.910	0.918	0.924
120.8	0.903	0.915	0.924	0.931	0.937	0.942
120.9	0.933	0.941	0.948	0.953	0.957	0.961
121.0	0.964	0.969	0.972	0.975	0.977	0.979
121.1	0.997	0.997	0.997	0.998	0.998	0.998
121.2	1.030	1.026	1.023	1.021	1.019	1.017
121.3	1.064	1.056	1.050	1.045	1.041	1.037
121.4	1.100	1.087	1.077	1.069	1.063	1.057
121.5	1.137	1.119	1.105	1.094	1.085	1.078
121.6	1.175	1.151	1.134	1.119	1.108	1.099
121.7	1.214	1.185	1.163	1.146	1.131	1.120
121.8	1.255	1.220	1.193	1.172	1.155	1.142
121.9	1.297	1.255	1.224	1.199	1.180	1.164
122.0	1.340	1.292	1.256	1.227	1.205	1.186
122.1	1.385	1.330	1.288	1.256	1.230	1.209
122.2	1.431	1.369	1.322	1.285	1.256	1.233
122.3	1.479	1.408	1.356	1.315	1.283	1.257
122.4	1.529	1.450	1.391	1.346	1.310	1.281
122.5	1.580	1.492	1.427	1.377	1.338	1.306
122.6	1.633	1.536	1.464	1.409	1.366	1.331
122.7	1.687	1.580	1.502	1.442	1.395	1.357
122.8	1.744	1.626	1.541	1.476	1.424	1.383
122.9	1.802	1.674	1.581	1.510	1.455	1.410
123.0	1.862	1.723	1.622	1.545	1.485	1.437
123.1	1.924	1.773	1.664	1.581	1.517	1.465
123.2	1.989	1.825	1.707	1.618	1.549	1.493
123.3	2.055	1.878	1.751	1.656	1.582	1.522
123.4	2.124	1.933	1.797	1.694	1.615	1.552
123.5	2.195	1.990	1.843	1.734	1.649	1.582
123.6	2.268	2.048	1.891	1.774	1.684	1.613
123.7	2.344	2.107	1.940	1.816	1 720	1.644
123.8	2.423	2.169	1.990	1.858	1.756	1.676
123.9	2.504	2.232	2.042	1.901	1.793	1.708
124.0	2.587	2.297	2.095	1.945	1.831	1.741
124.1	2.674	2.365	2.149	1.991	1.870	1.775
124.2	2.763	2.434	2.205	2.037	1.909	1.809
124.3	2.856	2.505	2.262	2.085	1.950	1.844
124.4	2.951	2.578	2.320	2.133	1.991	1.880
124.5	3.050	2.653	2.381	2.183	2.033	1.916
124.6	3.152	2.731	2.442	2.234	2.076	1.954
124.7	3.257	2.810	2.505	2.286	2.120	1.991
124.8	3.366	2.892	2.570	2.339	2.165	2.030
124.9	3.479	2.977	2.637	2.393	2.211	2.069

Summary of causes and control of

Cause of spoilage	Bacteriological findings
Under processing	
1. Contaminating organisms of exceptional heat resistance	Usually only a single species is present. In low- and medium-acid products, the organism is invariably a spore-former
2. Excessive microbial contamination	Ditto
3. Faulty retort operation (a) Air or water causing localised underheating	Usually a single species is present but in severe underheating, two or more types may be found
(b) Inaccurate instruments	Ditto
4. (a) Heat-penetration retarded by excessive weight of solid in solid-liquid packs	Usually single species only
(b) Improper alignment of cans in retort	Ditto
(c) Cans too tightly stacked in retort. (This applies mainly to square or rectangular cans)	Ditto
Inefficient cooling	
Cooling period too short or water temperature too high	Usually thermophilic souring
Seam leakage	
1. Seam defects	Usually mixed infection but single spore-forming species may be present when cooling water is chlorinated. Cans occasionally sterile but have no vacuum
2. Excessive contamination of cooling water	Usually more than one species of organism, including non-heat resistant types
3. Handling wet cans (after processing) on infected equipment	Ditto
4. Straining of seams during processing	As for seam defects
5. Damage in handling	Ditto
Pre-process spoilage	
Delay between preparation of material and heat-processing	Numerous organisms are present in the food on microscopic examination but cultures are sterile

microbial spoilage of canned food

Appearance of can	Control or remedial measures
'Blown' or normal (i.e. 'flat-sour')	1. (a) Locate source of organism and eliminate or reduce contamination (b) Adjust processing conditions
Ditto	2. Reduce contamination by sanitary measures e.g., cleanse plant, use clean raw materials, hold materials outside temperature range for growth
Ditto	3. (a) Eliminate air or water by thorough venting or drainage
Ditto	(b) Check agreement between temperature and pressure reading
Ditto	4. (a) Check weight of solid component. Check weight during filling operation
Ditto	(b) Stack cans so that convection currents are not impeded (e.g. cans of asparagus stalks should be stacked vertically)
Ditto	(c) Loose stacking to permit free steam circulation
Normal	Cool rapidly to below 37.8° C
'Blown' or normal. There is often leakage of liquid contents through seams of 'blown' cans	1. (a) Service seaming machines and check seams regularly during production (b) Check weight of seam lining compound
'Blown' or normal	2. Chlorinate cooling water
Ditto	3. (a) Instal can drying equipment (b) Pay attention to can runway hygiene
'Blown' or normal. With severe straining, expansion rings on can ends show 'peaking'	4. Reduce retort pressure more slowly or, if necessary, pressure cool
Seam or body near seam shows indentations. Cans may be blown or distended	5. (a) Eliminate rough handling (b) Check that tinplate is of correct substance
Cans are usually 'soft swells' or 'flippers' but may appear normal although contents are sour	Eliminate delays in manufacturing process. If necessary, rapidly chill or cool pre-cooked ingredients.

Index

www.ingramcontent.com/pod-product-compliance
Lightning Source LLC
LaVergne TN
LVHW011529200926
848484LV00001B/209